Live Music Productio

This book presents the days of live music production in the UK spanning the late '60s to the mid-'80s, when rock music was enjoying a meteoric rise in popularity.

The author, Richard Ames, will take you on a true behind-the-scenes journey of discovery. You'll learn who the people were, where they came from and how they went on to pioneer the first companies that would become the lifeblood of a unique industry. The interviews contained in this book record and present the raw stories of a few of the original innovators who set the stage for their performers but also for the hundreds of technicians who would tour the world following in their footsteps. The pioneers presented in these interviews share with the reader countless candid anecdotes that convey how their curious enthusiasm, energy, dedication and general can-do attitude was the driving force behind the creation of the many companies we know of as commonplace today. The book presents interviews that span varied aspects of live music production including lighting, sound, rigging, staging, trucking, bussing and catering. *Live Music Production* captures a piece of social history that promises to inform, entertain and delight.

Richard Ames has been a tour manager and tour director in the UK live music industry since the early 1970s, a profession in which he is still involved to this day. He considers himself incredibly lucky and privileged throughout his life to have been asked to help some great bands and their crews travel around the world and entertain their fans.

Live Music Production

Interviews with UK Pioneers

Richard Ames

Routledge
Taylor & Francis Group
NEW YORK AND LONDON

First published 2019
by Routledge
711 Third Avenue, New York, NY 10017

and by Routledge
2 Park Square, Milton Park, Abingdon, Oxon, OX14 4RN

Routledge is an imprint of the Taylor & Francis Group, an informa business

© 2019 Richard Ames

The right of Richard Ames to be identified as author of this work has been asserted by him in accordance with sections 77 and 78 of the Copyright, Designs and Patents Act 1988.

All rights reserved. No part of this book may be reprinted or reproduced or utilised in any form or by any electronic, mechanical, or other means, now known or hereafter invented, including photocopying and recording, or in any information storage or retrieval system, without permission in writing from the publishers.

Trademark notice: Product or corporate names may be trademarks or registered trademarks, and are used only for identification and explanation without intent to infringe.

Library of Congress Cataloging-in-Publication Data
Names: Ames, Richard, interviewer.
Title: Live music production : interviews with UK pioneers / Richard Ames.
Description: New York, NY : Routledge, 2019.
Identifiers: LCCN 2018014632 | ISBN 9780815373575 (hardback : alk. paper) | ISBN 9780815373728 (pbk. : alk. paper) | ISBN 9781351242974 (ebook)
Subjects: LCSH: Rock concerts—Production and direction—Great Britain. | Rock concerts—Stage-setting and scenery—Great Britain. | Sound engineers—Great Britain—Interviews. | Stage lighting designers—Great Britain—Interviews. | Stagehands—Great Britain—Interviews.
Classification: LCC ML3534.6.G7 L58 2019 | DDC 781.66/143—dc23
LC record available at https://lccn.loc.gov/2018014632

ISBN: 978-0-8153-7357-5 (hbk)
ISBN: 978-0-8153-7372-8 (pbk)
ISBN: 978-1-351-24297-4 (ebk)

Typeset in Times New Roman and Gill Sans
by Apex CoVantage, LLC
Printed and bound by CPI Group (UK) Ltd, Croydon, CR0 4YY

In Memory of

Elvira Fletcher

Copperplate engraving © Sydney Fletcher

Who bravely transcribed most of these interviews

Contents

Foreword by Harvey Goldsmith, CBE   ix
Preface   x
Acknowledgements   xii

PART 1
Pioneers of Lighting   1

1. Brian Croft   3
2. Jon Cadbury   32
3. John Morris   51
4. Richard Hartman   58
5. Annie Pocock   80
6. John Coppen   95

PART 2
Pioneers of Sound   117

7. David Hartstone   119
8. Bryan Grant   135
9. Tony Andrews   152

PART 3
Pioneers of Stage Design   167

10. Ian Knight   169

PART 4
Pioneers of Full Production Services   189

11. Roger Searle   191
12. Barbara Pendleton   208

PART 5
Pioneers of Rigging — 223

13 Jon Bray — 225

PART 6
Pioneers of Trucking/Outdoor Staging — 243

14 Roy Lamb — 245
15 Del Roll and Ollie Kite — 271

PART 7
Pioneers of Bussing — 301

16 Len Wright — 303

PART 8
Pioneers of Catering — 321

17 Sandi Grabham — 323
18 Val Bowes — 334

PART 9
Pioneers of Travel Agency — 343

19 Mike Hawksworth — 345

Appendix: Richard Ames' Early CV — *361*

Foreword

It all started in the '60s. Young people wanted a say in life and began to rebel against the post-war austerity. The rock scene began in small clubs (I opened mine in January 1966) and spread like wildfire.

At first, bands pretty well took care of themselves, with the help of a few friends. They would drive the van and help unload the instruments, etc.

Then there became the need for managers, as the agents had too much control. Club owners and Student Union Reps. became promoters.

Suddenly, a business began to form, and alongside it all the trimmings that artists required to make their lives easier.

An industry developed that included loyal road crew, road crew chiefs, tour managers, stage managers and production managers. Alongside the staff came specialist sound companies, lighting companies and of course trucking companies.

Set designers appeared along with lighting designers and so more ancillary companies came into being.

Finally, the icing on the cake . . . on-the-road caterers were required to feed the ever-growing team of people. After all, an army only works at its best when it's fed well and regularly.

Specialised travel agencies and insurance companies soon followed.

Dedicated merchandisers, selling official goods to beat the touts (a problem that still exists today) expanded the team even more.

I started a number of these businesses, beginning with renting out Manfred Mann's sound system when he was not on the road.

I was fed up with curly sandwiches at venues and no hot food, so I started 'Toad in the Hole'. I took over Brockum merchanding in the early days and made it a worldwide business. Trinifold (Bill Curbishley, Tony Smith and myself) became a household name in rock travel headed by the wonderful Mike Hawksworth.

This book tells the story of those pioneers. I have worked with or employed virtually everyone in the book, and they are the unsung heroes of our business.

Harvey Goldsmith, CBE
February 2017

Preface

I thought about putting this book together many years ago, as I've always wanted to tell the story of how this extraordinary industry that I have spent 40 years of my own professional life in came about—and what better way than to recount the history through the words of the people who were there?

These are the early days of live music production in the UK, from the late '60s to the mid-'80s; where the people came from and how they went on to set up, or work with, the first companies that became the lifeblood of a unique industry; their stories in their words.

Recorded here are the words of just a few of the 'pioneers' on behalf of the hundreds of technicians that have toured around the world and made the shows happen. In their youth, the enthusiasm, dedication, energy, curiosity and can-do attitude was the driving force behind the creation of the many companies that encompass the varied aspects of live music production including lighting, sound, rigging, staging, trucking, bussing and catering. They are but a few of the people that were 'pioneers' in their fields; there are obviously many others, who I have not interviewed, and who are as equally important to the story. There are also several people I interviewed who for lack of space have hit the cutting-room floor and not made it into this edition, although they will all have been given honourable mentions for their integral roles.

I have worked with many of these 'pioneers', and throughout the last decade I have interviewed them in an attempt to capture the memories of these talented people who became the 'behind-the-scenes' backbone of UK live music. It has been an honour to have worked alongside them, and it has been a pleasure, all these years later, to listen to them recounting their stories.

Over half of the 'pioneers' still work in the music business; others have retired and some have sadly passed away since I started the interviews. I sat with them and recorded their individual stories, letting them talk about their part in the formative years of this industry. Of course, it's important that something is written down before those who helped to create it in the first place disappear forever, and history becomes distorted through hearsay and speculation. Recollections can become a little hazy, and for those instances where the stories may have become a little foggy I can only apologise and hope that no offence is taken if facts, dates and chronology are imprecise on occasion. All I can say is that fading memories will affect most of you at some point!

I hope that this book as a piece of social history will inform, entertain and delight those who were either there, or were on the periphery of, or are of an age when rock music was on its meteoric rise. Just as equally, I hope that this will become supplementary reading for tomorrow's students so they can see how the foundations of this remarkable industry were forged with the 24/7 strength and spirit of these 'pioneers'.

Richard Ames, 2018

Acknowledgements

I can't quite believe this book has taken me more than 10 years to complete, but it has, and finally I can stop asking my partner, Carrie, to "Please take a look at this!"

In 2007 I bought a dictaphone and some software for Elvira Fletcher, to enable her to transcribe my meetings with these interviewees. When Elvira passed away in 2010, I progressed to using my iPhone and sent my mp3 interviews to India to have them transcribed. But the work didn't really start until I had something in writing to proofread and edit. I couldn't have finished this book without the dedication and help from Carrie.

I must thank every one of the interviewees, who gave me their time and memories that now hopefully will never be forgotten. And, of course, thanks to Harvey Goldsmith for not only writing the Foreword but also for believing in me and giving me some amazing work opportunities at the beginning of my career.

Huge thanks go to all those who let me use their photographs to bring pictorial life to the stories, and their individual names are credited with the relevant images.

Special recognition must go to Rick Burton and Reverend Barker, who have created amazing websites dedicated to the history of the Rainbow Theatre and UK Festivals.

Thanks to Sheila Searle for helping me edit the introduction to each section of this book, and to Ruth Gladwin for generously giving up her time to help me understand all of the elements that I was ignorant of, like clearances and copyright!

A big thank you to Sergio Zavaleta, who in doing his PhD in '60s live sound at Edinburgh University spent many hours emailing questions and getting answers from David Hartstone. Sergio has shared his research into IES with me and allowed me to enrich the chapter with additional information.

Finally, I'm extremely grateful to my editor and publisher, Lara Zoble, for acknowledging the importance of a social history record of live music production and encouraging me to complete it.

Part I
Pioneers of Lighting

Cockney Rebel soundcheck early 1975

In the late '60s bands played more and more live concerts, and it was due to this constant touring that the bands and their crews decided they needed to indulge their audience in a visual as well as an auditory experience. Just switching the house lights on and off was no longer enough. An expertise began to evolve on the road, particularly in three London venues that spawned some great production people: the Institute of Contemporary Arts (ICA) in Pall Mall, the Roundhouse in Chalk Farm and the Rainbow Theatre in Finsbury Park, formally The Astoria Odeon.

Brian Croft spent a great deal of his working life as a director, or partner, in a succession of lighting rental companies. His early history was rooted in theatre before eventually becoming technical director of the ICA in 1967. Live music came to the ICA and, at the same time, The Rolling Stones' American production manager, Chip Monck, came into Brian's life, and it was the beginning of a highly successful relationship.

At around the same time that the ICA started radically changing its programming, another institution was being created up the road in Camden in a place called the Roundhouse. Jon Cadbury also came from a theatre background and was taken on as a carpenter, then later, in the heady acid days of the late '60s, became the venue's production manager. Jon is now one of the account directors of the UK's largest lighting production company, PRG (Production Resource Group) and he remembers, with slightly scary accuracy, some of the detail of the early concerts at the Roundhouse and also, a little later, at the Rainbow Theatre.

The Rainbow was transformed, from a cinema, into the first rock concert venue with full production facilities, in 1970 by a bunch of Americans who used to be the technicians for the Fillmore East in New York City. Brought over by John Morris, who had been the manager at the Fillmore East, the story itself is fascinating, and over the next 3 years created many future enterprises in the live music rental market.

There came and went, in those few years, a crowd of people that, in all honesty, became the cream of music production. I haven't spoken with them all by any measure, but 'The Innovator' Richard Hartman, a man who has become synonymous with innovation in structural lighting design (and was one of those original Fillmore guys), describes in great detail what transpired at the Rainbow.

The man who carried on the production responsibilities of the Rainbow in its second coming was Dick Parkinson. Dick created one of the initial lighting rental companies called Rainbow Lights, which rivalled that of Brian Croft and John Brown's ESP (Extra Sensory Projections). Whilst he was production manager for the theatre, Annie Pocock became his business partner and John Coppen his general manager, both of whom recant their memories of early touring and production values. Annie was one of the first, if not the first, girls to work in UK live music production, starting her career as a 16-year-old 'runaway' and follow spot operator in the Rainbow.

1 Brian Croft

Brian Croft
© LSi magazine www.lsionline.co.uk

January 29, 2007 with Brian Croft at His Home in Ealing, London

So Brian, from the beginning, please.
I was born in London, and I went to a school in South London called Alleyn's College, which was actually, technically, a public school but almost entirely scholarship as far as I can remember. A 'direct grant' grammar school, that's what it was called. I only ever knew one boy whose parents paid a fee. It was courtesy of the Clement Atlee post-war labour government that my brother got in there; he is four years older than me. He went there immediately after the war in 1945, and I guess I went in 1949, and it was actually a very good school. I mean we had serious doubts about it then, but when you get older you realise that actually the teaching was fantastic because most of the staff had good degrees, many from Oxbridge. It was a day school; the only thing that made it a 'public school' was that the head went to the Headmasters Conference for historical reasons. It was a very extraordinary school, because it had all the trimmings of being a public school, but many

of the kids were rough diamonds from Walworth Road, Camberwell Green and Bermondsey. They were there because they were very bright, so it was an extraordinary school, real untypical.

My housemaster there was Edward Upwood, who was quite a reasonably well-known novelist in the 1930s, a friend of Christopher Isherwood and W.H. Auden, and because of him, there was a young guy who'd left Oxford, been in the Navy and came and got a job there as an English master. His name was Michael Croft, no relation, and he started to do really big-scale plays. Shakespeare really, really properly, massively long rehearsals, and they were of a very, very high standard. He was only in his late twenties, and he was doing these (plays) at school in the summer term. They would do large-scale Shakespeare, and he got the national press there. They would come, because he had been a journalist.

He went to Oxford late because he had already done Naval service in the final years of the war, and then he went to Oxford. He was a late student, and then he became a journalist and then a teacher. Anyway, so he was a huge, I now realise, massive influence, and changed our lives completely, and I, who had never wanted to act, but wanted to be involved, would always do 'walk-on' parts. I mean I was like the Regimental Sergeant Major of Macbeth's army, and I realised I had this kind of talent for organising how things should be done backstage, and I became the stage manager really.

In 1956 Michael left and went back into journalism, a national I think, *Observer* maybe. He had worked for the *Manchester Guardian* as a theatre critic etc. Then there was such a demand from the boys that he actually formed the Youth Theatre, 50 years ago. We have just celebrated our fiftieth anniversary of the formation of the Youth Theatre in 1956, The National Youth Theatre now.

It's called The National Youth Theatre.
Yes, it changed its name about 30 years ago when it became a national organisation. Anyway, he formed this company, and it was really for kids. There were a lot of very good actors at the school; there were people who are now in the profession: Julian Glover, John Stride, David Cameron, not that David Cameron. A lot of kids of my generation who are now actors, Simon Ward, he's a bit younger than me, David Weston, R.A. Hampton, Colin Farrell. They are all still in the profession with various degrees of success. To cut a long story short, the Youth Theatre grew out of Alleyn's, and we put our first show on the year I left school. I was probably going to go to Art School, because all of my friends went to Camberwell Art School, I mean all my best friends who I am still friendly with went to Camberwell Art School. I was pretty good at art, I did it at 'A' level, and I was a bit more academic than them in terms of history of art, and Michael said to me one day, "Why don't you go into a drama school?". I hadn't thought of it at all, but you know how you do things on impulse when you are young, you never weigh it up do you, so he got me a grant from the LCC to go to the Bristol Old Vic Theatre School.

The LCC?
The London County Council, precursor of the GLA (Greater London Authority). London County Council ran all of the education in London, the inner Boroughs

and then the outer Boroughs, Bromley, Croydon, Barnet. . . . There was a woman called Maisie Cobby, who was the 'Inspector', she managed to wangle me a grant to go to Bristol. Normally they would only give it to the London drama schools, for LAMDA (London Academy of Music and Dramatic Art), for RADA (The Royal Academy of Dramatic Art), but I wanted to do stage management, and the courses were terrible there according to her, so she said, "Go to Bristol because it has got the best reputation".

So I went to Bristol and did one year.

What year was this?
1956/57. The year I was at drama school it was the Suez Crisis and the Hungarian uprising, where the Russians put down a Hungarian revolt with the tanks; anyway, I went to drama school.

What is it called?
Bristol Old Vic Theatre School, BOVTS. It's still there, still producing good people. It was attached to the Bristol Old Vic Theatre, so once a term you got to work on a real production at the Theatre Royal, Bristol and it was a good company. Peter O'Toole was in the company and people like that; Alan Dobie, really good actors, high standard. There weren't many notable people in my year; Brian Blessed probably is the only one that anyone's ever heard of. Anyway, going on from there I went into the theatre as a stage manager. I went to Perth Rep, and then I went to Stratford on Avon for two years, two seasons, where I was top of the tree, as assistant stage manager.

That would be before the New Stratford Theatre.
Yes, that was before, it wasn't even called the Royal Shakespeare Company; it was called the Shakespeare Memorial Theatre Company, and it was run by a really nice old guy called Glen Byam Shaw. I did the last year of Glen Byam Shaw and the first year of Peter Hall. It was real high-class stuff. I went to Russia for a month with them, first ever cultural exchange—not exchange but we visited. We took three plays to Russia with Michael Redgrave, Dorothy Tutin, Richard Johnson and Geraldine McEwan.

Stage managing?
I was the assistant stage manager. I never wanted to act. I have acted, it's where everyone starts, it's where you have to start; doing props; being on the prompt copy; doing sound in its infancy. I did that and after two years I left, I went to Canterbury Rep where I met my wife, she was an actress, and then, all the time I kept going back to the National Youth Theatre because they always did a big production in the summer and Easter holidays.

At the school still?
No, the first production in 1956 was at Toynbee Hall in Aldgate; by 1960 we were getting quite big and playing seasons in the West End, and I was getting paid to go back and be their production manager.

Did they have a home theatre?
No, they didn't have a home, they had offices. Nice offices in Eccleston Square paid for by a grant from Gulbenkian,[1] Arts Council money, Department of Education money; they were reasonably well funded and they had a small staff. Michael was the Director, he had a full-time secretary, male secretary, Dave Fournel, who I still see and we all used to go back and do the big shows. They were actually on Shaftesbury Avenue at that time. We did the Queens for two years, did a modern dress version of *Julius Caesar*, which he had originally done at Alleyn's in the open air outside the cricket pavilion; it was really revolutionary, with huge armies, swirling armies and real guns, real blanks, because we had a very strong cadet force at Alleyn's and no health and safety!

Anyway, in between, I would go back and do professional jobs, and then I would work with the Youth Theatre whenever possible, and we did lots of tours. Toured Holland, Germany, Italy. Half the year I was working as a paid employee of the Youth Theatre being a production manager, which meant overseeing building of sets, overseeing the lighting, putting all the ingredients together, and we were on Shaftesbury Avenue and doing big tours. We represented England at the Berlin Festival in 1961, so that was like the '60s, and then in the middle of the '60s I got to actually work for them full time for three years because Michael got a grant from the Gulbenkian Foundation and had three employees for three years. I had just got married, so that was like 1963/'64/'65, so I had a permanent job working in Eccleston Square all the time, and then going on tour and doing lots of big, big things in the '60s.

At the end of the '60s when that fund ran out, this is the key turning point in my life. I was very friendly with a guy called Michael Kustow, who was a theatre director I had known from Oxford. He and a good, good friend of mine, who was in the Youth Theatre, Geoffrey Reeves, they both did good English or History degrees, and then they both did post-graduate degrees at Bristol University to become theatre dramaturges, the academic side, the intellectual side of theatre. Anyway, Michael Kustow, quite out of the blue, got offered the job of running the ICA, the Institute of Contemporary Arts, which was a little private kind of very exclusive group of art collectors, based in Dover Street in Mayfair. A drinking club, and the chairman was Sir Roland Penrose, who was a wonderful, wonderful man, who was Picasso's biographer. He was a friend of Picasso and was married to Lee Miller, who was an American black-and-white war photographer, and all part of that 1930s Surrealism movement: Man Ray, Marcel Duchamp, Dada, Max Ernst and so on—Roland founded the ICA with Herbert Reid.

Anyway, Roland was smart enough to realise that they had to grow up, so they were offered this building in The Mall where it still is, a huge building, government building I think, and they got Michael Kustow in. The ICA had always been fine art and poetry basically, and he had the foresight to bring in a theatre director to be the artistic director of the ICA, which was very frightening for them and for Michael. Well he asked me to go and be his tech guy, and so I was the technical manager of the ICA from when it opened, in 1967, through to 1971. It was this fantastic time, this wonderful, wonderful time; we had these hugely successful exhibitions where people were queuing halfway down The Mall to get in. It was '67/'68, it was Flower Power, stuff happening at the Roundhouse, and because it was that era we had to, not just 'do art', it was principally fine art painting and sculpture, but we had a little cinema there where they showed experimental movies, and we started to do theatre. We did a play written by Adrian Henry, the poet, and we built a temporary theatre at the end of the gallery for that. By this time I had taken on an assistant, one of the kids from the Youth Theatre at LAMDA, that's the London Academy of Music and Dramatic Art, and I took him on as my assistant, and he was into music and he said "We have got to do music". This was a multicultural centre, anything went, you could have concrete poetry, you could have nudism, pornography, whatever, you know, everything was art so we had to have music, and we started doing concerts in the gallery.

At the opening party, Mark Boyle,[2] he did the lights, the Mark Boyle Light Show; he was actually an artist, he is dead now but his family still do it, its extraordinary. Anyway, it's a long story but he also did liquid light shows for Jimi Hendrix and Soft Machine . . .

Pink Floyd?
No, that was Peter Wynne Willson,[3] who is still around. But no, Mark Boyle, he was an artist, a very temperamental Scotsman, and he had a very dynamic wife. They were very pushy but they were good, and they used to do these liquid light shows, they did Soft Machine and then Soft Machine got a job as a support act on a Hendrix tour. This must have been around '67/'68, sort of round that time, and therefore Hendrix used his lights, too.

Can I stop you for a minute? Your assistant you brought into the ICA was called?
John Brown. I brought him in, and he was into the music, he was that much younger than me.

Were you not into music?
Well, I was into jazz really. I wasn't really into pop music very much, but I always liked The Stones. We did some good things at the ICA. We did big concerts with Julie Driscoll and Brian Auger, Centipede, and we also did very experimental modern music, Philip Glass and all that 12-tonal stuff, weird you know, anyway we did everything, everything. At the opening party, which I guess must have been in '67, we had a big opening party, and this was in the art gallery, a fully functional art gallery with Picassos and Andy Warhols on the wall . . .

This is at the ICA?
Yeah, it was huge, one massive gallery, it was probably 400 feet long, a big old gallery; it's broken up now into small rooms but it had a clubroom upstairs, and a restaurant and a cinema, which is still functioning, and it was a very, very exciting time. At the opening party we had The Nice, Keith Emerson throwing daggers into his Hammond organ and the Bonzo Dog Doo Dah Band, so that's the kind of

Brian scratching his beard looking at Jim Morrison at a press conference
© Russell Burrows/John Brown

thing it was—it was like old bohemian meets the revolution, you know, whatever was going on, Flower Power—blah, blah, blah. At the same time Jon Cadbury was with Joe Boyd up in the Roundhouse doing their thing up there, and he and Ian Knight and Hugh Price, what was that thing called up there . . . at the Roundhouse, Implosion was it?

Yes, it was.
Anyway, so John Brown and I started doing liquid light shows in our spare time, not that we had much spare time, we used to do gigs at the weekend, mainly at colleges. We would do Commemoration Balls and May Balls and things like that, and they used to book big bands. We did The Who with about four liquid light projectors and no other light (embarrassing), the Crazy World of Arthur Brown, all those acts that were on that circuit, and we got more and more into it. John got particularly into it.

We did a press release concert with The Doors in the gallery; in '68 the ICA was the place to be. Jim Morrison did this press release and played, I think they did half a song! Then along came a band I had never heard of called the Chambers Brothers, and they had a lighting guy, production guy called Chip Monck.[4]

They came from the States, and I met Chip, and he was just a whirlwind of energy. No, I had never met anyone like him, and he was a bit awe inspiring, he kind of made you nervous a bit, but that was good you know, and we just did this one-day show, we built a stage in the middle of the bloody gallery overnight, you know scaffolding, and he and I became very, very close in the course of the two weeks.

The next week he said, "This now works very well", and made a phone call and the next week we had Country Joe and the Fish, who was also at the height of his fame, doing anti-Vietnam stuff, and we were part of the revolution there.

It was all terrifically exciting, and in the meantime, every day we opened the doors at 10 o'clock and thousands of people would come in and see the paintings, I mean we had a series of exhibitions. The first one was called The Obsessive Image, it was about figurative art; we had loads of things, themed exhibitions, anyway everything was a marvellous blur.

In 1969 we were fitting up a show, and I suddenly realised on a Saturday afternoon that I hadn't got any crew left. They were all supposed to be putting this painting exhibition up, and they had all disappeared, and I wondered where the hell they had gone, and they were all at The Stones concert in Hyde Park.

So I didn't think anymore about Chip or anything like that, and then just before Christmas in 1969 I got a phone call in my office, I walked in, and a woman in there said (my office was down in the gallery) "I've just had a phone call from The Rolling Stones. Could you call them back?". I said, "You mean the magazine?" She said, "No, I think their office, they said The Rolling Stones office".

I got this call from someone I didn't know, saying "I don't know what I am doing here, but I've just been asked by Chip Monck, in America, to look you up because he needs a lot of stuff doing".

What had happened is that Altamont had so freaked the band out with the death of that kid, that they said, "We want to go back to our hometown and we want to play a couple of ordinary gigs in London", and they had called the office in London and said "Organise it".

Well, they were to put two gigs on, one was at the Lyceum, which was a dance hall, a Mecca dance ballroom, and the other one was the Saville Theatre, which was a theatre in the top half of Shaftesbury Avenue, its now a cinema complex, a multi-screen complex, sort of halfway between Cambridge Circus and where the Shaftesbury Theatre is, and by remote control I had to get this stuff together to do the lighting. I guess I had done a lot of lighting in the Youth Theatre, but only because I was the production manager; we didn't have lighting designers as such, theatre didn't for years and years. The lighting designer in mainstream theatre didn't really come on stream until the mid-'70s, till then it was just done by the director and the chief electrician. He would say make it a bit blue over there, and the chief electrician would just do it.

Where was the desk?
The desks in the theatre were always built in . . . no, I beg your pardon, manual desks were backstage, big old levers up in the gantry. Just at that time electronic boards were starting to come in, Strand Electric boards, and they were always located front of house, and that's not even true, first one I saw ever was at Bristol Old Vic, the Theatre Royal, and it was in the wings, it was in the place of the old board, but it was a kind of weird console, with levers and weird stuff, like an organ keyboard, electronics . . .

Where did it come from, do you know?
Strand Electric. Was a big company, gone now, finally.

Sorry, back to The Stones.
Where were we? ... So this was all a complete blur, this was just before Christmas, and Chip arrived but his stuff had got held up in customs. He didn't have his intercom, so he got Theatre Projects, who I used to use, they were a new lighting company and a rival to Strand, he got them to build him an intercom in three days.

Why did he want an intercom?
Because he used to 'call' the show. That was an absolute innovation, there were no cues, he would call the show on a headset to the board operator and the follow spots, and his calling was amazing. He would call the follow spots just like they do now, but he was very precise, very military, and he picked up what he knew about touring, big-scale touring, from the 'ice show' business and the circus, Barnum & Bailey's, Ice Capades, Holiday on Ice ...

From America?
Yes, he was American, Chip was a New Yorker, Bostonian actually but living in New York, and he had done this 1969 Stones tour of America, which was the first tour that I can remember that really had production of any kind, and that was the one that finished at Altamont. So he arrived and we did these two shows by the skin of our teeth with The Stones in a smallish theatre.

What were the lights like, what were they on, trees?
We had them on theatre bars; they were old theatre lights.

They weren't PAR lights, were they?
Didn't exist. Did not exist. We used Strand Patt 243s, 2K Fresnels with ridged lens, that big, 10-inch or 12-inch Fresnel, 2K bulb focusable but soft-edged and very heavy. He used Century 2Ks in the States, sheet metal not cast, half the weight. Here I got all the stuff together, that was my job, I got all the stuff together for him from Theatre Projects and hung it and focused it, me and two other guys, Roy Lamb was one of them, a kid from the Youth Theatre. John Brown and Roy Lamb were both people who had worked for me, had been amateurs in the Youth Theatre, and I spotted them and got them into the profession, first of all at the ICA.

Roy Lamb of course went on to be ...
Roy Lamb is now production manager for Bryan Adams, Plant and Page, and The Who, did a lot of work for The Who. Before that he was Edwin Shirley Trucking, and he was also lighting crew on several Stones tours that I did subsequently, and he is still doing it. So anyway I think that was it; I think there was me, Roy Lamb and John Brown, were the crew, with the theatre crew, and we hung these heavy old lights up and focused them, it was all mad, mad, mad, the shows were good.

So Chip got an intercom. Did he have a desk?
The desk was run from the wings, but strangely enough Chip didn't operate the desk himself. Other lighting designers used to operate their own desk and call the

follow spot cues, but Chip called everything (to the board operator and the follow spots). And he didn't say "Go cue one" or "Go cue eight", whatever, he would read out the circuit numbers.

A few days after The Stones gigs, Chip's stuff cleared customs, and we went on to use it for a Crosby, Stills, Nash and Young tour starting at the Albert Hall. Now I got this desk for him, I got it built by Electrosonics in Woolwich, and it was a very primitive two-scene pre-set, but it was the first of its kind. It was actually the forerunner of all future lighting desks, inasmuch as these were the first use of portable thyristor dimmers in Europe, and I got Electrosonics to build a rack of thyristor dimmers. Before that they were all mechanical desks—rheostats; resistance dimmers. All this thyristor lighting was coming in, but they were only using it at that time for things like automated house lights, so you push a button and the house lights would dim, or exhibition halls or foyers for hotels, that sort of thing. I could see that it was coming because in America you kept hearing "thyristors, thyristors are happening".

So this is instead of 'on' or 'off'?
They weren't on or off, they dimmed electronically, thyristors had just been invented. Effectively, before that to dim a light you had to do it through a rheostat, which is just a way of decreasing voltage. Now we were in that transition period, I got them and went into it headlong. I had almost no notice to do this CSNY show.

They built this desk, they built a dimmer rack. It had no sides, it was just a standard instrument 19-inch frame with the racks mounted in there, and the multicore was wired on a terminal strip, which I had to screw in wire by wire each show.

So we had done these two Stones shows, and then Chip said, "Do you want to do a CSNY tour starting at the Albert Hall, and I think three days in Scandinavia?" and that was it, culminating in Stockholm Concert House, Sweden. I said, "Yes, fine". I took a few days off from the ICA, went out to the Albert Hall, and Roy Lamb told me how to wire. Roy wasn't going to come with us, it was just me, so he showed me how to wire up this multicore using a terminal strip, and it was like orange and white goes to one, green goes to two, solid orange goes to three, blue-and-white stripe goes to four—blah, blah, blah—on stage at the Albert Hall about an hour before the show with a little screwdriver. I have never been an electrician, I never have been, never was and never will be.

Roy should have done the tour really, it was really a nerve-racking time. There was this multi-conductor cable; first use to my knowledge of a multi-conductor cable, before that it would be a separate wire. The reason that the boards were always on the gantry in the wings is that all the wiring was hard-wired back to the board in conduit, but this was the first use of multi-conductor cable for control. We still didn't have multi-conductor for power, that was still individual cables about that fat (Brian indicates a circle with fingers about one inch thick) that were taped together to make looms, and they would drop off every two feet, and there would be an individual, great big 15-amp connector to each lamp. There were no multicores, no multicores for power, but there was a multicore for control.

So we took this show from the Albert Hall to Europe, and then they all went back to the States. That was Leo Makota, the tour/road manager, and Steve Cohen (one of the many Steve Cohens) was the LD (not the one who did Elton John and Billy Joel later). San Francisco-based they were, bikers you know, Leo was a biker. They all went away, it was like a mad, wacky experience for me. Then I went back to work at the ICA.

'69 still?
January 1970, I think, we did Christmas and New Year's Eve or something at the Lyceum and then about a week later, I went out with Crosby, Stills, Nash and Young. I just took everything in my stride, you know what it's like, I wasn't that young, I mean by this time I was 30, over 30, and it was just a job really, but it was kind of amazing inasmuch as the energy and the money, not personal money, but the fact that you could spend the money and no one was going to ask questions, you know what I mean, that's what attracted me to the whole rock-and-roll business.

It was experimental really.
Yes, it was like that, and that's why rock-and-roll would never have grown out of legit theatre 'cos legit theatre is incredibly well controlled, budget conscious, rather conservative in many ways in those days, technically anyway not acting wise, but it was just this energy. I don't know if it was coke-driven or what, which it certainly was most of the time, and I was on the outside of it going, "hey I'm just an arty guy you know". I was married with three kids by now.

Anyway, what happened is Chip was such an incredible powerhouse dynamo, but he had absolute zero sense of getting it done in a timely manner, he would fail absolutely, he would produce the world but he wouldn't produce it on time, which is very strange because his sense of timing when he is calling the show is extremely precise, but he didn't get the big picture!

There's a famous thing I used to tell all my people right up till the end when I retired, every now and again they always had to be told this, "You have to get your shit together at the right time. It doesn't matter if it's only *just* at the right time, but it's got to be together at the right time, because if you don't, at quarter past eight tonight Mick Jagger is going to walk on stage in the fucking dark, and you are all going to get fired. You know what I mean, 'cos the show ain't going to get postponed because there's no lights, it might get postponed because there's no sound, possibly, but it will happen in the dark with two follow spots and then one's out of a job, so whatever you do, even if you cut half the rig, you have got to be ready at show time, 'cos show time is showtime".

Now that came from being in the theatre where the discipline is very, very strong. There are absolute rules in the theatre that 35 minutes before the show everybody has to be in the theatre, all the actors even if they are not on for another hour. It is called the half, and 35 minutes is the half, 20 minutes before is the quarter, the five is 10 minutes before and then beginners is 5 minutes before, and that's it, and those are unbreakable rules whoever you are, whether you are

Laurence Olivier or whatever, and that I had and Chip didn't, and we made a fantastic partnership.

In the summer of '70 he asked me to put an English crew together to do a major European Stones tour, and I took four months off from the ICA. I don't know how they let me get away with it actually, obviously I was unpaid for this time off, and I had a substitute, someone deputised for me, eventually he got my job in the end, that guy, but it didn't matter, by that time Michael Kustow had spent many millions of pounds of the Arts Councils money and had a great time for over three years. Anyway, I went off and did this long tour, and my crew was Roy Lamb, Robin Murray and Jon Cadbury it was going to be, but Jon backed out for his own reasons so we got a guy from Theatre Projects lighting warehouse, Paul, to come with us, and we were four and there were about half a dozen Americans who had all done Woodstock.

All lighting as well?
Yes, these were lighting people.

So there are 10 lighting people?
Yes, big crew. Three trucks, three artics, first tour I have ever known with artics.

Do you remember who the trucks were from?
Yes, for this 1970 Stones tour, someone had done a deal with Hertz, a sponsorship deal maybe, and the logistics were done by a Hertz employee, and they used their own drivers. Earlier that year though on the CSNY tour, Artist Services did the trucking. Don Murfitt, who is dead now, and his partner Jerry Slater.

Don and Jerry rented an artic and drove it (double drivers, tight run) from London to Stockholm and then Copenhagen. Then in 1973 they had grown, and they did the Stones tour as a company, they didn't drive.

Right. I worked with Don, I tour managed Adam Ant, and he was the manager.
Yes, Don was a lovely guy, real fearless chancer. Here's a funny story for you, you probably won't publish it. This is much later, we had put some shows together for Robert Paterson—do you remember Robert Paterson? He was a show impresario, he was upper crust, he had a beautiful house in Notting Hill, big, big house, and he had wanted to put on big open-air shows in the vein of the Freddy Bannister shows. He was a sort of competitor to Freddy was Paterson, a lovely guy, a real gent, bit of a drinker but a real gent. So me and Paul Staples, who was a set designer, and Don Murfitt used to go looking for locations all round the country. We went to Hickstead, which is a show jumping place, which wasn't suitable, we went to a motor racing track at Thruxton, halfway down the M3, we looked at that, aerodromes. I can't remember what year this was, it would be around mid-'70s, so anyway every night we'd finish up reporting back to Robert in his big house in, opposite Prue Leith's restaurant in Notting Hill, near the Portobello Hotel. If you remember being there, you weren't.

Anyway, we would go back and report to him, and it would be late by then, be about 11 o'clock at night, and he had this lovely upper-class assistant girl, nice woman, and just as we were leaving, going down the steps, all of a sudden this girl came running down the stairs after us, and she says, "Oh chaps, chaps, I wonder if you could possibly help me?" Don says, "Yes, what's the matter darling?", and she says, "I just need someone to help me close the safe", and Don says, "Close the safe, darling, close the safe! I've fucking opened a few in my time . . .". I just collapsed, and she was completely above it all . . . Anyway, where was I . . .

Lovely, God bless him . . . Can I go back just a step to The Stones gigs in London after Altamont? You talked briefly about getting the gear from Theatre Projects.
Theatre Projects was a new hire company, they were an 'everything' company. They were founded by a young Richard Pilbrow,[5] bit older than me, I saw him a couple of weeks ago actually, he is in his seventies now, and he founded the company. He now lives in America and still runs Theatre Projects, he has just retired actually, Chairman Emeritus or something, ex non-executive chairman, and they are theatre consultants who consult on building new buildings 'round the world, they are consultants to the architects on building places of entertainment. He lives in Connecticut, and they have a big office in London.

He's English?
He is English, very thoroughly English, a man of the theatre; he started off with what's now the Royal Exchange Theatre in Manchester. I can't think what they were called then. He was one of the very first lighting designers, and he formed the company, and I think his wife had a bit of money.

And he rented to theatres?
They did sound and lights, and they rented equipment out of a tiny little warehouse in Neal's Yard, which is now full of trendy soap, aromatherapy and all that stuff and trendy media places, but there was a nice, old-fashioned Victorian warehouse with trapdoors, ropes and pulleys, and you used to have to pull up into Monmouth Street in your truck—can you imagine that now, Monmouth Street?—load it and no one would bother you, no parking metre attendants or anything!

Anyway, they were brand new and they took on Strand; Strand had a huge rental facility in Kennington Oval, massive, and all their stuff was very old, and they dominated British theatre, and along came Theatre Projects, young bloods, and took them on, and of course my natural inclination was to use them, and I have been a friend of Richard's ever since.

ICA used them?
No, it was as an independent lighting production manager I used them, we used to rent our stuff from them whenever we did shows on our own with John Brown who was my assistant. Come to think of it we probably did rent from TP for the Adrian Henry play about Apollinaire.

Did you get the liquid lights from them?
No, liquid lights were slide projectors, single slide projectors, two-and-a-half-inch square slides, we used to rent those from various camera shops and things, down Tottenham Court Road, and we used to return them covered in shit, and they used to really complain like crazy, and we would say "Well we are going to get them back again next week, don't bother to clean them".

About liquid lighting, can you tell me a little bit about them?
Well there were two kinds, there was the sandwich kind where you had two pieces of photo glass, and you had this sandwich in between, and the heat made the bubbles explode, and that was what Mark Boyle used to do, we got that from him, and the dyes you used were paint dyes, glass paint. You used to have to throw them away at the end because they would just crack, they would go hard, brittle, but the other liquid light shows were done differently.

There were two companies in the States, one was called the Joshua Light Show, which was Josh White, who is now a big TV producer in the States, and Joe's Lights, which was San Francisco based.

John, my assistant at the ICA, got very friendly with those guys, Joe's guys, and they used to do a different method in which they got big convex, or is it concave?, bowls. They were like the fronts of big clocks, big grandfather clocks, a shallow glass dish with good glass, and into that they used to put oil and dyes and water, and they would sort of swirl it around on an overhead projector so the light was coming up through this thing that they were manipulating, and then beamed out onto a screen.

I mean the problem with all light shows is it was fairly short-lived, and it needed to be short-lived because the light sources were not bright enough to compete with any kind of stage lighting. As long as the hippies were happy not to see Dave Gilmour—no, it wasn't him it was Syd Barrett then—as long as they were content not to see the band. . .!

I always had my doubts about the value of liquid lights because it was an art thing in a dark room and it was fantastic; the Mark Boyle shows were art, he did the Soft Machine because the Soft Machine were an arty band, but because they then supported Hendrix, he was doing a main tour with these lights, and everyone remembers that.

Probably go down well with trance music now.
Yes, it's perfect for people dropping Es and stuff, exactly, but for a live show it was a non-starter, we never went anywhere near it with The Rolling Stones, that's why we had to have real light. Joe's Lights, I think, were resident at the Rainbow at Finsbury Park, it had a big gantry at the back, and that's when Richard Hartman first came along; those guys, Chris Langhart, Richard Hartman and Joe's Lights, they always had back projection. That was an offshoot of Fillmore East and West.

The Fillmore down on the Lower East Side, which is where Joshua Light Show started I think, Joshua was New York and Joe's Lights were San Francisco.

Anyway, we were at the ICA, and we had a company called Extra Sensory Projections. What happened is John said, "Look I want to do this rather than stay here at the ICA", so he left. Well I was married with kids, and I had responsibilities, and I was on a salary. He left and got a little shop for ESP in the Wandsworth Road. Later when we moved to Blackfriars, a guy called Dave Cohen took over the lease of the shop. Dave started one of the original lighting companies, which became Zenith[6] eventually; he brought in Paul Turner and Jon Cadbury, all originally from the Roundhouse. I didn't know Dave that well, but he was a nice guy, and I think that he started or became Zenith, and Paul Turner bought him out.

So you were doing The Rolling Stones in Europe?
We did the Rolling Stones for a three-month tour of Europe, major, major venues.

The equipment, where did that come from?
Mainly from the States, yes it was, all from the States, shipped by sea, Century 2Ks, Super Troupers, and we all ran the spots every day, and we put the show up every day. We had this massive aluminium superstructure to erect every day, to hold the lights and some drapes, used for outdoor gigs and indoors (about 50/50); didn't need to rely on roofs indoors and riggers. It looked great, a big shiny aluminium 'statement' 80 feet wide and 40 feet high. It took a while to put up every day, and we had to haul all the Supers up onto spot positions, and we ran the spots ourselves because there were no English-speaking spot operators in those days . . . There were no experienced stagehands in those days either, just schleppers. Even with a large crew of ten, we were exhausted. We carried our own forklift truck and butane fuel; can you imagine that? And a ramp to load it on to the artic, and those trailers were full height, not step frames.

Did The Stones own all this equipment?
No, they had none—well they did own the aluminium superstructure, custom-built by Access Equipment, architect designed, and the drapes and stuff. The lighting was rented from Chip, and Chip rented it from Altman, which was a Yonkers-based company, a Broadway lighting company, and they bought the stuff on Chip's specification, on the basis that they would rent it out to The Stones for three months, six months, whatever. So we did that big tour, and we all ran spots, and there must have been—well, how many were in the band, five?—so two on Mick, no, he had Gladiators, which was an even bigger version, bigger housing, powerful, so we must have had six Supers and two Gladiators.

All flight cased?
The tour started in Malmo, Sweden, but we rehearsed in Copenhagen because the overall promoter was a Danish promoter, Knud Torberjensen of SBA, Skandinavian Booking Agency, and we rehearsed in the old Forum, which is like an ice rink in Copenhagen. We built all the boxes, bought local wood, got a local carpenter to build them, and I remember the whole crew painted them a bright, bright yellow, school bus yellow was the paint.

In '73 though they were all flight cased by Showco, who nobody had ever heard of; they were another company in Dallas that built very good fibreglass cases.

So you built the flight cases?
The local carpenter built the flight cases. Anyway, it's a long story, we did football stadiums, we did some indoors, some outdoors, arenas, a big long tour. That all over, everyone went away, and I went back to the ICA. Then in 1971 we did this amazing tour promoted by John and Tony Smith of English dates, and we put everything we had, everything, sound, lights, backline, into one 16-ton, 22-foot straight truck, rigid, driven by a good old truck driver, Jonny, who drove The Stones' mobile recording studio truck.

And we played, this is the amazing bit, we did a tour of England, did about 10 dates with a £1 ticket price.

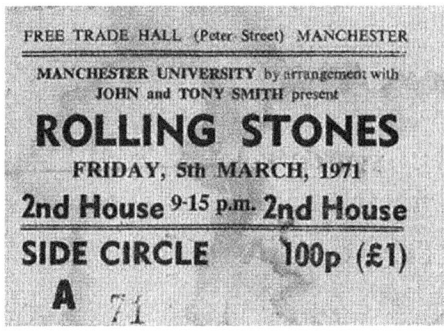

Old Stones concert ticket, 1971

Was that a lot at the time?
No, it was fuck all, I mean £1, because the boys they were like "we want to get back to the roots", they always wanted to play clubs; we actually did Greens Playhouse in Glasgow, we did Leeds University, we did the De Montfort Hall, Leicester, we did the Roundhouse in 1971, and when that finished, I decided the time had come to leave the ICA.

I went back to the National Youth Theatre, and we opened the Shaw Theatre in Euston Road, and I was there from 1971 to 1973 as the production manager, because we had a professional company as well called the Dolphin Theatre Company which did straight plays. We did some Shakespeare, the Youth Theatre in summer seasons, the good old days, we did Zigger Zagger, big hits like that, and then in 1973 John Brown called me (he was my partner), and he said "Are we ready to go, it's all taking off, can't you leave and I'll guarantee your salary?". So I did, I left the Shaw in 1973, and we formed a company based in Wandsworth, and we changed our name, our partnership was called Extra Sensory Projections. The note paper that we had was all multi-coloured, and it cost so much money we daren't ever type any letters, and if you did you couldn't read them because the typing and colours all melded together.

So we changed the name and became ESP Lighting Ltd, and that was it, that was our first company and that was 1973. Then immediately I got asked to go to Australia with The Stones as crew in the spring of 1973, February and March, and then I came back and went back to the Shaw. By this time I had already done three things away from the ICA.

Now that I had been to Australia, it was like, I have to get out of the theatre and into the rock-and-roll business; it was partly money, but it was partly the fact that these guys had the balls to spend the money not just on our own wages as it were, but on three artics. It was an act of faith really, and people like Chip commanded that kind of trust you know, and I thought this is powerful stuff. So I really got into the business unlike most people that I remember talking to; I didn't get into the business through the music, I got into it through the opportunities presented by these kinds of budgets and the scale, just the scale.

We went into Australia on our plane with all the gear on it, and I used to fly on the plane, a 1948 Super Constellation, bloody death trap this thing; it didn't have a toilet, it had a bucket in the corner and one row of seats bolted down amongst all the freight.

All the crew on it?
No, about three of us went, the rest of them went schedule, the rest refused to go on it. It flew out of Honolulu, it's alas another long story in itself.

Were they jets, Constellations?
No, four engines and each engine was like 18 motorcycle engines back to back in a big circle, blue flames shooting out the back it was so noisy. We did a 22-hour flight from Honolulu to Auckland, and you couldn't speak to anybody, you couldn't sleep, and if you wanted to pee you had to pee in a bucket; it was freezing cold, there was no heating, and we were only about 200 feet off the waves most of the way, no point in going higher evidently!

No!
Apparently not, just further to come down. Anyway, the mad American crew, a drunken pilot, a long story that.

Let's hear the story from the beginning then, tell me about the Constellation aeroplane 'cos that sounds mad.
Well what happened was, at the very last moment the Japanese shows got blown out because Mick Jagger had his visa application turned down, so he couldn't go to Tokyo.

It's a stupid case, the Japanese ministry of the interior, whatever it is, turned his visa application down on the basis of an English drugs bust, I think there was some marijuana bust, but because of that a huge chunk of the budget got wiped.

1948 Super Constellation
© Ron Cuskelly

The earning income of Japan shows?
Yes, the whole of Japan. We did a Nicaraguan benefit for free with Bill Graham at the L.A. Forum, then we did three days in Hawaii, very good shows. People who know remember them as being The Stones at their hottest, with a mad and crazy American 'up for it' audience, in a 20,000-seater and hot, hot shows.

I stayed in Honolulu, Patrick Stansfield goes with Peter Rudge and Co. down to Sydney to re-organise the tour, and all the time we are telexing backwards and forwards to cut stuff; we have got to cut the budget, cut the weight. In the end Patrick organises a plane that's coming from Florida because its cheap, it's the cheapest deal he can find.

So I am in Honolulu with Jim Gamble, Jim Chase and a few others, and I stayed at my post. Jim Gamble and Jim Chase were Tycobrahe Sound Co, they owned it and ran it, one was the front engineer and one was the monitor engineer. Tycobrahe was a very good sound company.

The rest of the people, like Chip Monck and all that lot, went cruising 'round the Pacific and down to Tahiti on 70-foot luxury boats, owned by rock stars and drug dealers, you know, and I'm going out to this hut at the airport every day to receive the next telex from Patrick. Backwards and forwards. They had weighing scales and I'm weighing every box, this is all in a freight handlers' Quonset hut who are charging us for storage, and it keeps changing. The manifests keep going backwards and forwards, and the plane is later and later and later, and we

Jim Gamble's business card for Tycobrahe
© Jim Gamble

have got an opening date (for a show) in Auckland, and it's getting nearer and nearer. Eventually the plane turns up, and it's this 1948 Super Constellation, petrol engines belching out flames.

It comes in and it's very suspect, grounded sort of temporarily by the FAA, and this is one of the reasons they were late, they had stopped in Ontario, California, and had a kind of hole cut in the roof for some navigational instrument, which was almost like a sextant. It was pretty primitive 'cos they didn't have the correct navigational instruments to fly the Pacific. The captain we subsequently found out was a drunk, the first officer wore a Stetson and 'nuttin will trouble me boy', and the navigator was Rick Ricotelli. Eventually Jim Chase says to me, "I'm not flying in that plane".

So they wouldn't fly in it, and some of the others wouldn't fly in it either. David Noffsinger said he would come with me, he was one of Chip's guys. Chip had been sent home because of complications with his visa.

So we are at the airport, we load the plane, and then there's a huge question mark about what the maximum take-off payload can be with maximum fuel. This thing only does 150 miles an hour, and we have got to get to Auckland via somewhere; Samoa it turned out in the end, American Samoa, Pago Pago it's called. So the range and everything is calculated, and they wouldn't let us take off, and then the second officer came to me and said, "What is the payload?". In the end we had this kind of mantra, and so when the FAA asked, "What is the payload?", we would say "11,000 pounds just like on the manifest" but we knew it was over. Why I am getting on a fucking plane that is. . ., I mean, the show must go on, okay, but we were facing possible death.

In the end, we simply took seven Super Troupers, the sound system and the backline, everything else we had to leave and send back by boat to L.A. with the lighting gear. It was okay because basically all the Australian gigs were outdoor, and a lot of them were pretty much daylight gigs, they were evening shows but it was very light, didn't get dark until 10 o'clock, so we got by with just the Supers.

How heavy was it?
I don't know, 50% more or something, I didn't understand it all, it was all to do with the amount of fuel this thing had to take on, which was hundreds of gallons of fuel. Eventually we had loaded it, and it was all strapped down we were ready to go, but the local FAA won't give us permission to go, we hadn't got permission. Then very early in the morning there was a change of shift, and this jeep takes the guys who had been supervising the airworthiness of the plane back to the airport, and the captain said "Is everyone on that's going?" and I said "Yeah". I was sitting on the step behind the cockpit, quite a big cockpit you could walk around with a navigation table in it, and the doors are open, and suddenly the pilot just guns it, and takes off with the door open and the little jeep is coming out to meet us to tell us we can't go! Guns it up the runway and we take off, of course it's quite a big airport, Honolulu, a nice long runway.

When we get to Pago Pago we have to refuel, and we take off from there, and it took, seriously took, 20 minutes to gain any height at all, we were literally only metres above the sea, we went out over the sea on our way to Auckland.

Was he doing that deliberately?
No, the guy turned round and said, "When we come back I sure hope we don't have to take off the other way"; we looked round and there was a mountain range behind you, you see, you couldn't have taken off. If the wind was blowing the other way we wouldn't have been able to get off the ground.

We had to go straight to the gig in Auckland, it was that close. But the Constellation was unbelievable, it was cold, no heating, no soundproofing, couldn't speak to anybody, couldn't read really or anything, but it became a big thing, and we had a tongue painted on it when we were in Sydney, and then we flew all 'round Australia. Of course the band, Keith, Bobby Keys and Jim the trumpet player, all came on the plane. When we are halfway across from Perth back to Sydney, Keith says he has had enough now, "I've had fun, drunk all the beer" and all that, but you look down and there is nothing but desert, well you can't stop here Keith, it's a long way, its like 12 hours! So that was a bit of a nightmare. There's me up till now I'd been doing a lot of straight theatre, and it was almost like an out-of-body experience seeing these mad frontiersmen and hippies, and I'm part of it, and I'm risking my life for the glory of The Rolling Stones. It's like madness when you think about it now, but we had some great fun, and the important thing about that tour was when I talk about entourage now—the whole group, crew, band and road-ies everything, everybody, tour manager, probably 22 people—we would all go and have dinner together, that was what was nice about it, you would all sit down and have dinner together because it wasn't an unmanageable number, whereas it's hundreds now in the touring party.

So then I came back from that and went back to the Shaw. A few months later I left to join John down in Wandsworth Road, and then the next day got a phone call saying could I put together the 1973 Stones tour. This was from Pete Rudge because they had fallen out with Chip big time because he not only had no sense of time, big time, but he could never get his budgets together, and his accounting, he wasn't really profligate, but it was just that he didn't produce the accounts in time. Peter just used to get so furious about the total lack of budget control, and so Peter asked me to put the tour together, my company, so I joined John Brown and within days we were doing a major Stones tour and we did everything in 1973.

You sub hired?
Tycobrahe did the sound, they had been to Australia with us, and we did the lights and all the structure for the ground-supported lights.

I invited two Americans onto my crew, both great engineers. One was Bruce DeForest; we were friends from the '70s Stones tour (Chip's Woodstock crew) to look after the ground support. The other was Richard Hartman, who was already living in London (Rainbow Theatre), to look after the mirror. The remainder of the crew were my guys, and from then on we were full-tilt boogie.

I mean this is how it goes; when I get down to Wandsworth Road I'd say, "What's on the books, John?" and he said, "Well at the moment we're doing . . ." and it was something like David Bowie, Elton John, something for The Who, a band called Led Zeppelin, and I would say, by the way, The Stones too because we've just got the gig.

When you look back at it, it was mad, mad, mad, and we didn't own any gear, we rented the gear, we would rent the gear from Theatre Projects or whoever, Bob See,[7] we rented a lot of stuff from Bob See, in fact he gave us the David Bowie gig. We would rent stuff from Bob See, and we would be the London end for See Factor. We did Bowie for him, and then we had enough money to buy some equipment. I think those Honolulu shows were the first shows I remember ever using PAR cans. I think they were the first use because I think Chip had previously used them on the 1972 tour.

So PAR cans came out of America?
They were film lights, they had existed quite a long time, it's an 8-inch, do you know why they are called PAR 64s?

No.
Only the Americans could do this, the measurement across the face of the bulb is counted in one-eighths of an inch, not metric—they don't know metric, they still don't know metric. One-eighth of an inch so it is 64 one-eighths of an inch, i.e. 8 inches, and there is a PAR 56 and a PAR 36, and these are parabolic anodised reflector lights that were still being used. They were introduced into the entertainment business in the film industry just to bash out white light, and they were housed in this can, and it was only that deep (Brian indicates a depth of about 8 inches).

From the nozzle, from the housing to the front about 8 inches. so you couldn't put a gel frame in front because it would melt, so what Chip did, and I think he was the first to do it, he had top hats made that went into the gel frame and extended that distance by another 8 inches, they travelled separately in a different box, and you would put them on every day, and they clipped in the can and the gel went in the front of this extension, the 'top hat'.

So it didn't melt.
To give that extra space and more ventilation as well, and there were quite a lot of holes round where the two bits joined. Those top hats were built for Chip by Ronnie Altman of Altman Brothers in the Yonkers. Very soon after that it developed into a purpose-built PAR can as we know it, and Strand did the first ones, they were quite robust, quite heavy. Then Altman built lightweight versions for a fraction of the price out of sheet metal rather than cast. I am flabbergasted to think how many of those things have been manufactured. I'm told by Marshall Bissett, a friend of mine in L.A., they still get them built in mainland China now, and they still bring them in by the hundreds, and they still sell them, so there must have been hundreds of thousands of these, and where are they all? I mean maybe millions of them!

So really that was a first of the glory days. Also Chip pretty much pioneered the use of trusses as well, and Hartman built them. Richard did the '73 European tour with me.

Did they have trusses then? Steel trusses, I think, not aluminium.
We had, yes, a truss, a big deep truss that went up, pushed up from the floor level on two huge things that were hydraulic, called Galloway Rams, and they were actually for loading stuff onto aeroplanes. It was a massive hydraulic ram on a big old base that you couldn't push, you had to drive it, and it had a battery and a little remote, you know turn right, turn left, rather like a tank you know, move that wheel and not that one, and you could move this thing into position.

This was the base?
You put one of these on the ground at each side of the stage, but on the floor of the ice rink or wherever you were. If you had a proper stage, you couldn't use them. You had to have a load-bearing floor, and you had to be able to drive onto that floor, in other words, they could only really go where a forklift truck could go because they probably weighed a ton, you couldn't pick them up or anything.

Galloway Rams came from the States, and they were very expensive pieces of kit from the aircraft industry, that's the sort of thing Chip did, you see. This was left over from the 1972 tour, he did a lot on that tour that I wasn't involved in, I was still at the Shaw Theatre. Your stage would be 60 feet wide, so your Galloway Rams were parked off each side of it, and they would push up to a height of 25 feet, and so the span between them must have been 70 feet, and that beam was a big box truss, and then at the front we had the famous mirror, the tilting mirror, Mylar mirror panels that were mounted onto its own truss that Richard Hartman built, and it was bolted together—radio antennae, that's what it was, it wasn't welded, you actually built it out of bits of metal that were bolted together, and it was actually a triangular truss.

Like Meccano?
Like Meccano, yes, but it was about 15 inches each side, equilateral triangle, and that would take the weight of this lightweight mirror which was on little servo

motors, and it tipped, and then the Super Troupers were on the floor behind the amp line shooting up into the mirror and back down onto the act.

So you never had to cart the Supers up into the house, and that was Chip's idea, you see he did innovate, he was a great innovator. It was a fantastic idea, and it was all based on practicality. If you are touring in theatres you don't need to take follow spots because they have them, and as it is now if you are touring, even in sports arenas you don't take your lights with you because they have all got them now, they have all got Supers and others, but back then they didn't, so you toured them. The effort of getting those up into a high position up in the seating and onto a tower was enormous. This is the kind of vision he had, he went, "why, we'll stick them behind the act and put the mirror up instead", and of course it was a great effect as well because you could tilt it during the show and the lights would hit it, and then it would tilt and go back, a great swathe of lights would rake up the audience and then back again during a sax solo or something. I remember I used to cue it, "Rip this joint", I think it was. We used the same one in 1976, it was such a good idea Peter Rudge wanted to use it again.

One of the first special effects, lighting special effects?
Yes. Then in 1973 one of our really earliest customers was the Moody Blues, this is all between Stones tours, we had Cat Stevens, actually quite a lot more, like Elton John, early Queen, Uriah Heep, some stuff for The Who and for Zeppelin and the first Knebworth in '74.

And all the time you are still renting the equipment in.
Yes, but buying stuff gradually as we could afford it, buying to rent out. Actually we were importing PAR cans by the hundred from Altman and selling them on to our competitors at a profit until they figured they could buy them direct from Altman. And we were making our own dimmers and desks—Paul Ollet was making them for us in the basement. So we had a lot of equipment by now. We decided we didn't want to be a manufacturing company (too much capital required, and our heads were into rental and full production), and we let Paul take his designs to Avolites at no cost, and they became the dominant industry standard, quite possibly a wrong decision commercially. Then in 1975 I went over with Patrick Stansfield, who was the production manager, and we did a whole long, long, long Stones tour of America. I was the stage manager, and that was with the lotus stage, the shape of a lotus flower, but in the big venues they had a hydraulic version where the stage leaves moved up and down.

They did Earl's Court, I saw it there.
Yes, we did Earl's Court. The only place in Europe we did it was Earl's Court. In America we did New York, which was Madison Square Garden, L.A. and Chicago, only three venues where we did the mechanical stage, the rest of the time we used the same shape stage and it was heavily raked, but it was static, you had a subworld underneath the stage because it was very high at the back. Because it was raked, the back of the stage was like probably like 12 feet off the deck, the front

of the stage was probably 4 feet off the deck, so you had this enormous rake and all these internal staircases and things up to the back, the amp line was partially buried. That was revolutionary stuff, which is commonplace now because the tech world is phenomenal. When you go and see a Mark Fisher design show now, it's all built-in, all the little guitar areas and everything.

You know it's beautiful, it's like a little city under there really, hydraulic lifts, mechanical lifts, extraordinary now. I saw the Stones in Mexico last year and had a tour of the stage, and it was fantastic. I went to the opening night in Boston actually. I am nothing to do with anything anymore because I am retired, but I still manage to get to one or two of the gigs because I know all the people.

In 1975 we did this massive American tour, and then 1976 Peter asked me to put a cheaper show together. So we went back to the '73 design with the mirror and a different set. Everything went very well except at the end of the tour Mick started to blame us for being cheapskate, me and Pete, because it was like "we have got to have something new, man", you know, which hurt a bit because actually me and Pete together made it come in on budget, whereas Chip, and to a certain extent Patrick, managed to make it come in short. On the American tour, I think we grossed 11 million dollars, we spent 12, so we had to go and do two more huge open-air shows in Buffalo just to have some money to give to the band.

What did they think of that?
Well, they didn't like it, but when you do the reverse and say, "Okay, we're going to do a European tour on budget" and in '76 I did everything for a fixed price, I did sound, lights, crew hotels and transportation and trucking all for the one fee through my little company, and that included renting Bill Harkin's stunning orange tensile roof structure and stage outdoors. All that was a big risk, and we probably didn't make any money on that, but we got a lot of kudos out of it.

And bought a lot of gear probably.
Yes, and we got ABBA out of that because they literally came to the show and said, "We want to see the same thing". Thomas Johannson phoned me and said, "Can you come out to Stockholm?", and I went there for the day and within 40 minutes they said "Yes okay", so we did about four years' work with Abba just because they were so impressed with the Stockholm Stones show. They wanted everything the same, different design of course, so that's the kind of kudos we got out of it, that was '76. Stones at Knebworth was '76 as well.

So, John Brown. Remind me how did you meet up with John Brown?
Well, in the Youth Theatre, he joined the Youth Theatre from Lancing (he went to Lancing), then LAMDA, and we got on like a house on fire, and I still see him now. He's not in the business anymore, he has got his own company that does purely commercial application of holograms for security things, little things on cigarette packs, that little thing that's on a credit card, he doesn't do those but he does similar stuff, it's to cut out counterfeit cigarettes, coffee, any kind of product.

So when he said to you, come on down to Wandsworth . . .
He said we are now making enough money to pay you what you need.

So what did he do?
He was managing director and I was executive director.

John Brown at Wembley, 1975
© Michael Freeman

So he left you at the ICA.
Yes, he left because it was all getting too much because he was doing way, way, way more of this liquid light show stuff than he really had time to do, and we had a third partner, Alistair Robertson, who was the genius, the inventor. I think he became a teacher in the end. So they left the ICA and started to turn the light show company into a real lighting company, and I would do work for them and stuff at weekends and so on, and I would do the odd bit of touring. During that time we used to do the odd UK gig for Crosby, Stills and Nash, Joni Mitchell, Laura Nyro . . .

Do you recall in that period other lighting companies?
Yes, Rainbow. When The Rainbow closed down, Rainbow Lighting sprung out of what was left there, Dick Parkinson and John Coppen, and they were called Lairhurst.

John Coppen, that's right, and Richard Hartman?
Hartman was never part of Rainbow Lighting; he had his own company called TTR, and then when Lairhurst went bust he bought them; in fact Rainbow was never called Rainbow legally because there was already a film lighting company called Rainbow, and they couldn't use the name. Lairhurst was the off-the-shelf company that operated as Rainbow. Richard Hartman had a company called TTR, Theatre Technical Research, and they bought the remnants of Rainbow. They were competitors of ours but friendly competitors, and then Zenith came along, but they were the three main companies. Then of course there was TASCO in the same period in the '70s.

We did our first Knebworth in '74 with Freddy Bannister. I was stage manager with Ian Knight, and he has been around a long time, he is a set designer really but he and I were co-stage managers for the first Knebworth.

Did Ian Knight do the Wings set in 1979 that we worked on together?
Yes, I think he did, he worked a lot with Showco. Anyway, in the '70s it was mad, we were very, very busy. John Brown, my partner, decided that he couldn't keep up, he didn't want to keep up with the pace of production, live production it was just too stressful in a way. When I was off with The Stones, he was doing all our other acts including Elton and The Who.

John was really just a very good businessman, but he also had a much better understanding of electricity and stuff than me, he wanted everything to be proper, not proper but organised and precise and definable, and he had this automatic business brain, and it was all a bit too hit and miss for him.

In '76 after the big Stones tour and after we got ABBA we had an offer from Rikki Farr, which was a sound company called Electrosound, in the Marshalsea Road in Borough, and he offered to buy the company. I went with it and John didn't. Rikki had serious backing at this point from a City institution, a merchant bank.

John took the money, and I took some money and some shares and went on and had a job. I went on to be the managing director of the London operation, and Rikki went immediately to L.A., even though it took 6six months for me to negotiate the deal for the purchase of the company, which wasn't Rikki's style at all, he would go "Yeah mate", shake hands on it and it's done, but then of course you can't do it like that, you have to have lawyers.

By the time we had signed the deal, Rikki had gone and already bought Tom Field Associates in Boston, TFA, which is the lighting company that had done The Stones tour with me in '75. A great company. Tom Fields was the boss, theatre guy, Tim Mahoney, John McGraw and Rick O'Brien.

So within the course of a year Rikki bought TFA, and they opened a brand-new place in Los Angeles and then eventually moved the Boston operation to Los Angeles, and all these New Englanders all moved to L.A. It happened so, so quickly, because Rikki had a nose for the business. It was like the Brit invasion, do you know what I mean? It was like we dominated the American rock scene really with Zeppelin and Genesis and The Who and The Stones.

TASCO were there at the same time.
TASCO went to L.A. as well.

Terry and Paul.
Yes, that's right, and of course we were all very near in England, in London we were just 'round the corner, they were in Lambeth we were in Blackfriars. So we sold ESP Lighting to Rikki, and he dropped the ESP name. We did the lights, he was a sound company, and the reason that he was keen to buy us is that for the '76 European Stones tour I had used TASCO for the sound and not Electrosound. The show you saw at Earl's Court where the sound was appalling was TASCO because they had got a delay system, and they didn't know how to use it. I felt sorry for the guys, the front mixer was Davy Kirkwood, crew chief was Keith Bradley, who now effectively manages Elton John on all of Elton's live stuff.

Do you have no history?
All written down? No. We were doing a lot of bands for either lights or sound and sometimes both. TFA had its old clients including Chicago, Beach Boys, Earth Wind and Fire and so on. Rikki had Rod Stewart and was picking up acts in L.A. including Foreigner, Rush, Boston, Styx . . . In London we had all my big acts like Neil Diamond at Woburn and Dylan and Blackbushe Airport Festival for Pat Stansfield (we even built the roofs for those), and we were picking up new acts such as Elvis Costello, Ian Dury, The Pretenders . . .

From '77 when I moved into Marshalsea Road we got rid of the lease on our other building in Glasshill Street. Rikki had gone, moved to L.A., I was running England, and Tim Mahoney was running Boston, and then he moved to L.A., and we had everything going for us, we were doing all the major acts both sides of the pond. We were now named TFA Electrosound, Marshalsea Road, in a big old warehouse, it was nice, lots of office space, workshops and then quite a lot of high ceiling space on the ground floor, terrible loading bay though, ordinary residential street, blocks of flats and things. We were there from '77 till '82. Rikki was managing The Tubes, he had a company in America called AIM, Artistic International Management, really trying to make his brother Gary into a rock star, put a lot of money into that, we were struggling but we had major, major tours, this was '77/'78/'79, we had all those great acts.

That's what happened really, wasn't it, in the late '70s, the production companies that had made it to the late '70s, basically amalgamated and crossed over and were running the sound and lights?
There was only TASCO and us frankly. I think that TASCO and us were the only combination, the only ones that did sound and lights both sides of the Atlantic.

In '82 TFA Electrosound's parent company went bust, which was an investment company called Norton Warburg, and they went bust in a big, spectacular way, and lots and lots of lawsuits, the shit hit the fan with a whole group of companies.

Pink Floyd opened restaurants on canal boats in Islington, lost a fortune. Tony Stratton Smith lent them money, then there were horseracing stables, it was all completely mad, you know. It went bankrupt in a very big way and out of all that, actually Andrew Warburg, poor old Andrew—nice enough guy, not a crook but just foolish really—went to prison for four years. Bristol-based company, they were licensed deposit takers, which means they borrowed money from rich people and wasted it on rock-and-roll bands and management companies, but they spent 3 million pounds of the Floyd's money. They were just crooks basically, but Andrew wasn't, he was just in charge; all the ones who were crooked scampered and hid under his coattails, but he was the boss, he was the chairman, so he took the rap.

Very sad, anyway, at that point it was a crisis time, and my company, which was TFA Electrosound, was the only company in the entire group that was saleable, and I had lots of talks with the receiver, and we carried on working during that period (in fact, we did a whole Springsteen European tour in receivership), and after a few months they sold us as a going concern. Basically Zenith made a bid for us because they had an investment guy as well who was going to buy them and us. Our sound guy was a good friend of mine, Bill Kelsey, and he went and asked the accountant, a really nice accountant from the old Roxy Music, Brian Eno days called Sam Alder; he was Mark Fenwick's accountant. Mark Fenwick was a manager, he managed Crimson, Roxy Music. Anyway, Mark Fenwick's accountant, Sam Alder, put in a bid, a consortium bid for us. Zenith bid and Theatre Projects bid and we went with Theatre Projects in the end. They weren't the highest bid, but they were the only ones that I wanted to go and work for. They all said, "We are only buying it if we have you and the staff and the crews", and I said, "Okay, in that case it's Theatre Projects".

Let me ask you briefly now, so who was *your* biggest influence?
Chip Monck.

Yes, of course. Now what do you think has been the biggest influence in the industry during that period?
I think what has changed most since I was on the road, I mean not since I have been involved in the business, but when I was actually out there in the front line even up till '76 with The Stones, I would say the biggest influence has been in the logistics of transportation; sleeper buses were absolutely critical. Sleeper buses solved the problem of exhaustion and made everything possible.

I once on a Stones tour did not see a bed at all for nine days. I asked my people to do the same thing, and I remember people coming up and saying, "I don't want any more coke, I want to go to bed", and it was what you had to do to do back-to-backs 'round Europe.

I was logical about it, I mean I'm not a stupid guy, you plan it out, and you think the only way we can do it, is for every four people in the crew to have a Hertz top-of-the-range car picked up locally. We did fly every time there was a day off, but when you did the back-to-backs you finished up, four of you, driving a big Mercedes for every four or five people, so there were maybe 10 of those going from town to town because this was the only way to get there. You did a load out and finished at 2.30 in the morning, you had to load in at 8 o'clock, and it was a 4-hour drive, and you couldn't do it any other way, so we drove.

Now short of having a chauffeur for each car, which means you lose another seat so there is only three of you, one person had to drive up the bloody Autobahns in Germany, and they hadn't slept for nine days, so sleeper buses are absolutely critical to the business and proper handling, forklift trucks, proper trucks, ramps, computers, mobile phones, all that stuff. I mean when I go visit a production now I am flabbergasted how totally fucking together the likes of Jake Berry and Opie are on their tours.

They have these immaculate offices with fantastic girls who are friendly, stay awake, know the business inside out, the production assistants. Now they have to cope with a lot, perhaps unnecessary bollocks from the band's end, the logistics on the band has gone completely nuts inasmuch as the touring party is vast and everybody has an assistant and everybody has an assistant's assistant and there's a baggage man and assistant baggage man—that is all fucked as far as I'm concerned, it's all unnecessary. I think that sleeper buses, if there's one single thing, made the biggest difference.

Who do you think has been the biggest influence in the business?
From the technical side of it, the boys work hard, extraordinarily hard, and what they achieve in one day, which is to virtually build a town, you know, in the middle of a baseball stadium in Boston, is stunning. The logistics of leapfrogging the steel, what you can and what you can't, and what's critical.

I would say in that respect one of the major influences on the business from that level was actually Bill Graham and Michael Ahern. Bill Graham took the touring business onto a different level. He was the overall promoter of long stadium tours with giant revenues, guaranteeing the artists their piece of that.

Michael Ahearn was the production manager for The Stones at that time, and he bought into Bill Graham's concept. After the helicopter crash that killed Bill Graham, Michael Cohl took on this same concept. Patrick Stansfield was another one. That's all part of the vision and, of course, the budget control got a lot better. It cost a lot of money, but at least it wasn't being pissed up the wall. Talking of people, I think Mark Fisher had the most incredible effect on what a show looks like.

Please tell me a story to finish off.
Right, so in '74 we were doing the Van Morrison tour. I was doing production, and I was the lighting designer, and the reason I was the lighting designer is they didn't pay much attention to lights. We rehearsed in Manticore[8] actually, and I found out that in fact Van Morrison was extremely nervous, temperamental and would rather play in the dark, he hated lights, so you gradually took any lights out front off him, including follow spots.

So I found myself lighting the only pop group I have ever lit, and it was Van and he hated being lit, he wanted to play in a glimmer, it was actually a big challenge. He used to come off to the side of the stage and say, "Man, keep the lights down". I was running it from the side of the stage and couldn't run it from the front because he kept coming and wanting to talk. Anyway, we get to Dublin, we do a double-header 6 o'clock and 9 o'clock, or 8.30 or something like that, and I don't think that he really had been told, and he goes on and does the early show does a bare 40 minutes and walks off. He played well, really did well, good band, terrific brass section, at his peak really, but really nervous and skittish. So he goes off and would not go back on stage, so we put the house lights up, and of course the audience refused to leave. His personal road manager was a guy called Ed Fletcher, Fletch, and he finishes up down the front of the stage, jumped into the audience, and starting laying about him and having this fistfight with the audience. Whereupon the doors at the back of the hall burst open and the Guarda come in, loads of them, about 50, and are arresting everyone, arrest Fletch, clear the house. Everyone decides they have got to go because there is the second show. By this time Fletcher is in the local nick, and he tunes guitars and everything, I mean he is a roadie, so I find myself going down the nick and bailing him out with my own money. I thought, wait a minute, what is wrong with this business, you know? I am lighting a man who doesn't want to be lit, and I have got a road manager who is down there smacking people in the audience who paid good money to come and see the gig, and I am having to use my own money to go and bail him out. It was an absurd tour really but a lovely gig.

So that was my introduction, my first and last time I ever lit anyone, but I was also the production manager. When I was in the theatre I only used to light plays because after you put the thing together and everything was working you thought, well yes, I might as well do it myself, might as well light it, but I never had any great ambition to be a lighting designer. Having said that, I wasn't a bad lighting designer for Shakespeare and stuff, but my experience with Van Morrison . . . well!

Brian, thank you very much.

Brian Croft, now retired, was awarded the industry's highest honour for Live Event professionals in the US at the Parnelli Awards in 2001. (This from Parnelli website—"Since 2001, scores of our highest achievers and most admired innovators have been awarded the Parnelli. The award recognises pioneering, influential professionals and their contributions, honoring both individuals and companies. It is truly the "Oscar®" of the Live Event Industry".)

Notes

1 https://gulbenkian.pt/en/
2 www.boylefamily.co.uk/boyle/about/index.html
3 www.peterwynnewillson.com/
4 www.chipmonck.com/
5 www.richardpilbrow.com
6 www.samuelsonfilmservice.co.uk/wp-content/uploads/2013/04/SamScene-Take-2.pdf
7 www.seefactor.com/history.html
8 www.brain-salad-surgery.de/manticore.html

2   Jon Cadbury

Jon at Richmond Park circa 1973

April 27, 2007 with Jon Cadbury at PRG, London

It's nice to meet you, Jon. I do remember the Roundhouse very early on in my life, which is where you started in this business, isn't it?

Yes, that's right, I had been working at the Chichester Festival Theatre, and this was when the National Theatre was there, Laurence Olivier's company. Lindsay Anderson (who had been directing at Chichester) befriended my best friend, a guy called Jeff Torrens, and offered him a job at the Roundhouse in Arnold Weskers' new arts centre. He had set up a movement called Centre 42, which in fact Brian (Croft) was also involved in, which in its crudest form was taking the arts to the masses, to the working class. There was a trade union resolution called Resolution 42 made in about '65 or '66, to set up touring arts festivals going 'round trade union halls and working men's clubs, with poets and playwrights and artists and so forth. I know Croftie did a fair bit of that; he actually went on some of those tours. They decided they needed a base, and exactly how it came about I am not sure.

A group of businessmen and Jennie Lee (Minister for the Arts, '67–'70 Labour administration) were involved in finding the Roundhouse, which had been a bonded warehouse owned by Gilbey's, the gin manufacturer.

And originally a train turning point.
Exactly, the trains used to come in, turn around, and be shunted off into any one of 24 bays to be worked on. It had been built in 1847, I think, right at the beginning of the railways. Within about seven or eight years of its construction they changed the gauge of the rails so the bays that were used in the entrance and exit

Front entrance of the Roundhouse
© Roger Morton

were too narrow for the wider gauge, so it became pretty redundant. Then, it was just a warehouse space and, as I say, Gilbey's, I think, had it through the late 19th century into the 20th century.

When I went there it was in its very first version, there had been a few things happening there. They had a thing called the Dialectics of Liberation in 1967, which brought in some of the American beat poets and various intellectuals to discuss where they thought the modern world was going. It was the Vietnam time just beginning to kick off at that point, and these people were free thinkers, so there was poetry there and some music, although not a great deal. I think UFO was running in Tottenham Court Road around that time. I think I remember going to a UFO night in what became the Open Space Theatre in Tottenham Court Road in about '66/'67.

What was UFO?
UFO was effectively a club for hippies to come together and have a psychedelic rave, that's effectively what UFO was. They did them at Alexander Palace, they did them in Tottenham Court Road, and I think they did a couple at the Roundhouse in '66, maybe early '67.

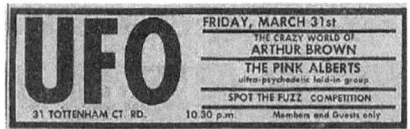

UFO newspaper advert, 1967

When I pitched up to the Roundhouse in '67 with my mate Jeff, it was an empty shed with a wooden balcony, wooden columns, cast iron girders, very similar actually to what it is today, but the balcony has gone. It has recently been completely refurbished, and a very nice job they have made of it, but it was pretty bare back then, there were no lavatories, there was no power, so people would bring in temporary power.

A complete 'in the round' temporary seating arrangement that we put in for Joint Stock Theatre productions of 'Epsom Downs' and 'A Mad World my Masters' in 1977
© Roger Morton

I remember going in there the first time, and there was a group called Exploding Galaxy, a dance group rehearsing in the middle of the 80-foot-diameter centre circle. Really, I was a pretty straight theatre guy you know and quite bowled over by the power of theatre. I was thinking 'I am going to work in the theatre'!

Can we just step back? How did you get into Chichester? Where did you go to school?
It was a holiday job. I was brought up in Bosham, where my mum ran a children's home, so yes it was a holiday job working with the stage crew and working in the props department.

Great holiday job?
It was a great holiday job, yes, it really was, and it made me realise that the theatre was rather more rarefied because Chichester is quite a conservative town. The actors were entertained by the Bishop, and they were taken out by the local wine merchant, and they were feted by the Cathedral worthies and lived in very nice residences that were made available to them, and I thought 'that's a really nice life working in the theatre'. Actually there were a lot of very famous actors in the seasons I was working there. People like Ian McKellen, Derek Jacobi, Maggie Smith and Robert Stephens; it was just before the National Theatre moved into the Old Vic. Laurence Olivier took that company and moved it into the Old Vic, so it was quite an exciting time in terms of where theatre was going.

So as I say October '67 was when I first went to the Roundhouse and started working. Lindsay Anderson and Arnold Wesker were two prominent writers and directors there, and it felt very exciting. That first year there was some theatre, but it was very much a mixed economy. Arnold Wesker wanted it to be pure art really, he wasn't particularly interested in having contemporary music coming through, and he really wanted it to be a theatre base for experimental theatre. Jennie Lee and Harold Wilson (then Prime Minister) felt that it needed some sort of commercial director or administrator. and they brought in a man called George Hoskins who had worked for the Egg Marketing Board apparently, very successfully around the world.

So, George came in and decided rather than just wait for grants and subsidised theatre to come through. he started looking for activities where he could charge for the use of the building. The very early regular concerts at the Roundhouse that I remember were put on either by Middle Earth by Paul Waldman and David Howson (who were the two people who ran Middle Earth), or by Blackhill Enterprises[1]—Pete Jenner and Andrew King, who managed the Floyd at the time, and they were doing things like the early free concerts in Hyde Park. I don't know if you recall that, but there were quite a lot of free concerts in Hyde Park we worked on.

I remember The Stones, I was there.
The Stones, I was there too, that was the summer of '69. I do believe Blackhill put that on. It was the year Brian Jones died, wasn't it, and the butterflies just the most wonderful psychedelic moment really. There were other concerts in the park

because Pink Floyd were there in '71 I think; Canned Heat played there and Blind Faith played there.

There were some very good free concerts, but it was a scruffy stage. Charlie Watkins from WEM (Watkins Electric Music) used to come up with his speakers in the back of a flatbed truck and unload his speakers and set them up around the stage. But it got a little bit uncomfortable; the Hells Angels started muscling in on security on the later ones, and that sort of changed the whole vibe really in a big way. Those were sort of early, out of the Roundhouse, musical experiences brought about by the team who were working at the Roundhouse.

What was Middle Earth?
Middle Earth was a hippie club in Covent Garden; I don't know exactly where it was because I never actually went to the Middle Earth club when it was there (43 King St), but it was like a development from the Flamingo, the Speakeasy, those sorts of venues. Very specifically, *International Times* ('it') were very involved in working with these clubs, so it was very much that underground vibe. John ('Hoppy') Hopkins ran 'it', and it did have that sense of 'this is something very new and very different going on here'.

The concerts in the Roundhouse were really done using the in-house lighting system. There were lighting people like Jonathan Smeeton creator of Liquid Len, who would come into the Roundhouse with oil wheels, and they would hang white sheeting up between the columns and back project onto it from the balcony so that you had these 24 columns backlit and these extraordinary psychedelic light shows going on.

There was a company called Alpha Centauri, which has become A.C. Lighting (and as of 2008 A.C. Entertainment Technologies), who now work out of High Wycombe as a major distributor of entertainment lighting and sound equipment. There were three or four different companies doing these psychedelic light shows, and the stage was just lit with conventional theatre lights.

As I say, Middle Earth and Blackhill were the first major promoters, and they were always in competition with each other to be the promoter for the particular show.

And what were you doing specifically?
At that time I was assistant carpenter. The beauty of the Roundhouse was that we were all able to do something of everything. My job was carpenter but effectively I used to do work on the doors, on security, behind the bar sometimes, in the catering area but my full-time job was carpenter for the place. That developed into eventually becoming production manager for the Roundhouse.

At this time we had a series of rake seating which was timber rostra, timber-framed rostra, providing a sort of rake that the chairs had to be cleared from every Saturday night (after the theatre performances finished). The music shows I think were pretty much always on a Sunday, and so the stage had to be built depending on what was required. The place had to be completely re-jigged overnight Saturday for the Sunday and then overnight Sunday before the Monday. I was doing a lot of those turnarounds, and then generally nominally being on the

Middle Earth newspaper advert, 1968

security team but actually just having a pretty good time on the Sunday during the shows.

Those shows developed into Implosion, which became a regular thing. Implosion was set up largely to fund Release, the organisation that had been set up to help people who were having a bad time on drugs, essentially as a support organisation for people who weren't dealing with their drug problems. Caroline Coon and Rufus Harris ran it, and they teamed up with Ian Knight, Hugh Price and Jeff Dexter, who had come out of the Middle Earth and Blackhill world. Ian at that time had just finished managing the Soft Machine, and his background was art, really he was a graphic designer. I am still in touch with him; he lives largely in Thailand now, but he still works in the business. He still does Rod Stewart's shows; he designed the current tour Rod is doing. (Author's note: Ian sadly passed away in March 2010, but an interview with him appears later in this book.) Hugh Price went off to work with Gerry Bron in the Bron Agency.

Which was next door to the Roundhouse, wasn't it? A new building?
Yes, that's right, which came after about three or four years though. It was brand new and they had a recording studio (Roundhouse Studios) in the basement with an echo chamber built into the cellars of the Roundhouse. An extraordinary tiled echo chamber that was piped through to the studio so they could get these strange sounds for some of their acts.

Back to Implosion, was it a regular event?
Implosion was a regular event, yes, and it probably ran from about 1969 or 1970 through to about '72 or '73.

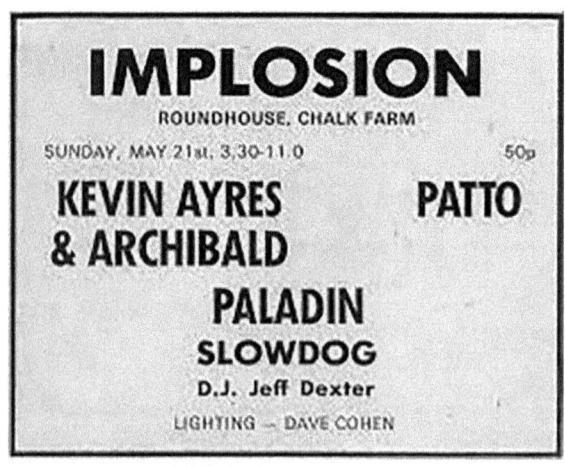

1972 Implosion poster

I think I saw the Floyd there.
The Floyd, certainly. Blackhill presented The Doors and Jefferson Airplane there.

Tell me more about Blackhill?
Blackhill was Peter Jenner and Andrew King. They had an office in Notting Hill, just at the corner of Westbourne Grove in Notting Hill, and they were managing a number of bands by that time. Marc Bolan was one of their acts, T Rex would play regularly. Jeff Dexter tells stories of Marc Bolan hanging out, just waiting so he could get on stage and play for a bit, and Jimmy Page hassling for a gig, and these sorts of extraordinary pictures of these world-famous musicians now, so it was really very exciting.

What I loved about it was this juxtaposition between weekend concerts and theatre in the week, theatre and other events. We did a television show there called The Roundhouse Forum on a Sunday morning. They wanted to recreate Speakers Corner, so people would come in and talk about whatever was on their mind and be heckled by an audience, a live 'speakers corner' type environment. Lasted a little while. Some of the big London orchestras would come in and rehearse in there, they brought some lights and they would put down a wooden floor in the centre, which slightly improved the acoustics. The acoustics were pretty dreadful for theatre, actually is the truth of the matter, because it was a brick wall building and quite a lot of wood. It was never great for spoken theatre. That was one of the reasons why it was absolutely necessary to develop into other areas. We did ice shows, we did circuses and we did some very exciting theatre.

We had the Living Theatre from New York come through, Nicol Williamson did *Hamlet* there with Marianne Faithfull playing Ophelia directed by Tony Richardson. Just before he made Tom Jones he did *Hamlet* in the Roundhouse. Andy Warhol came and did *Pork* there, which was an extraordinary show. Kenneth Tynan did *Oh! Calcutta!*

Photo of hole in the wall
© Roger Morton

'Last Girder' was from 1971/'72 when the cellars between the underground service tunnels were excavated and the old wooden gallery was replaced by a new steel and concrete gallery at the instructions of the local authorities, who were not happy with 140-year-old timber beams, and wanted everything ripped out and replaced. This was done as an in-house building project headed up by Robbie Simpson and crewed by many of the Roundhouse regular show and changeover crew.

I saw Marcel Marceau there.
Yes, I remember Marceau being there. *The Sunday Times* (and National Film Archive) put on an exhibition called Cinema City, in September/October 1970, and a range of people attended from Liz Taylor, Richard Burton to Harold Lloyd, so it was really interesting. There were films shown every day, a repeat of old films, new films, experimental films and massive film luminaries coming through to give a talk to the audience, in fact Taylor and Burton got heckled off the stage by among others Sylvester McCoy, who went on to play Dr Who, because Sylvester was working in the box office at the time.

I remember one of the stewards in those early days on a regular basis was Harvey Goldsmith with his partner who came in to run some of the stewarding called Michael Alfandary.[2] I don't know if you remember him? Michael was Harvey's first partner, so I knew Harvey way back then, and he was a very nice guy.

I can't for the life of me remember how I got involved with Harvey, but I ended up freelancing and doing a lot of work through his office in the early '70s. He helped me up the ladder a lot.
I mean that is exactly the vibe.

Were you ever involved in the Reading festivals?
I have a horrible recollection of the first Reading festival. I was asked to get involved by this guy whose name escapes me completely! It was '73/'74, something like that, I still had the truck maybe even earlier, maybe '72. I had that truck for about three years, but the job, this I hadn't understood, was taking all the plumbing for the lavatories and all of the pipework. It was fine taking it there . . . '

. . . Oh no, I know what you're about to say!
Bringing it back was just . . . I think I sold the truck really soon after that, but I got rid of it so I have always had a bad feeling about the Reading festival!

The Roundhouse was really a very interesting and exciting place to work, and I stayed there until '73 when I left the full-time job, having been able to go off and do odd festivals. We worked on things like a festival in Krumlin[3] in Yorkshire that was in 1970. There was a festival in Bickershaw[4] also in 1970 where the Grateful Dead came and played, and The Dead rehearsed in the Roundhouse beforehand. There was another free festival called Phun City[5] down in Worthing, which was with Keith Albarn and Ian Knight involved, I think.

By this time I had got a truck, bought a lorry in '70, so very often I would take the lights. We would go down to Strand Electric, just south of Vauxhall Bridge, they had an outlet there, and they hired lights out. Ian (Knight) arranged the hires, and I put the lights in the truck, and we would take them to the festival, set them up and then I would stage manage the shows, well nominally stage manage. Very often, by the time we got to the actual shows the stage management was really nominal at best; in fact at Bickershaw I recall the person who actually did the stage management (although it wasn't intended) was Neil Warnock (much later founder of the Agency Group).

Bickershaw Festival 1972 levering vehicles out of the backstage mud!
© Jon Cadbury

Touring stages for these festivals were almost always scaffolding constructions usually without walls and usually with some sort of plastic roof. The lights were just borrowed from individual theatres or, as I say, hired from Strand Electric. Sound systems I do remember Charlie Watkins coming 'round with his truck full of WEM speakers on a regular basis and, quite early on, Dave Martin was supplying speakers. He did his first installation in the Roundhouse actually with the first Martin speakers.

Was yours an artic or was it a fixed wheel truck? A seven-tonner?
That was a seven and half tonner. Jeff (Torrens) and I were going to go off and tour Europe with it, and Ian said, "Well why don't you just come and take the lights to this festival for us and do the festival and then go off on your travels?". I hadn't actually worked out that if you take the lights out there, then you are probably going to have bring them back again. Of course, we didn't think about things like carnets, so we got on the ferry at Harwich, got off at the Hook of Holland and customs impounded everything! The guys who became Mojo Concerts, Berry Visser and Léon Ramakers, eventually sorted it out; they were the people who promoted that festival (Holland Pop Festival 1970).[6] They paid some sort of bond that got the lights in and got them out again. I actually took the truck to Schiphol airport to collect the Floyd's equipment when it came in and had the entire Pink Floyd's equipment in this seven and half ton truck . . .

With a carnet?
No, no carnet, no, I had to do a deal with the customs agent at Schiphol, which was basically, "I have got to get this to the site you know, they're the headline act on this bill!".

So my deal with this customs officer, who was a young guy who was into music luckily for me, was that he would release the equipment and he would come to the site with me as long as he could collect a bond. So I went to the site and told the organisers and Steve O'Rourke, that we couldn't actually unload the truck. I said I wouldn't let it out of my truck until the customs officer had got his bond, which was probably exceeding my brief, somewhat, but we got there. Everyone was sort of passing the buck to someone else to pay the bond, and I said, "Well, I'll have to take it back to the airport then. That was the deal I made with this guy so if you don't sort it out . . .". So they did pay the bond and it did happen, The Floyd made an album with all of their equipment lined up in a great photo on an aerodrome (Ummagumma back sleeve). That was *that* equipment, and those were the roadies who were dealing with it. It's extraordinary.

Around that time I had met Brian (Croft) from the ICA. There was a network of underground venues around London, the ICA was one, the Roundhouse was one, Charles Marowitz had the Open Space Theatre where he worked with Thelma Holt, and Jim Haynes ran the Arts Lab in Drury Lane. *International Times* were publishing, and they were all creating this movement, this underground movement.

'it' logo
© www.internationaltimes.it/archive/

Was it a free newspaper?
I don't think it was completely free; it may have started off free, came out around the time that OZ was first published. It may have been a bit before OZ.

What happened was a sort of network of understanding between these different venues—'this show doesn't fit the ICA but it would fit the Roundhouse' or 'this show isn't right for the Roundhouse but it could go to the Arts Lab' or 'this show's too big for the Open Space so let's put it up at the. . .'.

Brian came to the Roundhouse with a show called The Hero Rises Up by John Arden about Nelson. I met him probably '69 maybe early '70, and he had just hooked up with The Stones, so he offered me a tour and said, "Well why don't you come out and do this tour, we're going off 'round the world, and we are having the staging built out at Zip Up scaffolding factory, out near Reading. We are putting the system together out there, so come and join the crew. There will be 12 of you, there will be eight Americans, three English guys and me and we will be taking this around the world".

With The Rolling Stones?
With The Rolling Stones, yes. I got there and there were two guys called Robin Murray and Roy Lamb and I, we were the three English people in this group. The LD was a guy called Chip Monck, who was a bit of a legend in his lifetime. I spent three or four days there, and I didn't quite feel comfortable. I had become production manager at the Roundhouse before I'd left for this, and I'd been put into a very junior role . . . just to "do this . . . do what I tell you" by the Americans. Then it also became clear that we were earning exactly half their salary. Roy, Robin and I were earning exactly half the US salary, which I thought was unreasonable, and when I questioned Brian about it he said, "Well Americans have a higher standard of living than English people" so I just said "okay".

I mean if he had said, "Look, you haven't done this before, you are inexperienced, that's the reason", that would have actually been more truthful! At the same time my wife or my partner who I had met in the Roundhouse was pregnant with our first child, so she said, "Well listen, it's great, it's going to be really exciting but if you do it, I won't be here when you come back because I can't really have a relationship if you are going to be travelling 'round the world, it's not going to work", so that was a fairly clear-cut moment, and so I thought okay, I won't do it, and I backed out of the tour.

But the tour obviously happened, and it was probably one of the first where they took a stage, the lights, the sound system and everything. It was inventive touring because the early tours, the early rock-and-roll shows I saw were all done as variety shows more or less.

The first show I went to see was Bobby Vee. I bought some tickets for Cliff Richard in about '64, and then I went to see The Beatles in '63/'64 on tour at the Portsmouth Guildhall. It was a variety show, there was a comedian, a magician, The Springfields closed the first half of the show and The Beatles closed the second half of the show. That was how touring worked, it went into venues with the existing arrangements whatever they were: lousy sound system; lights on all the time; nothing happening and then the band came on and played, that was how it was.

Around that time, 1970-ish, was really when you saw the beginnings of a touring world as such. Apart from theatre, which travelled huge amounts of equipment 'round on big theatrical tours, concert touring wasn't really there as an industry.

Did you have in-house concert lighting at the Roundhouse?
No, we just used the existing (theatre) lights, just what was there. Whatever was used in the week we would put different gels in so there would be some colour in the lights.

No PAR cans then?
No PAR cans then, no, the first PAR cans I remember seeing were that same year probably in the autumn.

Ian (Knight) said, "The guys who did Woodstock are coming across to open up a theatre in London. They are taking over the Astoria at Finsbury Park, why don't you come in and work there and help set it up and get it going?". So a group of us who had been at the Roundhouse went across there, I think I went first, and met Richard Hartman, who was the advance guard from this group of Americans. He had actually taken a job in the theatre—they were still showing films. I remember distinctly when I first got there it was still showing Carry On films in the afternoon and evening prior to closing down and the Rainbow organisation taking over. Richard wanted to get started because he was just a firebrand of energy, and he did a job as an usher so he could be in the theatre. In the daytime prior to the theatre starting, he would go in and I joined him, going into the empty rooms that eventually became the dressing rooms and the backstage rooms. I remember they were full of pigeon shit, just full of pigeon shit because the roofs had been leaking and birds had got in. We spent weeks just clearing this pigeon shit out and painting the rooms white. I remember there was a poster from the early '60s of a variety bill which was headed by The Beatles of a show there up in the gallery that someone had stuck on the wall, I wish I had kept that. But a group of us, as I say, had gone across from the Roundhouse, including a guy called Paul Turner who was one of the lighting guys from the Roundhouse.

I remember that name.
He started Zenith Lighting with Dave Cohen. They borrowed lights from the Roundhouse to go and do shows. I think their first big customer was Emerson, Lake and Palmer, and they started off by borrowing the equipment and then they started buying equipment. Both Brian and Paul tell me that in those days, bands who wanted to take equipment on the road in the early '70s would phone up and book months in advance, you know, "We want 12 of those six-lamp bars" and "We want two of your Genie towers", but that's rushing on a bit . . .

At the Rainbow, the Americans who came in, the guy who fronted it was called John Morris, who I had met before this, at the Holland concert. He was looking after The Byrds, a sort of tour manager for The Byrds. John had been at Woodstock. He was one of the guys on stage at Woodstock when things were getting pretty crazy with Michael Jaffe. A guy called Chris Langhart was the technical director; Richard Hartman was the all-round production man and myself. The crews at the Roundhouse were a lot of freelance people, a lot of people travelled through the Roundhouse, so we had a sort of team of people who could do scaffolding, or could do electrics, or could do sound. We all migrated to the Rainbow, maybe half a dozen of us, and became the stage crew at the Rainbow for the first few shows. The first show was The Who, I remember; I have a programme from it and there is an article by The Who's lighting man, Roger Searle. I remember that opening at a show at the Oval where The Who and The Faces double headlined, and the earliest merchandising I remember was pink T-shirts with two girls kicking their legs up, I suspect maybe it was there.

So what were you doing at the Rainbow?
I was stage manager there. I was the first stage manager. I worked there for about three or four months leading up to the opening in 1971.

What happened that first time 'round was these guys who had done Woodstock: Jaffe, John Morris, Richard and Chris Langhart, a guy called Bruce Bile, a guy called Mark Morris and a guy called Bob Goddard, who has a company now in New York making DMX fixtures and very clever—Goddard Industries I think they are called out of New York—these guys were actually creating a rock-and-roll venue, and they were building equipment for it, which is where the PAR can came in. I remember Richard making the square PAR cans, and that was the first time I ever saw them. I remember Chris Langhart experimenting in his office with what was a little flying saucer of some sort, I don't know how it worked but I saw it fly. They had a camera built into the front of the circle that was suspended effectively so it had completely free movement, almost like a gyroscope-type situation, and their idea was they would film these concerts, sell the imagery and that would be another string to the bow, but they didn't have a great deal of money.

Do you recall who owned the Rainbow or who let them in, in the first place?
It was Rank, one of the big entertainment organisations. The company they set up was the Sundance Theatre Company, and it was really funded out of the proceeds of concerts. They were booking these major acts through every weekend for two or three nights. The Who were their opening night act.

The Americans were doing the booking?
The Americans were, whether they were booking directly, I don't know. They were dealing with the people who set up the early agencies; well not the early agencies of course, there have always been agents, but the people who were specialising in rock-and-roll. Mountain played there, Joan Baez played there, Quintessence came and did a show, and there were a range of bands.

But that particular lease of life came to an end with the Mothers of Invention. They came to play three nights, three sold-out nights. On the first night, a guy jumped on stage at the beginning of the set, grabbed Frank Zappa and threw him into the pit, broke his leg, broke a rib and paralysed his arm. As I say, I was stage-managing the show that night, and I was up on the electrics floor, which was 12 feet up from the stage on stage right with this Grand Master, which was the big old sort of house lights desk. Up with the stage lights, just taking down the house lights, and this happened almost immediately. I saw Zappa go into the orchestra pit, which was a 14-foot drop, and I looked down and he was broken up. It was obvious that was the end of that. I had to go down on stage and say to people, "I am sorry but the turn has been injured and there is going to be no show tonight . . .".

You had to tell the audience?
Yes, which was not a comfortable experience. Meanwhile, this guy who had done it was being beaten up by Zappa's roadies backstage very severely, to within an inch of his life, and that effectively brought that first era of the Rainbow to an end.

Why?
Because there was no money. As I say, they were paying the running costs out of the proceeds each weekend, and that was it, there wasn't substantial funding in

the background; it really was these guys who had worked with Bill Graham in America, had done the Woodstock festival and had an idea . . .

Came out of the Fillmore East.
Exactly, came out of Fillmore East and wanted to create one in London. Essentially the rumour of what it was supposed to be was Fillmore UK.

So how long did that last?
Six, eight weeks, that was all. They tried to keep it going but the company folded. I remember Richard stayed.

And that was the Sundance Co?
Sundance Theatre Company they were called. I remember we did stay on a little bit afterwards. I had a child by this time, and Marianne came and worked in the box office at the Rainbow. We needed to earn regular money, so part of the deal of working these extremely long hours in the build-up to it was, "Don't worry, when we are open we will be paying you properly, blah blah blah . . ." and this all sort of went by the board.

So after two or three months, I think probably after about three months, I had had enough and I thought, I am not going to stay here anymore. I don't know exactly what happened. I know the initial business organisation had collapsed, and I know Richard stuck around. I do remember the receivers came in and pulled out anything that was left. The people who knew that it was going into receivership, the day it went into receivership, trucks arrived at the back door, and people who had been working there were coming and taking things, you know, taking whatever was worth having out just before the receivers arrived. We had built a sound system, Richard had instigated a workshop in the basement where we built the monitor system for the theatre, we built raised staging to improve the sight lines, we built a huge amount in there to actually make the theatre a viable proposition and all hand-made.

Richard was building those cans, Langhart was building the lighting control systems and Arthur Max was in there, he was Pink Floyd's very first LD, he was there as one of the lads on the roundabout. So people were building and innovating. That was the first time that I remember specialised equipment for the music business being created in this country, and it came out of the Rainbow. Immediately after that Paul Turner, with Dave Cohen who had been a fellow electrician at the Roundhouse, set up Zenith Lighting and decided there was an industry here. They started buying early lighting desks, early dimmer systems, packaged lighting equipment, and so they were running Zenith from about '71 I think. Then I went back and did more work at the Roundhouse; there was a big building project going on there.

So, Zenith may have been the very first lighting rental company?
Possibly, I don't know the chronology. I think it was probably before Rainbow Productions. I sort of lost touch with Richard after that, and as I say then the

Rainbow did have a second lease of life, but all of the Americans had gone. John Morris, Michael Jaffe, Chris Langhart had gone but Richard Hartman had stuck around, Richard had stayed and had hung in there.

So where did you go off to?
I was really freelancing then for three years doing festivals, doing building work, doing the Sunday changeovers every Sunday between the concerts and the theatre, so I was around the Roundhouse a lot or working out on music festivals or doing whatever I could to make a living.

My experience of the early festivals, as I say, there were half a dozen or so that we did around the UK, it was a completely hit-and-miss situation that you may or may not get paid. I thought, 'I can't make a living at this, if I go and work for two or three weeks setting up a site and running a festival, I want to get paid for it'. Usually we did, but the last one I remember was Fun City down at Worthing. MC5 came over and played, I recall, among many other bands, and the only money I made was by storing the beer for the guy who was running the catering in the back of my truck. £20, so it was sort of okay!

Then in '73 the Roundhouse said, "Come back as the production manager". So I went back from '73 to '76, by which time Implosion had more or less stopped, well by '74 Implosion stopped happening. I think by '74/'75 George Hoskins, who was still there till early '76, got very ill. He had really been the person who was a respectable administrator for the place, and everything else about it was not considered particularly respectable. By the time he got ill, there was a movement to take it over, there was a movement by the staff at one point to just take it over and to actually self-administer it, but it petered out.

I had learnt an awful lot, I had learnt a lot about dealing with local authorities. One of the biggest lessons was not to sit in meetings pretending you understood. If you didn't understand, stand up and say, "I don't know what you are talking about, can you actually go through that again". And so it was a great learning curve for me the GLC (Greater London Council), the architects department, the district surveyor and the fire officer, and I took it fairly seriously the second time 'round. It was good times but not quite as vibrant and exciting as the first spell in there; the first spell really was anarchic and a lot of drugs and a lot of 'close to being out of control' situations but a lot of fun.

I couldn't see it going very much further without a drastic change. You need the backing of institutions to keep a place like that going, and we had no business sense, no business knowledge, and I certainly didn't have the vision to see quite how we would get from closing the doors, locking out the straight world and moving it forward. I think that it was in the first three years there that movement happened, but as I say, by the time George got ill and really had to cease looking after it, by '76 it was clear that it was going to change.

Then a woman called Thelma Holt came in to run it, and she was very much focused on, "I want good theatre here. If we keep having concerts on Sundays and theatre in the week, you can't create viable sets so you can't fund the concerts. It doesn't raise enough money to fund the changeovers and the change rounds, so really you have got to say it's going to be theatre". She went for Arts Council subsidies and subsidised companies coming through, and there was some very good theatre there for the next few years. Then it closed around '83.

Camden tried to set up a Black Arts Centre for a while, gave it 3 or 4 million pounds, which sort of vanished into the ether without very much being done, and then it pretty much closed down until fairly recently. That's the Roundhouse story really, and now its back, it's a great venue again.

So, you obviously got back into it at some point.
Well what happened was in '76 I left again, and with Marianne and our two children decided to drive across the States, well actually we left in July of that year and travelled around England. That Autumn I had come into a bit of money so we thought well, let's have a break, visiting various friends around the country, and then about October/November we decided to go to the States and spend three or four months. Take the kids to Disney World, buy a vehicle and drive across America. We had some friends who had sailed to the Virgin Islands and they said, "Well just stop off and stay with us for a week or two before you begin your travels". We got to the Virgin Islands, and they were living on a houseboat in St. Thomas, it was just completely glorious. That winter, the citrus crop in Florida froze, they had a dreadful winter in the States that year, and we were thinking, this is beautiful here, and this is really idyllic. So we spent three months in the Virgin Islands and ended up having just a week or ten days in America.

I came back at the beginning of '77, and then I was working pretty seriously with fringe theatre companies. A company called Joint Stock (Theatre Co) I worked with for three or four years, and I also did some teaching at a place called the Drama Centre, which was a drama school in Chalk Farm. Joint Stock was a company set up by Bill Gaskill and Max Stafford-Clark. David Hare wrote for it, and Howard Brenton wrote for it, so it was pretty serious theatre that had grown out of the Royal Court. I built sets, toured their shows, production managed their shows for them for three or four years. Then around 1980, Paul Turner said, "You know this business is really getting to be quite a big business, and I need some production knowledge, so why don't you come and work with us?".

Zenith?
This was with Zenith, yes, Dave Cohen had gone away by then, so Paul was running it on his own. I went and did a year in the warehouse there to get an idea of actually what the business entailed, you know, what the business was. Then I started working in the office taking on shows, looking after shows, and then it was taken over by Samuelson's in about '83, and then Brian Croft's company was taken over too.

So you all ended up at Samuelson for a while, didn't you?
Then TTR (Richard Hartman's company, Theatre Technical Research) was taken over, so yes, we all ended up at Samuelson's, which became Concert Production Lighting, well, Theatre Projects Concert Production Lighting. I went to the States

in '87 for them, primarily to look after The Cure—actually it was to make sure it went out properly. Richard (Hartman) was running the American office. There was Samuelson Concert Productions Limited and Samuelson Concert Productions Inc., but by this time Samuelson had sold out to Eagle Trust, and I went to America as I say, to do The Cure, and Richard was putting together the big Floyd tour.

I was going to ask you Jon, three questions really: who was your biggest influence in the early days?
Ian Knight without doubt because he gave me chances to do things, he gave me the chance to go to the Rainbow, and he got me involved with these festivals, followed very, very closely by Brian Croft.

Who do you think has been a big influence in the industry?
On the lighting side Brian Croft without doubt, and Richard Hartman as an innovator. If you talk to almost anyone in this business today, Brian would have touched their lives in some way over the last 30 years; he is a hugely influential man.

And the third one was: what do you think has been a big influence in the lighting world?
I would say the advent of business. The advent of it becoming a real business, that's what has developed it and kept it going. It's actually changed from 'making it up' into a very highly organised business, so that's influenced everything about it.

I must say I don't think I have seen smarter lighting company offices!
PRG[7] now are probably the biggest company of its type in the world. Set up by a guy called Jere Harris who was a scenic carpenter out of Broadway, he had a company called Scenic Technologies and started buying up companies in this business.

I don't think it's Dow Jones listed, I think it's funded by private equity. He has been very smart. This company in its current form has shown me how important it is to adhere to sensible business principles, i.e. just getting a job for the sake of getting a job is not always the most sensible thing to be doing from a long-term business plan type of view. As much as your heart might be in it, it is not necessarily what your head is saying. It's interesting that it's taken that long, it's only in the last three or four years it's really become completely clear to me. You know the people I am talking about—Brian, Ian, most of the people who I have worked with over the years—have been people ruled by their emotion, by as much as anything else. It's the mentality of 'their friends are doing the tour', 'my friends are doing this tour', so 'I am expecting to do it' or 'I have got to do it for this price' so 'okay I'll do it for this price'. It's not always great.

Jon, is there a funny story that you tell, is there one that comes to mind?
I wish there was. I don't have the sort of Brian Croft range of repartee! I can remember a story out of the Roundhouse, a show called Atomic Sunrise put together by some of the people who had been at the Rainbow, production managed

by Sam Cutler, who used to be the Stones main man. The Stones did the Hyde Park show, and it was a sort of mixed-media show, it was art, it was music, it was poetry, it was a complete mixed-media show, an evening event. We spent a couple of days getting the building decorated and set up ready to go, and then on the evening of the show, an hour or so before the door opened, a great big cauldron of punch arrived for everybody. All the people, maybe 150 people, had worked preparing the building for this show, so we were going to have a communal drink before we opened the doors and all have a wonderful party. Everyone came along and had a glass of punch, and then within about 15 or 20 minutes the rumour started circulating that this punch was spiked full of acid, so I had a couple more glasses just in case, and the rumour was absolutely true! About 150-odd people who had put this show together and were supposed to be running with it that evening were all completely off their heads! I remember at one point hiding in the store where we kept all the seating rostra because the rostra was open framed and you could crawl in underneath and just thinking 'I have got to get out of this, this is extraordinary, mad'. I mean the show happened, it went off and somehow or other everyone survived until the next day, no one died, fun was had, but the whole thing was done on acid. That was one story and that featured a couple of times.

There was also the night that The Faces were given a big cake ... it wasn't icing sugar on the top, it was sprinkled acid ...

Simon Napier Bell's book exactly describes that whole scene with The Faces and The Beatles and The Stones, and he did, they all did, got seriously into acid. I have never taken it in my life.
Well, I did a few times, and it was awe inspiring but ...

Didn't have a bad trip?
Got close to bad trips, you start getting very hung up on a particular train of thought so you can't get out of it. I had a couple of bad nights and I stopped.

Very wise, Jon. Jon, thank you very much indeed.

Jon is currently an Account Director (Theatre) for PRG UK, having worked there since 1995.

Notes
1 www.ukrockfestivals.com/blackhill-enterprises.html
2 Early show organized by Alfandary — www.ukrockfestivals.com/camden-Festival.html
3 www.ukrockfestivals.com/Krumlin-festival-1970.html
4 www.ukrockfestivals.com/bickershaw-menu.html
5 www.ukrockfestivals.com/phun-city-menu.html
6 www.galeriekralingen.nl/hpf/hollandpopfestival1970/festival1970/index.php
7 www.prg.com/about-us/our-people/profile/jeremiah-jere-harris/

3   John Morris

John Morris
© Barry Z Levine www.woodstock
witness.com

April 12, 2008 with John Morris on the Phone from the USA

Listening to these great stories from Jon Cadbury and later from Richard Hartman about the beginnings of the Rainbow Theatre made me more inquisitive about the reasons behind this invasion from the US. So I started to delve into people's phone books and finally tracked down both Chris Langhart and John Morris.

Talking to Chris Langhart at the beginning of 2008, I wanted to get to the nub of why and how the Americans came over to open the Rainbow Theatre. He told me it was John Morris who originally set it up and 'invited' him over to London in September 1971 to create a rock venue in London. John Morris, of course, already had an amazing history behind him, having booked all the acts for Michael Lang for the Woodstock Festival. John had set up a bunch of backers to pay for the crew from the defunct Fillmore East to move over to London and set up from scratch at the Finsbury Park Astoria, turning it from a cinema into London's first rock theatre. The cinema had closed down in September 1971, and the lease was taken on by John Morris and his holding company, Sundance Theatre Co., and for the first time was renamed the Rainbow Theatre. Chris got together a bunch of the crew that had been so much a part of the Fillmore East, and they all came to London. They created a full-on light show with the Joshua Light Company, built a sound system from mainly JBL components, got the crew from the Roundhouse to build a proper stage and began to run the venue in November 1971.[1] The opening shows were The Who, followed by Alice Cooper, Barclay James Harvest, Joan Baez, Curved Air, Family, America, Yes, The Byrds, The Kinks, Isaac Hayes, The Faces, Pink Floyd and many other famous names.

Over the phone I asked John to tell me more about the history behind the opening of the Rainbow and his involvement from an idea to fruition.

John, it's great to be talking with you. Let me ask you initially about the Crawdaddy venue you managed before you set up the Rainbow.
Well that was before the Fillmore East in New York. The guys at *Crawdaddy* magazine came up with an idea about running live concerts in a place called the Anderson Theatre, which was better known as the Yiddish Anderson Theatre because it was the home of Yiddish theatre in New York.

There was a guy called Tony Lech, who was a bar owner in New York, and Tony put some money into it, and that's where Chris Langhart and I met. Josh White, who was doing the light show brought Chris over, because he was teaching

Old Anderson Theatre poster

at NYU (New York University), and so what happened was that we did a series of concerts there.

Was this on 4th Street, John?
Yes, on 4th Street, well, between 3rd and 4th Street. The Anderson Theatre was the other side of the street to the Village Theatre. We got Bill Graham to come into town to look at the Village, because I knew Bill, I'd toured The Airplane and The Doors in Europe. Bill came and looked at it and we literally cajoled, tricked . . . whatever you want to call it . . . him into taking over the Village Theatre, which became the Fillmore East on the other side of the street.

Tony Lech was a bit strange, he was a semi-village heavy, he was a Lower East Side bar owner, it was a situation where Crawdaddy didn't have much money, Tony didn't have much money, but we opened up at the Anderson and did brilliantly. We had BB King opening for Janis Joplin.

What year is this, John?
This has gotta be around 1966/'67. Finally got Bill to say, "Do it". Bill and I put together a partnership to run the Fillmore East, which was Bill, Ron Delsener, Albert Grossman, Burt Block, who was Grossman's partner, and a guy who was involved with Loews Theatres, . . . and this partnership got together to do the Fillmore East in New York.

The Fillmore West was open, Bill had a reputation for that and the shows in San Francisco, and I had worked with Bill managing the Airplane. Ron Delsener, who was the major promoter in New York, called me up about a week before the opening, we had all the meetings in Ron's office actually, and Ron called up and said, "John, if you don't mind, I'm looking around, and there's all these heavyweights involved in this deal, and I don't know what's going to happen, and I was here before everybody else, and I have a feeling I'm going to be here after everybody else, so if you don't mind I'm going to politely decline to be part of the partnership". So he backed out, and he and I still laugh about it today, I still see him every five years or so.

We went ahead and opened the Fillmore East, and my job was basically the managing director, which was to get the thing put together. We rebuilt the theatre, we put in the light show, we built the sound system, and it was 90% Chris Langhart and his students and his friends. We did it and we opened up, and I ran it for six months to a year, I can't remember exactly. Then I got asked by the Airplane to take them and The Doors to Europe.

By then the Fillmore thing was together and working, and it wasn't as much fun, so I said "Sure". Which circles around to a guy called Ian Knight. Do you know a guy named Ian Knight?

Yes.
Okay, well Ian was working with George Hoskins from the Roundhouse, so we took the Airplane in there. I had seen four or five bands play there; those were the days in the Roundhouse when you had to watch where you walked because some of the boards in the floor were missing, but it's what inspired me after Woodstock

to come back and do the Rainbow. So Chris came over, I brought Chris with me and Hartman was there, and the amazing thing was, we had 22 Americans working there putting things together, we had a whole lot of Englishmen too, but 17 of the 22 Americans were people who had made their own way there and just showed up! People who had worked with us at the Fillmore or on Woodstock. It was just a fistful of guys who just showed up and said, "What can we do?".

Can I just ask how did you get EMI, Air Studios, John and Tony Smith to invest?
I worked my butt off; I knew some of the people already, not George (Martin) but a few of the others.

How did you know the theatre was there and available?
At the outset, after Woodstock, I moved to England, and I lived in London for over a year and looked at every venue that existed from the Brixton Astoria to Camden Town Rank. I loved England, I'd done a show in the West End five or six years earlier, loved England, always have. I'd be living in England now if I could afford to, I'm there once or twice every year. So I decided I wanted to do a Fillmore-type operation in London, and I started trying to raise some money for it in the States, raised some, and then I thought screw it, the only way to do this is to go to London.

I went and looked for a venue for a year and went back and forth with all sorts of people. In the meantime I had done a few concerts so I knew people like Chris Wright (Chrysalis), and I knew Tony Smith (the promoter) and a bunch of other people. Tony got interested in it, and Ian Malfeeney who was running Warner's at that point was very interested in it, really wanted to do it, and I believe it was him who turned me on to EMI.

1936 postcard
www.rainbowhistory.x10.mx/

It was a battle to raise the money, we didn't get tons but some, and raised £160,000. We got some money from EMI, some from George Martin and Air Studios, and from Tony Smith. George Martin had an accountant called Alistair Rainsford, and when we went under at the end he managed some kind of a deal with Rank where they tried to keep it open and kept it open a few months.

Top Rank owned it, didn't they?
Top Rank owned the building, yes, it was a movie theatre and it was only in fair shape. What was interesting is, Ian Malfeeney's father was a violin player and opened for The Beatles in the building in the '60s. I found a poster downstairs (at the Astoria) with his father's name on it and The Beatles, and gave it to him years ago, he and I are good friends. He, I think, made the introduction to EMI for me or put me in the right place.

The company was called the Sundance Theatre Company, wasn't it?
Yes, that's right, that was the name of the holding company.

Am I right in thinking you put a circus into the venue?
The day we opened up the theatre to sell the tickets for The Who, we had 10,000 people lined up around the building, literally throwing money on the floor as we were shoving tickets out the door; one of the Chipperfield brothers walked into the theatre, this is late October. He said, "Hello, my name's Chipperfield, would you be interested in doing a circus?", and I went "Yeah, sure".

What I didn't know was that to do circuses, you book theatre parties over a year in advance, you go to the ladies that book package trips and they book it, you don't just 'do a circus'!

So we struck a deal, and I was really excited, and it sounded good, and Mr Chipperfield was great, a real nice guy, and we died! We didn't get any audiences, because all the other circus things that were being done around the holiday time had been booked a year in advance by the kind of people who do package booking to schools and everything else. We got clobbered!

We literally had more lions and tigers on stage than people in the audience on some of the performances, and then the bird lady . . . the birds got loose and went up into the Spanish village over the proscenium. I mean, it was insane. The cops came after me at one point and said, "Your elephants are creating a hazard". They were being kept in a lot a couple of blocks away and walked to work and pooed everywhere; we had to build a ramp to get them on stage.

Was it the circus that financially put the theatre into liquidation?
Well no, it got us halfway there, maybe a little bit more, but when Frank Zappa was barged by that guy and knocked him into the orchestra pit, where he hooked his leg in the scaffolding and shattered it, and it was the first of six sold-out shows . . . that did clobber us.

We had to give all the money back (to the audience), and that sort of finished us off. I spent a week on the phone trying to raise help. Mountain came over again

Chipperfield Circus Rainbow Theatre poster
© Rick Burton

and tried to help out, but we did not have the back-up to be able to ride out the two of those things.

Yes, that's what buried us. Malfeeney was trying to get Warner's in at that point to bail us out as well.

The main idea with the Rainbow, which was naive on my part, but semi-worked (because I had done the Fillmore East with Bill Graham), was to bring in bands with the kind of production standard that we had created basically at the Fillmore East. It just got too expensive to get people to come all the way across the ocean to play. The English bands were pretty supportive, I mean, we opened with The Who, but it just wasn't enough to keep the overheads and everything running.

I do recall when we were in trouble financially, trying to keep it going, and I went to Chris Wright and his partner Terry Ellis, who I really detested, at Chrysalis who had Ten Years After and a bunch of others, and Terry put me through the ringer. One time he called me at 3 am in the morning and said, "Could you come to my office immediately?" and I said "Yes, okay", and I got over there about 3:30 in the morning, and he said, "I've decided not to put the money into your project unless I have complete control over everything" da da da da, and I politely said "F*** you", or not so politely, "You get me out of bed at three in the morning to try to horseshoe me into something that gives you all the plays, no way", and I left and I've never forgotten that meeting. The theatre went down at that point.

The thing about the Rainbow story, which is sort of sad, is that it quietly disappeared. You know when I left, we just couldn't keep running it, we were losing too much money, and in the end we went bankrupt, Rank took it back, and it sort of fizzled out.

John, thank you very much.

John currently is co owner/producer of 'Objects of Art shows' providing a full range of production services. From planning and layout of floor plans, to technical supervision, assembly and construction of art, antique and other shows.

Prior to co-producing The Santa Fe Show, John and Kim Martindale founded and co-produced The Antiquities Show in Santa Fe, which ran for 15 years. John has lived in the USA, England and the Caribbean.

Note
1   https://books.google.co.uk/books?id=Ng8EAAAAMBAJ (article from *Billboard* magazine, 18 December 1971, page 51)

4   Richard Hartman

Richard Hartman

March 20, 2007 Sitting Here with Richard Hartman at Stufish Entertainment Architects, London

Richard, let me start by asking you where were you born and where did you go to school?
Well probably the school that's relevant to this conversation would have been my last school, which was NYU School of the Arts in New York, and I say relevant to this conversation because that school was on the Lower East Side and was opposite the back door of the Fillmore East. Because of that proximity to the Fillmore and what the Fillmore East was in the US music business in the late '60s, i.e. an epicentre of new and psychedelic music, the school became a think tank for new technical ideas on how to present this new medium of music. A lot of the ideas for new and different theatrical things came out of that NYU group and were actually field tested at the Fillmore East at concerts.

The reason that this little niche was so important at the time is to be aware of what was happening around it. It was the time of Woodstock 1, Altamont and the Isle of Wight Festival. There was a desire that there should be something more to these grand events. The nature of what theatre presentation was conceived to be, prior to these events, was based on legitimate theatre in that, it was all about Lekos and 2K Fresnels and similar equipment that would do Broadway or West

End theatre shows. As a result of what went on at NYU and wanting to go a step up in lighting presentation at the time, we were developing other ideas. What were needed were other light sources to do big shows. We recognised, being avant-garde students, that there was a real lack of something for the music business that conveyed the intensity and vibrancy of what was really going on. Designers had to make do by utilising what was available. Nothing was really available for where we wanted to take design. Where we wanted to take it was a very simple concept; it was to make 'light' the 'emotional scenery'. It was the psychedelic era of the '60s; there was Joe's Lights, who were experimenting doing wiggly blobs of colour on a back screen in a theatre environment, but it didn't have the punch to match the music. You had designers like Arthur Shafransky at the Fillmore East, who became known as Arthur Max (Pink Floyd's designer), trying to find brighter sources and really trying to do something with music and light and make it almost a three-dimensional environment to match the music. Unfortunately, the Fillmore East and the Fillmore West shut down in the late '60s, so the ideas from the NYU think tank were never implemented to the full. I then came over to the UK as part of another project, which was to open the Rainbow Theatre in North London for the very first time. It was to be the next generation of the Fillmores. In that theatre we brought these concepts with us of how we were going to make it the next step up. It was to have the best sound and the best lights and actually be able to film record in it, which at that time was a totally unknown quantity.

Woodstock was recorded though.
Yes, that was on film, and it made all the bands nervous because they suddenly thought the road revenues were going to disappear if they had their stage act put onto any kind of recorded medium, but that's another issue. MTV was in its infancy, too. At any rate we went ahead and we re-equipped the Rainbow Theatre. Originally, there was a production group led by the tech of the time, Chris Langhart. In fact, Chris Langhart, out of the States, was also the leader of the NYU theatre tech think tank. And John Morris was a promoter type, who worked under Bill Graham at the Fillmore East and also worked on Woodstock 1. He came to the UK after the close of the Fillmore East, and with some other techs of East and West Fillmores and myself brought the new concepts into England.

So into the Rainbow Theatre we put 3200k PAR 64 cans. We had 400 custom-made units in there, which was more than anybody had ever used anywhere in one place to do illumination of music acts. There were several ideas that were tried out. One idea was four-quadrant lighting systems that maintained the light intensity level in any one of eight areas when the light colours changed, and from one direction to another. This made filming a snap. Unfortunately, that idea was a little too advanced for the time, and a pre-set control was developed. But the main drift was that we brought PAR 64s onto a stage in England for the first time. With smoke in the air, the light beams became the scenery and were electrifying.

Sound was also enhanced in horsepower. At the time English sound was based on WEM cabinets, which were basically guitar amplifier cabinets. To get a big, clear sound, people would stack up endless quantities of these things to make walls of sound, not really understanding how sound was created per se. It became 'size equals volume' kind of a thing but WEMs weren't very efficient. We brought in JBLs from the USA, and I think we were using H&H UK amps for our system. We really had a lot of it, so that with that much power and clarity the sound was crisp and clear.

The building was leased?
Yes, John Morris leased it from Rank. The deal went off with this venue in North London; it was one of two of the biggest venues in London at 3200 seats. Its sister theatres were what's now called the Brixton Academy in South London and The Sundown in Edmonton.

The Rainbow Theatre that I started with here in the UK brought in labour from the Roundhouse in Chalk Farm, which was the 'Rock Palace' of the day. Many of the Rainbow crew went off after it closed down and started their own lighting companies, especially after seeing what was put into the Rainbow.

At the same period of time as the start of the Rainbow a lot of transatlantic artists were coming over to play, and they saw the results of the Rainbow lighting system. There was a very small community of lighting designers who twigged onto use of PAR 64s as the lighting instrument of choice, especially the 3200Ks, so their use proliferated.

The 3200Ks were a movie light, they were not these 2900Ks they use now, which are a long-life kind of lower intensity lamp. Back then it was still the psychedelic era, and it was still about lighting intensity. You also had people using thousand-watt aircrafts along with PAR 64 3200Ks because they were absolutely the brightest things you could get. All the people who started with the Rainbow and the acts who came through it picked up on this vibe, and the 'Genie was out of the bottle'. That venue under that management went through a period of about four months before it closed. It was bringing in big US acts, a lot of British acts played there too and twigged onto what was going on with the lights and sound. I mean people like The Who and Yes played there, and of course, once they got into that environment, they saw a means to play the 20,000-seat venues so went away and got themselves a lot of their own equipment! This in part was because at the time there wasn't the rental equipment base that you now have to draw from when you do a big venue.

Think back to when they talk about The Beatles playing Shea Stadium. Nobody could hear them, as all they had was backline in the middle of the stadium. Everybody was struggling to expand technically, especially after Woodstock. Promoters' eyes saw dollar signs; they had to start using the big venues where the gate was secure, and suddenly the formula of 'intense lights and clear sound' started to spread all over the place. Like all they have got to do is get a big venue and then fill it with people and a rock star and 'boom' they have got a lot of money. Well yeah, but you have to support that with sound and lights. Suddenly demand started to be for not doing small theatres but doing 20,000-seat venues where you really did have to have that intensity of light, really did have to have the sound that was going to clearly fill a big venue. All of that started about the same time, very early '70s, using really high-intensity lighting that was portable, using sound that really could fill a stadium. That's why it went beyond WEM cabinets because they didn't have it. The push was on to develop all of this new equipment that could specifically do this scale of things. Everything that you now see in PLASA (Professional Lighting and Sound Association) which everybody takes for granted, was in response to people trying to develop specific products.

PLASA is the sound and lighting show at Earl's Court every year, that's constantly evolving with new theatre technology. All the equipment generated, connectors, cables, amplifiers, dimmers, lights etc. was all in response to this emerging market that came about starting in the '70s essentially as a result of festivals like Woodstock.

Suddenly there were a lot of acts out of the English market that leapt on this bandwagon. In the early '70s, groups were leaving here and going to the States to do big arena tours. All your big groups now are ones that went through that era and technically they developed or invested in equipment that started out from those simple beginnings at the Rainbow Theatre. The Chalk Farm guys reformed, and it became another lighting company called Zenith Lighting. It was down on the river and Jon Cadbury was with it. They serviced a number of different acts. Another group took the ideas and worked out of the Midlands at LSD, Light and Sound Design, and became the Birmingham group of people. They serviced another whole genre of acts that were in the North.

Now you are missing a big technical gap here that I would be interested to hear your version of because John Coppen told me quite a lot about the advancement from static lights held up by trees, static trees, hydraulic trees and then I think your famous towers, and at the same time trussing. Can you tell me how that evolved in your head, because that came through or after Rainbow, didn't it?
Yes, it did.

You were only at the Rainbow for a few months; what happened?
Well, the Rainbow, the way that was originally set up, its premise, was to try to get in big-name acts and to do it well. The problem is to pay for a big-name act and the overheads you need to be able to do two shows a night, and you need to have a liquor license. That's the model. The management struggled to get the liquor license, and they could never get the two shows a night. That was because a lady who lived opposite the back loading dock door complained to the council. Hence, the shows could never go past whatever it was, 11 o'clock, I think. When you back that time up, it means you could never do two shows a night. This means you never get your gross up enough to cover the costs. That's why that company failed after four months because they just couldn't get past the economics of having invested so much in, basically, gutting that theatre from an old cinema and filling it with lights and sound.

We also had our own generators to run the lighting because the LEB (London Electricity Board) was going to charge us horrendous amounts of money for the electricity to do a pop show, you know, because of maximum demand especially in November, December and January. All that went on in that place, but it just could not crack it because it couldn't get the second show. So it folded.

Then it had a reincarnation and some Chrysalis management came in. It was on that second generation that Dick Parkinson and Annie Pocock got involved in managing the technical side of what went on in the Rainbow Theatre.

Now, of course, at that point, it had some leftover equipment from the original inventory but not a lot; it had basically been sold off or it was hired and got sent back or repossessed or something like that, who knows.

At any rate the seed had already been planted, and so at that point Dick and Annie called their first company Rainbow Productions that came from the Rainbow Theatre and that group of people formed.

Meanwhile, the original others went off to Chalk Farm and formed the company that Jon Cadbury was a part of. Brian Croft was down with John Brown and that group of people at ESP along with Roy Lamb and Edwin Shirley. Later, Roy and Edwin would split off from ESP and form EST, Edwin Shirley Trucking. But let's get back to structure now . . .

When we first opened the Rainbow we used theatre bars to hang all this lighting on because it was basically a fixed installation. Now subsequent to that, when people started to tour the equipment around, there was a cry to find something that was portable, that you could get the lights in the air with. The initial thing that happened is trees with Genie lifts. These were single-masted, CO^2-driven Genies that would put up frames of lights like trees. Eventually one had trees and lights all over the place and everybody realised that this really wasn't what we needed because we just couldn't get enough lamp density, mass or whatever, in the air, and so we were all looking for other solutions.

I was trying to produce 'truss' because originally when the Rainbow first folded I went out on the first Wings tour, and the only thing we could get to put lights on was antennae trussing, and that's what we used. We put it up either on ground support Genies or it was hung from the building.

Three bars welded together in a triangle?
I think we actually had steel antennae section that was three bars welded together with web trussing—solid steel rods and very heavy. From that we said, "There has got to be a lighter way". That's when we found Grainger in Surrey, who was doing aluminium trussing. It bolted together, it was antennae section, vertical mast. I remember at that point I was doing freelance work and design, and I designed truss for Brian and John Brown at ESP that was triangular. It was cocked at an angle with footplates, and it held bars of lights inside it because we worked out that you could actually put it on its side. So the truss was really that and the lamps hung in it; we put frames on the end of it so that you could actually roll it around. It was kind of like the first pre-rig truss, because it had the PAR cans inside.

Were you making PAR cans as well?
Yes, in the beginning I was making square PAR cans. That's what I made for the Rainbow first time round. They were about two-foot-six or three-foot-long PAR

cans, and that's what controlled the beam spread. That's why you could take a Very Narrow or a Narrow PAR, put it into the end of one of those, and the can would actually focus it down into a really hot spot so you got rid of all that peripheral light spill. That's why the cans were so long because you can't focus a PAR.

It had nothing to do with the fact that it was easier to make squares than circles?
Well I always had it that with a square can it was easier to make it for me because I had a bench press, and also if you mounted the lamp in there the heat would go around the face of the light, up through the corners of the can and out the back, which meant you weren't cooking the gel. At the same time, technically gels had not improved in so much that they would take high-temperature beams of light. You would change gels about every other show because these things would just cook and burn the centre out of the gel, hence, that was another reason for making the barrels so long at the Rainbow; at that time you couldn't get high-temperature gel. Gels were all Cinemoid, which was an old theatre gel that worked in front of regular, cooler theatre lights. As I say technically all these things were struggling to catch up with what the demand was, and high-temp gels weren't there. Lighting incrementally was moving forward trying to improve, and trussing was too.

And before trussing?
We had Genie trees, okay. After trees you had Vermettes, which were a cheap and cheerful mechanical crank-up tower, absolutely dire and not very safe. Those things started to take up all sorts of trussing combinations, any kind of truss, either my truss or anybody else's version of a truss. About that time before '75 we had a fellow called Eric Pearce[1] come along. Eric Pearce had a lighting company called Showlites, originally called Keylites, in Bermondsey. He got a contract with ELO, and he also hooked up with a fellow called Mike Crisp, who actually came out of Strand. Mike went out to Norwich and he started a company called Tele-stage, and that was really the first piece of welded aluminium trussing that had any impact on the market.

That was in the early to mid-'70s, and actually that material was a mainstay into as far as the '80s. Now the other thing that prompted the use of trussing is that the big designers in the States were doing rock bands like The Stones. They were trying to find heavy truss that you could hang all these big theatrical lights on like 2Ks and things that really had some kick. So they got scenery shops in New York like Altmans to build scenery for rock-and-roll tours, and what they built were big, heavy trusses. I can remember doing a Stones tour in the early '70s, and we would have a truss that was about a metre high or so, and it had lamps in it. It had a lot of heavy cabling in it too because it was all 110-volt equipment. They would be on tour, and they would come over here with it, and of course people over here would see it, and that would kind of generate interest in truss and 'that's the way we ought to go but not as heavy'.

What was holding all this up?
Well all sorts of things. I can remember one client coming over with hydraulic rams, I remember doing a Cat Stevens tour with sets of hydraulic rams that would go 20 feet in the air. I mean, these rams were probably a thousand pounds each, which were really interesting to tip up on end!

So we are talking weight.
Huge weights but a really inefficient use.

Sorry, I don't understand a ram?
It's a big cylinder with another cylinder inside, you turn the oil on and it pushes the first cylinder up like the ones you see on the back of these 40-foot lorries that tip.

Like big Genies then?
Yes, but it was hydraulic, and I mean the cylinders were heavy.

Did Genies become hydraulic?
Not that I remember. I came up with using steel tube towers because they were relatively light, mobile and compact, and I could put 1-ton chain blocks at the top to lift battens or trusses or anything else. Having a stiff, strong tower for ground support did spawn other ideas. Trussing became lighter, hence why aluminium came into the mix.

This was steel truss on edge?
No, at that point we had gone into aluminium truss. The towers were thick-wall steel tube, the trussing was all aluminium.

When I first started in the early '70s, I might have used that Grainger stuff to do it, in fact I think I did, we had to have something that would take a lot of weight,

Grateful Dead stage 1978, Giza, Egypt
© Adrian Boot

Richard Hartman said, "What is shown in the photo is an aluminium bolted-together antenna tower made by Grainger and modified to raise and support truss. It was a forerunner. From this, the 'Hartman Tower' evolved as a steel 3" or 3.5" x 1/4" wall square section single-tube mast to hold a manual chain hoist mount at about 20–22'. I think the downstage towers are Vermettes".

and give us a base to do some kind of headlock, which we could then use chain blocks or something like that. Then you got into a period where lifting became the criteria, and it usually had to do with money.

The Americans came over here and started bringing CM chain motors. Those came in all sorts of configurations, well many couldn't actually afford them in the UK, and so another alternative was to use a Verlinde chain hoist. There were a whole bunch of companies that grew up using Verlindes as a way to go, it was out of France. That was a development that was progressing as you get into chain motors that would lift things rather than manual hoists to lift things.

So, going back to Eric Pearce, he got the money backing from ELO and made contact with this fellow Mike Crisp, and they drew up a welded aluminium truss. He then went and looked at where lighting was going. At the time everything was single hung or single connected, and it took a lot of time to rig. With the numbers of lights going up, to single connect them one would never complete them to do a show!

You would need lots of crew.
Yes, so Eric looked at incorporating into the ELO set, multicores, and hence the Socopex cable connections were multicore, and that became an instant success. It's the date of that ELO tour, which again is probably in the mid-'70s or something like that, which is a landmark. It was the spaceship tour. That was the thing, the spaceship landed, and when it went up, there was the band. That was I think the first time that I know of that they used multicores, multicore bars and aluminium truss and probably Verlindes. That combination started to come about in the late '70s because automation didn't come on the scene until about '81. That's when Varilite came out, and that's when you also had Pan Command come out.

So during the mid-'70s you had a rapid development of things that were trying to satisfy a market that was growing far faster than you could imagine. You had Altman sending PAR cans over here with the big bands, and people here looked at them and thought 'well we could make those'. Then you had companies like Thomas who started to make PAR cans spun from aluminium.

Where were they based?
Thomas were out in Pershore; it's where their plant was. Also you had, I think, a group up in Birmingham probably starting to make their own PAR cans, again spun. Eventually the Taiwanese started to make PAR cans too because worldwide the market just mushroomed as things took off, and the PAR 64 became the lamp of choice. They were doing rock-and-roll tours because of its horsepower, it was the only thing that you could make a statement with, you had to have something on stage that would make the stage brighter than the house lights in the hall, and that was it, that was the reason they went for that lamp and then the numbers of lights grew.

PAR cans didn't exist in theatre?
No. The PAR light is a movie light. Originally the Los Angeles movie studios used them.

To make night into day?
Well yes, you know how movies are made, right, you need a lot of light. Originally all the movie studios in L.A. were outdoors with sunscreens over the top because they used sunlight. They shot in the daylight so when you go inside a studio, you are trying to get really bright light, for film intensities. To do your outside shots inside, you needed something with horsepower, and GE developed a range of PAR lights that were 3200K for film called 'redheads', which were made by a Los Angeles lighting company. I forget their name now, but they made these little short, stubby things, and I first used them on a tour I did in '68 in the States because I was doing a movie tour called Medicine Ball Caravan with a French director called Reichenbach.

What's a movie tour?
Well we had four basic acts that we used, and then the record company provided a bunch of other acts. It was a movie about a group of people, hippies out of San Francisco, taking a journey across the States and eventually winding up on a farm in Kent with Pink Floyd playing on the back of two flatbed trailers back-to-back. That was sort of '69, and that was when Pink Floyd was just getting started.

And this is at the same time as you were in Tech school?
Yes, I went from Tech school, and then I went into that movie, and then the year after that Fillmore East closed, and I had the opportunity to come over to the UK to the Rainbow Theatre and worked there at the start of the '70s.

It was a bit of fate really that rock-and-roll lighting came out of movie lights rather than out of theatre lighting. The power was never in theatre lighting in the first place.
We knew the intensity was there; we just had to refine them, you lose too much in theatre lighting, it could never have enough kick to do it. People started trying to do rock-and-roll in Fillmore East using 2Ks and 4Ks, but they were just too big. You had to use Fresnels, which was the biggest thing you could get to throw that much light out, you just didn't have the instrument to do it, it wasn't until we latched on to the movie lights. That's what we were going to do at the Rainbow.

Now Roger Searle tells me that going to America with The Who for the first time he saw steel trussing and couldn't quite believe that it was actually going to stay up; he was very scared at the time because he had never seen anything like it before. He also told me about lighting boards and that they used to be obviously on stage because there were no multicores.
We had none of that stuff either in the States and, yes, you could have lighting boards on stage. Then, we had to develop the whole thing with multicores and connectors. The only one who could do connectors at the time was the military, and they were very expensive. What wound up being a real standard were tank connectors, which were these multi-pin things from Litton Avionics, really expensive, gold-plated pins, insertion type. It was the only thing that was bulletproof enough; everybody was searching for a solution that would not fall apart on the

road. In the beginning the bands were trying to get out there and do what they really wanted to do, but there were no real companies that could support them from a rental basis. It was a struggle for them because the parts were so very expensive to build up that fast to do a big band in a 20,000-capacity arena, so a lot of the bands actually bought equipment themselves. Hence The Who bought all their own equipment and wound up turning it into a rental company when not on the road.

Led Zeppelin used to hire it, and Roger and a couple of other Who crew members used to put it up for them before they created ML Executives.
Exactly, I mean a lot of groups in the beginning had to invest in the equipment because they couldn't find the equipment fast enough to do the tours. Then you got into tour managers and accountants and production managers who were looking at the cost of things and thinking, "we paid that much, to do this?!", you know.

They were out there busting their butts and not getting a lot of return on it. In part it was because they wound up paying for a lot of crew to singly hang lamps or some low-tech thing. The bands really wanted to find other types of equipment on rental, or they made or developed what would suit their needs so they could get away from the tour cost of having to invest in older technology.

Other markets were obviously opening up. In the European market Germany was the only one who had large halls. That became the mainstay of a European tour; a string of big halls that Hitler built in all the major cities in Germany in the '30s.

In the early '70s you would go to the capital of each country to a big theatre and that would be it?
Yes, most of them were just big theatres. The French promoter wound up having to create a larger venue by using that abattoir place in Paris, the Pavillon; the Spanish wound up using all the bullrings they could find. The promoters had to be really crafty to find big venues where they could secure the gate and then hence do the tours. Fritz Rau in Germany tied up all of the huge Messe halls for large concerts.

They were trade halls really, weren't they?
Yes, or they were velodromes without the bicycle rinks in them. Scandinavia had very few, and if they were anything, they were ice rinks, you had to cover the ice with fibre duckboards. In Italy there wasn't much, they were into velodromes and outside football stadiums, which were a challenge to do, so everything was kind of struggling through time. Equipment was struggling to catch up.

From what John Coppen remembers, you were always working in tandem with other people. Did you go off with Dick and Annie then when they created Lairhurst?
Well I am trying to think how I ever wound up with them because after the Rainbow I did a lot of freelance stuff. I worked with Mickey Tait at Yes, and he went to the States and opened up Tait Towers, and so at some point I must have tied in doing something for them. I don't know where I met them, but I wound

up needing a place to work because I was building lights, trusses, dimmers and eventually desks, and building some control equipment.

This was Lairhurst?
This was Lairhurst. I used to rent the equipment to them, and then they used to put in the Genies and the lights, the PAR cans and the towers and the trussing, although I built their trussing.

So you had the control system?
Yes, I would build their trussing for them, and then they would buy it and I would rent them the control equipment. I was kind of like their 'deep tech' during that period. Then I built towers, you know, the lighter-weight steel towers to hold up the trussing.

Did you have any formal education on this? I mean, it crosses several different skills you could have got three different degrees for electronics, construction and as a surveyor.
A bit of this and a bit of that in the '60s, and then I went into formal school. I have been in theatre most of my life since age 12 doing different kinds of theatre, so I had that as a background. Then when I went to NYU that was when the experience started, because of the philosophy to look laterally. Up until then it was just about, you know, flats and canvas and scene paint; after NYU it was about hydraulics and plumbing and pneumatics and pyro and electronics and amplifiers and more.

This is where it all came from?
Yes, all of that stuff, that's what the Fillmore East was all about. People were starting to record remotely there, you know, 16-track machines for sound, and sound was getting bigger. Transistors were just on the market. I can remember re-wiring amplifiers for Sly and the Family Stone because they kept blowing things up at the Jimi Hendrix studio and incidents like that. Back when everything was a single transistor and all the amplifiers were direct-coupled amplifiers. One transistor goes and it takes out the whole lot, and there wasn't any of this integrated circuit stuff that you know as ICs, they hadn't been invented yet. I mean, you had to look at parallel technologies of the time as to what was going on.

Were valves ever used for PA?
Absolutely, they were used for PA. In fact, a lot of the stage amplifiers, which are favourites of the groups, remain as valve amplifiers. Santana plays on a Boogie because of its sound. It fades, it doesn't clip and that's why I think all those people like the sounds of tube amplifiers because they don't clip.

But the PAs they went transistor very quickly, didn't they?
Yes, because they were looking for pure horsepower. If you go back and research Chicago, which was a big band out of the States, they toured a lot, and in their

touring they were one of the first to develop a large amp of two or five thousand watts. It was compact yet powerful. I mean the thing is it was almost all transformer; they went to all military transistors at the time because it was really about power handling. To do this they were pushing the limits, and they were trying to get it into a small box as well, they made custom amplifiers that were incredible because it was all about packaging, too.

I'm trying to think of the English ones now.
Well when we used H&H at the first Rainbow we stacked these up because they could only do 100 watt per amp, and there was a lot of air in the space to move.

At the time they were pretty good, but again it was about power, you either blow the amp up or you blow the speaker up because we were really trying to drive it, trying to get the efficiency up, you wanted to be in the feel of music too, not just hearing it.

So chronologically from the Rainbow through to Lairhurst you became a manufacturer, then a rental company to a certain extent that turned into TTR.
Yes, after Lairhurst folded in '80, what else are you going to do, right? Lairhurst folds and Dick and Annie decide they have had enough, so I caught up with John Coppen and Dave Murphy and Smoking Bill. We said, "Well what happens if we try it on our own and we do the whole thing ourselves?" So we take everything down there into Edwin Shirley's yard, I rented a big place for all my stuff, which I had to move anyway out of where Lairhurst was. The rest of it got liquidated, I don't know what they did with it, so we really had to start from scratch. At that point we started with welding aluminium trussing, but also at that time what was happening in the lighting industry was that automation was coming out, and automation is a lot heavier a light than a PAR can. So the ability to put that much automated lighting on a piece of Tele-stage, which was at that point the truss of choice if you like, was becoming increasingly difficult. I was still manufacturing equipment, I mean I was also making the first roofs for ESS (Edwin Shirley Staging) and getting into aluminium welding, because TTR had to make heavier-duty truss. I made everything you know, I made control desks, dimmers and eventually made heavier trussing. I made folding trussing, too.

Different companies used to buy off you or would you rent out?
Some did or I would rent out. The touring market was developing and required large amounts of equipment, so TTR continued to inter-rent even though it was busy enough on its own.

TTR went on until the mid-'80s, and at that point the impact of automated lighting was such that the lighting companies were becoming a black hole that you could continually pump money into. With automated lighting it meant that there was starting to be a shift in the budgets that went on tours, everybody wanted automated lighting.

I remember all of a sudden we have got two LDs out front...
Yes, you got your Varilite LD, then you got the conventional LD. Well the conventional lighting market went down the tubes in the early '80s because automation was taking more and more of the lighting budgets. Suddenly, the groups didn't like it that they were now paying £10,000 a week for lighting and £8,000 of that went to automated lighting and £2,000 of it went to conventional lighting and all the other equipment support that you had to have, ground support trussing, mains distribution, intercom all that equipment. The only way you were ever going to get around that is you had to own the automated lighting, but there wasn't any to be had because there was still patent things going on between Varilite and the rest of the world.

You didn't build your own version, did you?
No, I sold the company instead because I realised that in order to get into the automated market it meant you would need an enormous amount of money to do it without infringing patents. Varilite was already in there, they had already secured their place. One of the relationships I built, I suppose because of my American background, is that I became an English translator between the Texans at Showco and European tours that Lairhurst did. I became a marketable item for Dick and Annie to be able to use to support a lot of the acts that they got.

That Showco did?
Yes, so when a Showco act came over here, I would be the interface person that would smooth out all of the bumps that goes on between Americans and English techs because I had lived here for so long I could translate, and they trusted that; you know, I did the same thing for Santana.

Well Europe was pretty scary for the Americans back then.
Yes, we did Led Zeppelin, Bowie and we did any American act that they brought over. Subsequently, when TTR came about we continued to do those acts, and so that Showco link to the American acts became the mainstay of the company.

Because we had supported Varilite into the early '80s I realised that there was a certain life limit to this thing. The return was getting smaller and smaller, automation was taking more and more of the budget. Yes, a few more automated lights were coming out, but they cost the earth, and if I was going to stay in the lighting business it was just going to be too costly an endeavour.

Anyway, a company called Samuelson's suddenly got taken over by some people who decided that they wanted to, for want of a better phrase, 'stand in the mixer position and watch the stage artists' and think 'gee, that's ours', and they bought my company too, which was TTR. At the same time they also bought the company that Jon Cadbury was with, whose name is Zenith, and they bought the company that Brian was dealing with, TFA, and they rolled them all together.

Eagle Trust was the company that was buying you all out?
Eagle Trust, yes, they rolled it all together. I have to think back, yes, Samuelson bought TTR and then Eagle Trust took them over.

Nearly everyone worked for them at that point, didn't they?
Yes, if you look at the London lighting companies, Eagle Trust had cherry-picked the top three at that time. What I thought of as the three top lighting companies in London. I like to think of TTR as a top lighting company, although there was Neg Earth out there and Meteorlites.

There was Entec.
There was Entec, yes, I used to do a lot of rental to Entec as well. It all gets fuzzy back then. I got out because at that point I would have had to invest a lot of money that I didn't have in automation to keep up.

Do you remember TASCO's moving lights?
Yes, but it couldn't compare. It was an attempt to try and do something against Varilite and make themselves a sound and light company; all singing, all dancing, another Showco, but it never got there. Clearly Varilite had a real superior system at the time because the control was bullet-proof. I don't think they have even changed the control today, it was that good at the time, they have updated it but it was good from the start. Lamps were a bit iffy, but as the generations went on they got better.

Does Samuelson's still exist?
No, well not as such. I think one of the Samuelson brothers kept the film entity, which might still be operating today, but not very much. It was the younger Samuelson who was on this rock-and-roll acquisition quest and loaded up the house with all of these lighting companies.

Can I just take you back for a minute to your recollections of when outdoor staging started? Was it with ESS (Edwin Shirley Staging)?
Actually, there were a lot of people doing it early on, the Isle of Wight festivals or the Glastonbury festivals or whatever would have some kind of staging, and it was probably all scaffolding.

It was scaffolding, as I recall, I worked on The Who concerts at Charlton, Swansea and Glasgow in '76, which was the first time lasers were used in big arenas, to this day one of the most awesome things I've ever seen. Did you see any of those shows? They had mirrors halfway up the lighting towers of the football stadiums and bounced the laser beams from mirror to mirror.
You could do a lot more then without people running around worrying and looking at their insurance policies worrying about liability!

'What insurers?' I would think! I remember Harvey worrying because there was a guy climbing halfway up one of the stadium's lighting towers, and he was really worried about him falling. I mention that because I was looking after all the bands backstage for Harvey when I was very young, early in my

career still, and I remember the scaffolding companies coming in and building the stage.
Scaffolding was the way to go, in fact Edwin Shirley latched on to that and the trucking. Because they couldn't find enough stages, they did deals with GKN and wound up with GKN staging as the basis of their staging. Then they had to have roofs.

GKN?
GKN was a scaffolding company that make quick-form scaffolding, and that was the basis of Edwin Shirley Staging. ESS would rent in masses of amounts of GKN standard scaffolding as a system and then hire the crews to go out and erect it. They were one of the biggest and one of the first.

And the roofs at the time, what were they?
Well once they got into scaffolding they realised you could put scaffold towers on the sides of the stage with head blocks off the top on soldiers, and then you would lift up a big flat grid. TTR built the first flat-grid roofs for ESS because TTR was in their yard, and they knew that I was building trussing.

So you put motors on those flat-grid roofs?
No, they bought all the motors, we put the system on. A flat roof is just a big flat plane of trussing, a square grid or a rectangular grid, and it's lifted by motors that go overhead blocks on the side scaffold towers and go down to the bases where they are tied into the base structure of the stage.

And it would be tenting over that to keep the rain away?
Well ESS could never get that right, and so anybody who had an ESS roof always knew that if it rained you were going to get wet! No matter whatever they did, the roof would stay dry (because it had a big roof skin over it), but the PA towers to the sides of the main stage, there was no way you could get it waterproof, so it became like a 'Visqueen city' underneath it. Because of that, that's how Stageco made their name.

Hedwig at Stageco developed a staging system where he could totally encase the stage by the way his towers worked. Therefore, he could keep somebody dry in a Belgium rainstorm, I mean that was the selling point. Then he wound up starting to do a lot more shows for touring acts in competition with ESS, but obviously ESS have their dyed-in-the-wool clients, and they just will not change.

Again, it was about investment and about development of a product for an application. I mean, the number of monitor mixers who would sit in the wings and get absolutely drenched and their boards would get drenched under the ESS system. They got really pissed off, and that's when Stageco came along.

Who were Stageco?
Stageco were over in Belgium. Hedwig De Meyer, he developed staging over there that was tower based, it wasn't about using scaffolding, it was a specific tower that

was designed as a system. It was faster to put up, it was more compact to travel, and it was waterproof so you could keep a band dry.

And you personally didn't get too technically involved in the outdoor staging and roofing at the conception of it?
Not further than supporting.

Did Woodstock have a roof?
It was supposed to, it had some kind of a roof, but production ran out of time to instal a lot of that. We kind of had to deal with it as it was, it had a bit of a roof over it, but it was more like stretched and taut materials than an actual structure per se.

It looked nice-ish?
It was all scaffolding, you know, scaffolding for days because it was the only thing there was. No roofing and staging companies, it was all home-built. You get 'building scaffolding' to do the job of staging, and then you had to think of how to make something out of a product that was not really designed for it.

But, yes, I made stuff for everyone, you know, I had access to those engineers. Because we were going to make this type of equipment it had to work, and we had to deal with much larger numbers, big payloads and really start to be a construction project, so there were a bunch of engineers who started to evolve who could deal with theatrical things. They are still out there today as the basis of the engineering that goes on in this industry.

Did your towers go in as holding trussing on outdoor shows?
Maybe if they were under a roof they did. I never went into the business of building towers that large and being in the market for it, TTR was just a lighting company, and that's as much as I could deal with.

Stages are massive nowadays.
Well again, that's what I say, the stages are usually from Stageco. That's who's doing most of the touring of big stuff now. ESS still does some. The other thing that Stageco did was that it sped up the get-ins and get-outs.

Now promoters are booking outdoor concerts every other day with a system. You will get Genesis going out in two months' time, and they have got three sets of steel. It's like you can do all this leapfrogging of systems, multiples of the steel, then you start to get down to single sets of certain production items, and then all that has now been refined and designed such that it goes in on one shift. The truck rolls in and in 5 hours you can have a stadium show up on top of a subset of steel.

That's what we did with Supertramp.
They were basically doing an arena show outdoors where you could control the environment of the heavy elements by leapfrogging the heavy elements, and that's still the practice now pretty much, too.

I have asked the same three questions to everybody: Who was *your* biggest influence? Who do you think has been a big influence on the industry? And what do you think has been the biggest influence?
Okay, the three questions, the person who influenced me most was a fellow by the name of Chris Langhart, who was my teacher in NYU and subsequently he became a Tech Director at Woodstock 1. He inspired me to think laterally and to use all the information about all the materials that I could learn about in doing a project so I would say, without a doubt, he is one of those.

Do you still see him?
I haven't seen him in about four years. He is probably five or six years older than I am. He came out of Carnegie Theatre School in Pennsylvania.

And Chris got you into the Woodstock project; tell me how that happened.
Well I was doing a film in New York in the summer, and this fellow Chris Langhart contacted me and said, "Look, you know we have got this project that we are doing for a frizzy-haired kid named Michael Lang. They want to try to take a lot of the acts that were going to Fillmore East and do them outside and they'll film it". So he asked if I could come up and help him, get involved in this thing, and so I did; I used to do the film I was working on in the day, and then I would get transport up to this place in New York State. I would be working at night on his projects, and then I would come back down into the city and go back and do the film. I wound up building the artist pavilions backstage, all built out of telephone poles and some other miscellaneous bits back there. Then during the event I wound up trying to help them with the plumbing because we had real water problems.

For the toilets?
Well, when they put the plumbing systems in they had to buy 10,000-gallon tanks to hold the anticipated water. They put them all up on the hill, and that ran all the toilets and water supply at the top of the hill and down at the bowl of the theatre. They put all these stopcocks on the water that were not spring loaded, and so people would turn the water on and leave it on, or they would just break off the fittings or some fool thing. I was up there continually replacing them during the show, but unfortunately so much water wound up draining out that the whole place turned into a sea of mud.

Which is in the movie.
Which then proceeded to just kind of come down the hill to a sea of mud, and then it rained one of the days and that added to the sea of mud. Pretty soon the whole thing was just kind of a murky brown! (laughs)

Did you sleep over that long weekend?
I can remember sleeping a bit, yes, because I was up next to the helicopter pad at the back.

You had a tent or something?
I was next to the pad because that was actually how I used to get up to the toilets to fix them. There would be these helicopters coming in and out of the site with the talent and every once in a while I would go and requisition one of these things and have them fly me up to the top where the toilets were. Then I kind of abseiled into the toilets and try and fix them and then be pulled out and come back down, it was very gung-ho at that time.

It was just one of those events that happened, and it just had a life of its own. It was a very spiritual time. Everybody has their own story of Woodstock who was there, people who they hadn't seen in years suddenly they saw them at Woodstock, all that kind of stuff, but it was a life unto its own because no authorities could get to it.

I mean there was so much traffic going into this remote part of the Adirondacks that it got totally isolated, it was the first time they had shut down the New York State throughway in god knows how many years because there was so much traffic. People just drove until they could drive no further, and then they abandoned their cars and walked in, and so basically you had a situation where for a radius of somewhere between 10, 20 miles out from the event, the roads were totally blocked, you just couldn't go through with anything. The only way you could get people in and out was by helicopter, and it was a trip, it totally side-tracked the government, you know, caught the whole bureaucratic institution at the time totally off guard and actually scared the hell out of them.

As a result of Woodstock all the politicians suddenly were aghast that anybody could get four hundred thousand people or half a million people together in one place with one idea. So hence as a result of Woodstock 1 there were investigations, there were sub-committees investigating people and everything else because they thought it was a communist plot.

They got fearful that those are all potential voters and, 'My god, if they should have an idea that they want to do something and vote me out'. I mean, it was such a politically intense thing that nobody knew was coming when it happened, but here was this undercurrent of commonality of trying to go to a music festival that the frizzy-haired kid had read right, as now was the time to do it. It was after Altamont the big festivals with the Stones, three days of peace, love and music.

Had Mike Lang done anything before?
The frizzy-haired kid Mike Lang? I don't think so.

How did he advertise it? Was it just a local New York thing?
I don't know how, word-of-mouth spread it maybe.

Didn't he get *Rolling Stone* magazine or something to advertise it?
Nothing was that organised; it was through I think a music network that was developing. It went to the radical elements or it went to the student unions of the universities, because keep in mind what was going on at the same time. We had the Vietnam War, we had Kent State. Kent State, Ohio, that was where army reservists

who were called in to a campus to quell a demonstration against the war and they wound up shooting students. As a result, the backlash was that all the students in the country shut down all the universities and took them over. I took over NYU and we ran stuff out of NYU, we took over the computer centres and shut it down until the universities agreed with the students that they would capitulate. As a result there really was a communication network that had been set up to communicate university to university group about what was going on anti-war whatnot, and so hence when Woodstock came along it was at the peak of that kind of period. So you can see that politically, politicians who were in there and trying to fight for keeping the Vietnam war going, da da da, suddenly realising that this wasn't just a bunch of, you know, a few dozen protesters about the war, suddenly there was four hundred thousand to half a million people in one place all about anti-war, and it scared the shit out of them. The FBI were investigating everybody for a long time after that as a result.

Fascinating.
So that's the first question more than answered! The second question?

Who do you think has been the biggest influence in the UK industry, but it doesn't have to be a UK person or persons.
Well I think technically one of the pivotal points would have been the ELO tour and the introduction of aluminium welded trussing with multicore cable and Socapex connectors. I think the course of taking raw PAR lamps, transitioning and then going into huge arrays of lighting. This is when bands were going out with 12, 14, 1600 PAR cans, especially the heavy metal acts, because they were always good earners and had such a huge quantity of lighting.

When you are talking multicore, what are we going up to? We talking six at a time?
Six, basically six. Then you got into pre-rig truss, which is what Ronan Wilson developed at Meteorlites when he was doing Judas Priest or Def Leppard, somebody like that, because it all fitted into a sea container. From that you had people like James Thomas Engineering copying pre-rig trussing and building that out in Pershore. Pre-rig flooded the industry. Then you had people like LSD out of Birmingham, and Meteorlites who were always competing for the same acts. They were always kind of pinching them off each other, and they were using pre-rig truss whether they were building their own or using Thomas's or whatever.

What is the difference between pre-rigged and the triangular?
The lamps are inside the pre-rigged truss, which meant it was faster to deploy your lighting system. Ronan Wilson put them in the truss. Being the lighting designer he could do anything he wanted. First, you had Eric Pearce who introduced the multicores and the Socapex and the bars of six with the triangular welded truss, and then you had Ronan taking it through a pre-rig design. Both still happen.

Okay, that's the who, now how about *what* you think was the biggest influence?
PAR lights and automated lighting. Because automated lighting was the next stage.

Big gap between the two, too?
Yes, but it's just giving motion to the light. Automation lost in the horsepower, and I don't think to this day most automated lamps can match a PAR lamp's output.

One last thing I was going to ask, a road story, one that maybe you would tell over dinner. I quite often tell the story of when I did The Grateful Dead in front of the Pyramids.
So do I!

Harvey Goldsmith was asked to set up a European tour for The Dead, and I ended up getting the gig. I was setting up the tour when they just asked me to get all the equipment to Egypt.
Bill Graham took The Grateful Dead to the Pyramids and I, along with Lairhurst, provided the lighting.

I think the Dead's LD, Candace Brightman, might have asked me to hire you. I was the production manager.
Somebody hired us, but I wound up in that funky hotel next to the Pyramids.

The Mena House Hotel, I was there for a month with a fantastic view from my room!
I wasn't there that long. We had the Bedouins standing guard over the mains cable because the locals kept pinching the copper.

Photo from R. Ames' bedroom window, September 1978

Here is a rare photo of the man, his work pass for the Athens Olympics!

Do you remember the earthing system?
Not offhand.

Running water. A stake and running water.
Okay, I can remember being really sick there. Gippy tummy or whatever it was they called it but ever so thankful getting back into a British Airways plane.

ML did the sound and the trucking. Roger Searle helped me put it all together.
Yes, they had a couple of lads come in with our gear, and we put it up and then they had that bloody windstorm that blew it all down. We managed to get it all up again though and get it all turned on.

Did you see the show?
Yes. The show was basically purple, and they would go "Oooooh" and it would go from one shade of purple to another shade of purple, you know, "Aaaaaaaah". Someone was wandering around putting acid in any kind of liquid that people were drinking. They were all getting off on this and I thought, all right great, and at the end of the show they all got on horses and rode off into the desert to some bar, and in the middle of it there was an eclipse of the moon. I thought, 'that's a great lighting effect'.

Richard, thanks very much.
Well thank you sir. Good luck with the book.

Richard continues to work at the highest level for supply companies like Stage One or PRG Projects or Brilliant Stages in various capacities including Project Coordinator. Richard also involves himself in projects for Olympic-size ceremonies.

Note
1 www.parnelliawards.com/showlites.php

5 Annie Pocock

Annie Pocock

April 26, 2007 with Mrs Anne Cullum, Formerly Miss Annie Pocock, at the River Studios, Battersea

Annie, I know we met with Rainbow Lights, soon after you left the Rainbow Theatre. Tell me, where did you go to school and where were you brought up?
I was brought up in Northwood, Middlesex, and I went to school at St. Mary's, which was a very old-style girls' school, and basically I bailed out halfway through my 'A' levels. My father died when I was 13, and I think the guiding hand had gone at that point, so at 17 I did bail out of school and did the sort of modern-day equivalent of 'running away with the Circus' which, in my case, was to run away and work at the Rainbow Theatre.

Initially I worked as a follow spot operator, and then as a sort of trainee lighting tech, and then very swiftly, which seems absolutely ridiculous now, considering I must have been 19, I was the resident lighting designer.

There was a permanent lighting rig in the theatre that had been installed, Richard Hartman probably told you about that, when the American Fillmore crew came over to open the Rainbow with The Who as the opening concert. They installed, which was then a very new idea, hundreds of PAR lamps, in huge steel very homemade cans which were very difficult to manoeuvre. If you saw them now they would just look so amateurish, but it was the light show, the beam show in those days, which was a very new idea.

A lot of bands would come in with their own rig, which would sort of supplement the Rainbow's rig, and some bands would come in and just use the theatre's rig. Then the whole thing would be worked out on the day, and I would light it for the evening show.

Because a lot of bands didn't have rigs?
Yes, because a lot of bands didn't have their own rigs. A lot of single artists didn't carry huge amounts of crew. Donovan, I remember working like that with him and various other guys like that, single artists.

So who hired you?
There was a friend of mine, Dick O'Dell, he had a job there, and there was a bit of freelancing between the Rainbow and a sister venue they had at Edmonton (The Sundown). We used to go between the two, doing follow spot gigs. The follow spot gig was great because you got to watch the band and get paid.

Were there Super Troupers?
There were Super Troupers in the Rainbow. There were four Super Troupers, but the problem with them was that they were at the far, far, far back of the theatre. If you remember the Rainbow it was a very big auditorium, and that was a fairly scary business actually, just getting to them, because you had to go to the very top of the back circle, out of the door and up through, you know, really odd bits of roof to get into this room, which was totally cut off from the rest of the theatre.

They were never in great condition, the Super Troupers, and they were extremely difficult to operate. It was a very, very long way away (to the stage), and they weren't super bright, so you lived in terror of not being able to hit your target. If you blacked out and suddenly you had to come in (on cue) somewhere else, it was just absolutely terrifying because the room for error, of being in totally the wrong place by a really small movement, was enormous, very, very nerve racking.

You had to 'knurl' them by hand, you had to move the carbon rods together by hand, they weren't automatic, and I remember that being absolutely fraught. There was a 'knurl' knob; I don't know whether that was a technical term or one of the Fillmore guys' terms!

The thing we used to love doing was working the CSI follow spots that were in the Spanish village either side of the proscenium arch. They were much closer to the stage, and we were virtually hanging over the stage, and that was brilliant because you could be incredibly proficient. You were much more involved in the show, you felt that you were really there in the auditorium as opposed to being in this weird box.

The Rainbow was a weird building because there were so many parts to it, so many rooms backstage. I don't know if you can remember the backstage bar, which moved about to several different places?

In order to get to the lighting desk, which was hung under the balcony, you used to have to go through a series of rooms 'round from backstage, through a lot of rooms and in through a sort of 'hole' in the wall, very creepy, very weird. I think when the original Fillmore crew came over they actually sort of moved in. Did Dick Hartman tell you much about that?

No, he didn't, certainly not that people actually lived there. When was this happening, how long had they been there before you arrived?
I can't remember when they actually opened. I remember The Who opened it.

Okay, and were you there?
Yes, and a chorus line of dancing girls came on at the beginning with Keith Moon at the end of their line. I was working a CSI spot that night.

Were you the only girl there?
Yes, I was. The only other woman that I actually remember being around a lot was Jill Furmanovsky[1] (photographer). She actually has a studio and a series of outlets now, selling her back catalogue.

She's a real one-off. She turned up at The Rainbow, I think she was probably at college and she may have known Dick Parkinson or Mickey Martin, who was the other partner then, and said "Hi, can I take some pictures?" and I think Dick said "Yes, we can't pay you but you can take as many pictures as you like and you can be the official Rainbow photographer". So she photographed a lot of gigs. She would turn up for the shows and would either be in the pit or in front of the pit in the auditorium and she photographed. She must have a fantastic archive of early Rainbow days that are doubtless worth a fortune now.

Now apart from Jill, I don't think there were any other women involved. My friend Helen used to come and work a follow spot sometimes, she was at art college at the time. A few girls used to work front of house, in fact I think the only other woman I used to regularly bump into was Anne Bush, who at that time used to do various bits of promoting business on Mel Bush tours, but it was a very, very male preserve.

Jill Furmanovsky
© Nick Mason

Did the boys look after you?
Well, yes, it seems awful now when you look back on it, because there was a definite divide, and you had to make (as a very young woman working in that environment) . . . you had to make certain compromises about how you appeared to people.

Men who worked in rock-and-roll were extremely chauvinistic in attitude and divided women into two categories—you were either a groupie or you could be sort of, like, their 'sister' who you kind of had to look out for. Obviously I made the option that I would be a 'sister', and it worked out very well. People did look after me and people were respectful and protective, but I have seen at that time occasionally, other people in bands or whatever, who wanted to lead a more free existence and being given a very hard time. I think people found it very difficult, men found it very difficult, if you weren't categorised very clearly into 'groupie' or 'your sister', there wasn't very much in between.

Girls used to follow the bands, the groupie scene was in full swing then, it was pre-AIDS, the worst thing that could happen to you was having to take a few antibiotics and that was it!

With the productions that were coming through the theatre, I don't want to dwell on this particularly but now that you're mentioning it, did you see many other women actually working with bands?
Occasionally you would, it was a novelty. I remember Liza Minnelli playing the Rainbow strangely enough, and having a woman lighting designer who was absolutely terrifying, an American, really terrifying barking sort of "I want that there, I want this here", you know really wacky. I am trying to think who else was actually touring at the time.

At the Rainbow, do you recall any particular artist that was a favourite to light?
I was always very thrilled to get those sort of gigs where it was just you, you've got to do it, and that was always my favourite because it was a sense of progression, of achievement as you learnt a bit more about that. There were people like Roy Harper or Tom Paxton or Donovan who just turned up basically, and you spent a day trying to make something great in a short time.

I can remember more about the gigs than sometimes my involvement, because they all sort of blur into one. Dick would have a better idea of who we actually lit.

Apart from you, the first female I can recollect was when I worked for the Grateful Dead in '78 and their LD was Candace Brightman, and she first toured with the Dead in 1972.
The first tour I did after leaving the Rainbow was touring America with Rick Wakeman, I remember having my 21st birthday while I was on tour, and in a lot of venues it was difficult, you were regarded in the US, even more than here, with great suspicion as to what you could possibly be doing there.

By the union crews?
Yes, actually at one gig (with Dick Parkinson) I remember it was either Cleveland or Pittsburgh, one of those east coast union buildings; they actually stopped working because they didn't like me being on the stage around the rig.

There was a load of people in a big huddle, and I said to one of our crew, "What have we done? Why have they stopped working?" and he said, "Well, they are talking about you", and I was actually made an honorary member of the teamsters for the day! That was the way *that* was resolved.

There was occasionally trouble when people thought you were there as decoration, you had to constantly prove that you knew what you were doing. I remember being at another event actually, at The Palladium, and being on headsets while the resident lighting crew were also on headsets but didn't know that I was, they couldn't see me, but I was listening and it was like, "Well I don't know, she can't be more than 23. . . blah blah blah . . . oh no! Let the dog see the rabbit, let's have a look at what she can do . . ." so you had that kind of constantly, especially when you ran into older crew members, you know, 'she can't possibly know what she's doing'.

When were you born?
'53.

You're a year younger than me.
Am I? That's disappointing. I thought I was slightly older than you (laughs), you're wearing very well!

So are you (more laughs). I remember renting the Rainbow for two weeks in 1979 to rehearse the Kate Bush tour.
I'm going to wobble if you mention Bill Duffield . . .

So will I.
One of the things about my former life in rock-and-roll is that I lost three crew. At Rainbow Productions, we had a very faithful crew, we had crew that stuck with us through extremely impecunious times when no one was paid, but it was a very good life for people.

I don't regret the fact that a lot of those guys had a very good time that they might not have had, but the loss of Billy Duffield and Jim McCarthy and Andy Pollard, it was very, very hard.

When you joined the Rainbow Theatre[2] aged 19 in 1971, how did you end up being a director of Rainbow Productions?
Well it looked like the whole thing was coming down again, you know, the Rainbow was going to close again in 1975. I don't know where Dick had known Mickey Martin originally, I think there had been a bit of to-ing and fro-ing between the two of them over the years. Mickey was doing the sound if bands didn't have

L to R: Dick Parkinson,
Annie and John Coppen
© David Clark

their own system, as a full Martin Audio system had been installed in '72. I was doing the lighting and Dick Parkinson was the overall stage manager. But things changed, Mickey left and John Coppen joined us just after the theatre closed for good at our new home in Bethnal Green.

We used to do a lot of funny things at the Rainbow as well as rock-and-roll. There were always odd things going on, including Greek nights! It was a theatre that was never dark really. I wish I still had my diaries, you know. I have to try and piece it together bit by bit, and now when I think about it, there were months when we had something on every night. There was never a space between gigs, it was absolutely relentless the number of people that passed through.

When the Rainbow was finally closing, the idea just formed 'why don't we do this as a stand-alone business?'. If we are not going to be here at the Rainbow with the connections that we have made and the people we have met, with what we thought was a great idea about building equipment, then this could be a business for us.

Now Dick was a man with a bit of a plan, he always had ideas, unfortunately not backed up with an enormous amount of financial wisdom, which wasn't forthcoming from me either, so that was a bit of a bad combination. So we set out then to look for premises. I remember actually going to look with Dick. We trawled around Bethnal Green, which seemed like a good area, and found this warehouse in St. Matthews Row, Bethnal Green, near Liverpool Street. Brick Lane, just off Brick Lane, it has actually been pulled down now, the warehouse building. It was a great building, originally it was a transport company, and we shared that with Ground Control. I remember Ground Control came in, they supplied David Bowie's PA system in the Ziggy days.[3]

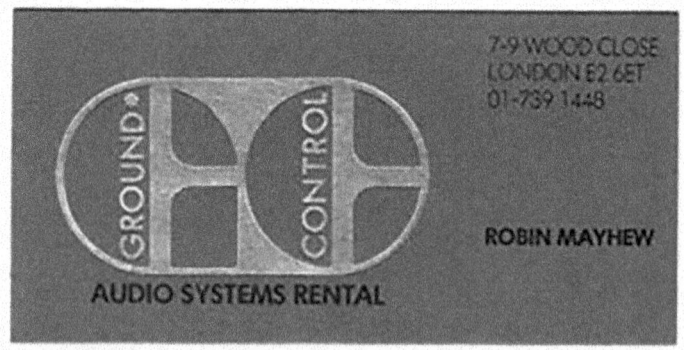

Ground Control business card
© Robin Mayhew

So you moved to Bethnal Green, and did Richard Hartman move in with you guys?
Richard Hartman had space there, but Richard at that time was living in North London. At some point he bought a place in the country where he had a big workshop and manufactured stuff. He used to come up all the time in his fantastic, I think it was a VW bus that he had painted green, and I mean literally hand painted! He was always tinkering with the idea of making equipment better and manufacturing better follow spots and better PAR cans. Although again, if you saw that equipment now, it did look incredibly, I suppose the most honest description would be 'homemade'.

But it was the first, wasn't it?
Yes, it was a great adventurous and amateurish business, incredibly unstructured, and we just used to make things and take them on tour sometimes with disastrous results.

As a director and LD of the lighting company, did you get involved with designing any equipment?
Richard Hartman was the driving technical force; we used to ask him, "Is there any way you can do this?" or "Is there anything available to do this?" I was never

involved in any kind of engineering of equipment. Richard was consistently doing that, he was always tinkering and modifying. When I think of some of the things, I mean, we were touring Australia with Rick Wakeman and flying out huge bars of PAR cans. Really huge, great long bars of PAR cans, in wooden casing, which basically didn't survive the journey; it seems incredibly naïve now. Gigs to play either side of Australia. I think we were going Perth to Sydney or Sydney to Perth and the stuff went in trucks. The roads were obviously pretty rough and whichever end the stuff turned up, I remember opening cases and just being horrified because it was just a layer of broken glass. It seems ridiculous now, but you know it was just very adventurous. A bunch of very young people trying to address a way of doing something that hadn't been done before and learning by experience and mistakes.

Did you go on the road before you moved to Bethnal Green or whilst you were still at the Rainbow?
I think the Rick Wakeman tour of America might have been before we moved into the warehouse.

Rick Wakeman, how did you get that gig?
You know I can't remember, we did have an association, a friendly sort of partnership with Mike Tait who was Yes's lighting designer, and when we did the Rick Wakeman tour we used Yes's truss. We took that to America, and it was a dreadful unwieldy piece of equipment that went up on triple Genie towers, and these triple Genie towers were in huge boxes.

What's the difference between a Genie and a triple Genie?
Three Genies in one box in a sort of structure that I think Mike Tait designed.

Gas controlled?
Yes, and there was some kind of stabilising structure inside. They were whacking great things, and it was something that you could only do in big venues because you need enormous numbers of people to lift these things.

The truss used to be put together on the ground, it was incredibly heavy, might have been steel, and then there was always the point, which was terrifying to me, where there would be a call for 'all crew to the stage', and you would have to have at least 12 people in a row to lift this truss into the coupling structures on the top of the Genie towers. My job as a sort of little monkey woman was to climb up on top of this triple Genie and guide the truss into the housing. You had to direct people as to what to do and that was always a really, really sweaty moment for me because that's when you got this kind of 'I hope she knows what she is doing moment, because we are all standing here' vibe and then when you get it right, "Oh, well done!".

I remember that being one of my early terror moments of which there were many. In retrospect it was a ridiculous idea, but I suppose that was the solution they came up with at that point.

No chain hoists in those days?
No.

Long live chain hoists! So first tour, Yes's lighting gear with the lighting desk out front or on stage?
Out front, it was always out front, which actually was a bad move, in my opinion; it separated you from the band. I always thought in retrospect that the people who had the best relationships with the bands they were working for always worked from the side of the stage, and were always much more part of the gig as far as the band were concerned than just being secondary crew. I always think bands have very good relationships with their monitor men, side crewmen.

So did you design from the side of the stage very often?
No, I was always out the front you know, it certainly didn't occur to me at the time, I loved being out in the auditorium. I can remember, I was 20, and on that American tour with Rick Wakeman we played Madison Square Gardens, and being out in the middle of Madison Square Gardens when that lights up, I just thought 'this is like being Queen of the World, this is a fantastic job'.

So multicores then?
Big multicores, yes, that was another one of my jobs. I remember the Richard Hartman multicore box, a lot of wood everywhere. You had this huge cotton reel structure that you used to have to lift out of the box and then reel it in.

It wasn't one cable, was it?
A few little extra bits for extraneous power lines for stuff but basically a multicore cable is just one cable, yes.

Okay, what happened after, this was '73/'74...?
'74/'75 there was a world tour through Japan and Australia.

Annie is now flicking through her old passport.
Yes, one of the few things I haven't lost, so '75 back in America again . . .

Who did you do after Rick Wakeman, do you recall?
I think when we started getting our own gigs, people like David Essex and Slade, who were of course huge in the UK and not anywhere else. A David Essex tour would be absolute madness, I mean a real sort of 'terror' in the auditorium and outside with the fans going mad everywhere, trying to get their hands on poor David. Then you would play the Bottom Line in New York and the clubs across America in a completely different atmosphere. All very peculiar but as a company we started getting other acts like Genesis, I mean the Peter Gabriel era, Genesis.

Did you ever go out on the road as head of crew but not designing?
No, I was only ever lighting designer/operator, which was a much more hands-on business at that point, I was never extra, always one of crew.

Annie as the LD (below Paul) on tour with Wings UK Tour, 1979

Didn't fly with the band?
No, a big mistake in retrospect... 'side of the stage, flying with the band'... ha! out in the front and the back of the bus!

Up first in the morning and last to bed. I always thought the lighting crew had the longest, roughest, hardest hours on a tour.
Yes, that's right—always in first and out last. I mean, I always felt terrifically responsible. I think one of the most stressful things was always worrying at the back of your mind about safety the whole time. Fortunately, we have never had any major incidents, never had trusses come down, anything like that, but we did have a couple of electrocutions.

Well you lighting people were playing with serious amounts of electricity most of the time.
Three-phase supplies, yes, the equipment had been rattling about the trucks, when I think about it now! I used to have two recurring nightmares, one was the truss falling, and my other recurrent nightmare would be to be sat at a lighting desk (I still have it now and I get very stressed, which is amazing), and the curtain goes up

and I look down at this huge lighting console and there is not a single mark on it, there is not a label anywhere. You know you mark it up with masking tape, probably not nowadays but in those days you used to mark everything up so you knew where everything was. The curtain goes up and there is nothing on it. Horrible! Stress! Nightmare!

Did that ever happen to you? The show starts and you freeze?
No, never, I think that was one of the main things, especially in the early days, it was always tremendous fun. We were very young, very resilient and didn't get exhausted. When I look at a lot of stuff now, you see pictures of people, and I think 'God, look at the crew, they're so old!'. I don't know whether that's true in your experience now, but I mean, we were all 18, 19, 20, up to about I suppose mid-twenties.

You know, there are a lot of people of our era still doing it, but at a much higher level than originally, so they are the bosses now. One would hope so after 30-odd years!
Still ploughing on. Yes, I do notice sometimes people are talked about as a 'legendary' lighting man or a 'legendary' soundman.

Annie, who was your biggest influence?
It's quite difficult to answer that question because I left and I'm not in the business now. It's very difficult. Without sounding incredibly pompous, I don't think there was. There wasn't anybody else doing it. It was difficult to kind of, you know, to model yourself on anybody or 'I would like to be like them' because there wasn't a role model to have.

Who do you think has been a big influence, not necessarily your biggest influence?
Who did I think was great? I think we always had this thing where we thought of the big American rock bands and their designers as being, well, you kind of felt in awe of, because they seemed to be so sorted and have it so together and such fantastic equipment.

You know all the people like the Doobie Brothers they seemed to have it all, so very slickly sorted out. I suppose Showco, I always thought that Showco—who we did hook up with in quite a loose way—you know, just seemed so much slicker. They made us look very amateurish sometimes, which I suppose we were!

Well they had those big buildings didn't they, and those outdoor arenas, their country so huge they could tour forever and never go to the same place twice.

John Coppen's biggest influence was Richard Hartman.
Yes, I would go along with Richard Hartman. I think Richard Hartman is an absolute one-off, an extraordinary character, and in many ways I am sorry

I haven't seen him for so many years; he is just, you know, unique. There were other guys as well, John Morris and Chris Langhart, they were very influential to me.

They also came over with the original Fillmore guys and Chris Langhart, in a similar way to Dick Hartman, seemed to me to have brains the size of planets! They were incredibly technically proficient, and the music was almost very, very secondary to them. They liked the way of life and they liked the milieu but you couldn't say, "What do you really like listening to?" because they would probably go ' "Well I like The Dead". John Morris, Chris Langhart and Richard Hartman, they could have almost ended up in another world you know, in theatre or any sort of venue, because what it was about for them was the innovation, the technology and that whole idea of creating a new way of doing things. I think John Morris and Chris Langhart were extraordinary, Chris was enormously tall, and he was really a great big bear of a person.

From a technical point of view, as a designer, what do you think was a big influence on the lighting world in your era? Brian, for example, said "sleeper buses".
I hated them, that was really tricky for me.

It allowed the crew to sleep between gigs.
I'd rather stay awake, and honestly half the time I think I did. Actually, I rarely went on sleeper buses, seemed to do a lot of tours where we flew or we drove, no sleeper buses, because there were no other women then. Later on, it was better when you got the girls coming in for catering, like Val Bowes and Judy Kelly, who worked for us at one point in the office, Flying Saucers and wardrobe, too. Then you could claim your female space on the bus, and people had to get used to the idea. But when it was just me that was really difficult.

Often I'd just have to cut myself off from everyone else because otherwise, I always used to have the feeling that everyone's not going to be enjoying themselves if they think I'm still around. I'm just going to check out, into my little bunk and disappear, while they're watching their porn movies upfront or whatever. It was uncomfortable and difficult, and I used to think 'I'm just going to have to sleep in my clothes, this is not working out well'. Well, I can imagine nowadays it's a whole lot more fun when there's the catering crew and other lighting people, but when you were that much in the minority, sleeper buses were just awful, solitary.

Ha! You're not going to agree with Brian on that one then (laughs)!
No, I'm really not, really, really not!

Did you ever go on the road with your husband, Ross?
Yes, the Roxy Music tour in 1981, he was doing sound and I was doing lights, it was very romantic.

Were you going out together?
We got married.

And then went out on tour?
I had gone to a scene builders to have a look at this set that we were having built for Roxy Music. Some of the time I was working for Bryan when he was being just Bryan Ferry even though he had Phil Manzanera with him, and then other times they were Roxy Music, and those two sort of crossover. Chris Adamson had brought Ross along to have a look at the scene builders, and that's when I first sort of set eyes upon him. We actually got together incredibly quickly; I don't know why we got married, it was a mad unfashionable thing to do at the time.

Where did you get married?
Chelsea Registry Office. We lived on Chelsea Embankment and walked 'round the corner. We were broke at the time, had to borrow the money for the license.

Was anyone else there?
We didn't want anybody there, and we were going to get two witnesses off the street. Then Ross's parents found out about it and they had to come, and my brother, and then Jimmy Barnett said, "I can't believe you are getting married and not asking me to come", it was like he was wounded.

Ross and Annie's wedding day

The internal politics and the touring unit are quite tricky, and if you have got a unit couple, people are not wholly comfortable with that. It was difficult, so we only did one tour, but we did do a gig at Perth, that huge stadium in Perth.

Was the support act Icehouse, do you remember?
I don't remember. Ross will. Why? Were you there?

I was the tour manager for Icehouse (aka Flowers), and we did some gigs with Roxy Music in Australia.
Were you there when Bryan did that big announcement thing?

I can't remember a big announcement.
Ross and I were at our respective desks and I was watching the follow spots, and the follow spot guy was really pissing me off because he wasn't doing what I was saying, and I said, "What are you doing?" There was like silence and the sound guy said "Okay, what's going on?" and suddenly the follow spots hit Ross and I at our respective desks and Bryan, who never says anything about anything, hardly speaks between songs. . . "and the next song is" or "thank you very much". I don't know if you have ever seen him at gigs, he doesn't talk much, but now he just says "Ladies and gentlemen, something extraordinary has happened in the world, blah blah blah, our lighting designer and our sound engineer, blah blah blah, and they

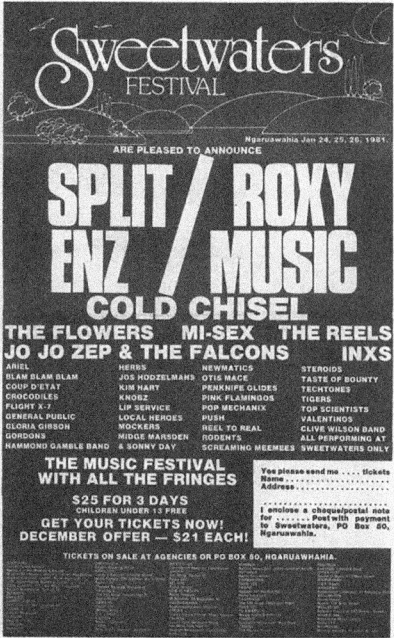

Sweetwaters concert poster

got married", and there was this like roar of applause, and for someone like me you can imagine the colour I was!

You just wanted to dig a big hole and disappear (laughs)!
And you know at that time an enormous number of people wherever we went, whatever we were doing, it was like "Were you those guys who got married?"

Did you go to New Zealand as well?
Yes.

A big festival?
Yes.

I am so crap at this. Were you there?
Yes, I was, I was with Flowers aka Icehouse, and we did the same festival, although we might well have been there on different days, but what a coincidence.

I think that's enough. Thank you so much, Annie, for your time.
Yes, that's quite enough, thank you very much!

Annie has more recently gone back to theatre production and has forged a late career as a tour and company manager with stints in the West End and touring the UK and America.

Notes
1 www.rockarchive.com/jill-furmanovsky-photographer.html
2 www.rainbowhistory.x10.mx/
3 www.robinmayhew.co.uk/pages/rmbiog.htm

6 John Coppen

John Coppen

February 26, 2007 with John Coppen at Cirro Lite Ltd, London

John, I met you with Dick Parkinson and Annie back in probably '74/'75.
Yes, Richard, 1975, I think, when you were tour manager with Steve Harley and Cockney Rebel. I think that was the first time.

And what was the lighting company called back then?
It was initially called Rainbow Productions and later changed to Lairhurst. There was a bit of a name problem, with ironically, a film lighting company, who were just around the corner from here called Rainbow Lighting.

Are they still here now?
No, they've gone now. There was a huge great wrangle about using the word Rainbow, and because we were both in lighting, we both fought it. Yes, we were in different industries, but we were from the Rainbow Theatre, Finsbury Park, which is where we got Rainbow Productions' name from, but Rainbow Lighting Ltd had been in business since about 1964, which predated the 'Rainbow Theatre', so unfortunately our name went. Nobody could come up with a good name to take over, so it was Lairhurst Ltd trading as Rainbow Productions. Lairhurst was the lighting supply company to the Rainbow Theatre in the very early days of the

second coming of that venue, and run by Dick Parkinson. Dick set up the company so that they could rent the lighting to the new owners of the Rainbow Theatre and supply the services.

A specially formed company called Biffo run by Chris Wright and Terry Ellis, who are also the bosses of the Chrysalis organisation, had acquired the lease on the premises. The venue reopened on July 1, 1972.

Did Lairhurst mean anything?
No, it was an off-the-shelf name. I think he went to the solicitors around the corner and got the name off the shelf and started trading. They needed a company to invoice for the servicing of all the lighting crew that were there.

Did you see many bands there before you worked there? I remember seeing Deep Purple.
I didn't see Deep Purple there, no. I did see Little Feat, but that was when Martin, my brother, was working there. I am trying to predate those days, but I am sure I have seen Yes, Pink Floyd and Traffic and probably Neil Young there, before being involved. I was working for Virgin Records at that time.

Can I go back, John, and ask a little about your family, your upbringing?
I came from a farming family in Colchester, North East Essex, and they were Dutch farmers from the 16th century, and they had land in London. There is a Coppen Road in Romford, named after my great-great-grandfather. It was a compulsory purchase after the First World War to house the Ford workers, and they moved from Whalebone Farm in Romford/Hornchurch area to Essex, North East Essex. My guess is they would have been farmers, market gardeners or market garden wholesalers.

Where did you go to school?
In North East Essex, Tendring and Landermere, and then on to college at Ipswich Civic College.

What did you do at college?
I did textiles, business studies, accounts; it was like an economics thing instead of doing A-levels. I opted to do something a bit more practical that would maybe give me a job, and it seemed like a way of getting out of school. I was there for two or three years.

Everything you do, everything you learn in life, it sort of pays its dividends later on, because without doing that I don't think I could have got involved in the way I did with Rainbow. I certainly couldn't have done what I'm doing now, it taught you about how to run a business. It gave me the basics of accounting and the basics of business studies, which was extremely interesting. Textiles, I am a qualified textile buyer, Richard.

The reason I am not in farming is because my grandfather gambled, so my father, after the Second World War, was a tenant farmer. His farm went from about 60 acres to something around 200 acres, quite a big holding, but he had always said to us, "Go and do something else. As a tenant farmer I can't hand it on to you, you might as well go and do something else", hence why I went to college.

I met a few mates there, and we decided to travel around Europe, bum around Europe for two years. We were working in hotels, bars, restaurants in Austria and stuff, and I came back after two years of that thinking I had better do something, and so I got a job at Brutus using my textile college connections.

Brutus shirts?
Yes, I worked for them for a couple of years, and then I got a job with Virgin Records. Actually, I almost got a job with Island Records.

Virgin Records when they were still small and down in Portobello Road?
Yes, absolutely. They had a base over in Avonmore trading estate, just off the Harrow Road, and I worked for Caroline Records, their export company.

What year was this?
This was '72/'73/'74.

And who was your boss there?
It was a guy called Nik Powell and Chris Stilianou, who was my immediate boss. Nik Powell was a partner with Richard Branson, and he went on to form Palace Pictures. They split up but Nik Powell was quite an instrumental part of the business end of the Virgin empire way back then. It was great fun working with them all, everyone was under 35.

When I first started, everyone was earning £35 a week, and the only difference was that the people with more responsible jobs had bigger record collections. While I was there, if people left the company, they would put a line underneath your debt. You see, everybody would have six or eight LPs a week or whatever and build up a massive debt, and when you left they just put a line through it.

Fantastic!
A chief accountant would have a 10-grand debt of records, and away it would go! While I was there the revolution happened; they actually changed the wage structure completely, and it was mayhem for about three months.

So you sold quite a lot of Tubular Bells then?
That's it, Tubular Bells, Hergest Ridge.

I'm tour managing him next week, Mike Oldfield, taking him to Spain.
I've got a feeling without Tubular Bells we wouldn't have Virgin Atlantic, we wouldn't have Virgin Railways, we wouldn't have any of that. That's where he (Richard Branson) made his money originally; it was a record shop in Notting Hill, Vernon's Yard, Portobello Road was the base.

That's where I went to be interviewed.
You went for a job there?

Yes, Branson asked me to tour manage Mr Oldfield's very first tour in '78 because he didn't tour when Tubular Bells came out. I didn't get the job, Mike thought I looked too young, and a friend of mine, Sally Arnold, did the tour.
They had a lighting and PA company of their own at one time, not when I was there but afterwards. I did a load of work with them, it was run by a French guy called Wiz but anyway that's by the by, they sold it all.

So you were in the export department, Caroline Records?
Caroline Records, yes, in charge of doing the licenses for the recording of any of the Virgin Records catalogue. So, for example, a record would be licensed for Norway to print say 25,000 copies, and maybe Sweden as well, or the records themselves would be shipped out, but they would be pressed here and sent out to Norway or Sweden or wherever.

Strangely enough at the time there was this massive business in exporting records from Britain around Europe, quite why it was cheaper to buy records from here I really don't know. I guess maybe it was because they couldn't get them, but it was a thriving business.

Fortunately or unfortunately there was a guy there called Rod Rickery, who became a bit of a friend of mine, and he was, I suppose, the business manager for Virgin. Believe it or not, it sounds odd now, but above or below the Virgin record shops that they were opening, there were spaces that weren't rented out, so they wanted to put other ventures into the spaces upstairs or downstairs. They decided to create a company called Virgin Rags and sell clothing, and I was kind of instrumental in opening the shops and getting it going.

I moved away from Caroline, because I had some clothing and textile experience so could organise Virgin Rags. We were putting sound systems in the Virgin Rags shops, decking them out, opening them up. Unfortunately, after about a year, it wasn't making money. There was no shop in London, they were all on the outskirts of London or other towns; it was a shame we didn't open one in London. Every shop was very different. What was selling in Brighton wasn't the same as what was selling in Edinburgh; you were buying all these different things, but it was very hard to make it work.

So Branson and Powell decided to put the chop on it and make everybody redundant. They offered me my old job back, and I said, "Yes, I'll think about it", but right at that time, strangely enough, a little turn of fate. Mickey Martin had left Rainbow Productions, and I was just talking to Martin and he said, "Why don't you go down and see Dick, we need somebody to run the office", and that was it, that's how I started.

Is Martin your elder brother?
Younger brother, and he's the reason I am in lighting. This whole bunch of people that you know, Billy Duffield, Mick Healey, Pip Robinson and Bill Martin, they

were all from St Martins School of Art or the College for Distributive Trades, doing point of sale there, along with Martin, my brother. Martin, with some twist of luck 'cos his artwork was so good, got in there a year or two years before he should have done really. He moved up to London and started living in college accommodation. He was too young to deal with it all. My old man was sending him money all the time, but he was a 16-year-old, in London, all over the show and subsequently got chucked out of college.

Which college?
It was at St Martins but it's the college affiliated with it, College for Distributive Trades, they do arty point of sale, exhibition designs, packaging designs, all those sorts of things. I think the whole thing was intertwined, and there were people he knew from St Martins School of Art and also from the College for Distributive Trades. Anyway, he got thrown out for not producing work or whatever and went back a day early to pick up his things and saw, up on the notice board, a job advertised: '*Wanted—fibreglass technicians at the Rainbow Theatre*'. So he went down to the Rainbow, thinking he would have to go back home to Colchester otherwise, and got the job with Dick Parkinson and Annie Pocock. And that was it, that's how the whole thing really started rolling. I think the fibreglass stuff he was building was for Yes, the band Yes. At this point Martin started to learn how to work a follow spot. If you learn how to work a follow spot, then you get some more money at night for operating the follow spot. Then they needed other people, and eventually his friends and mates from college all started getting jobs there. Luckily, most of them, I think, had finished their college work, and by then, enjoying being in the rock-and-roll industry. At the Rainbow at this time Dick Parkinson was running it, Mickey Martin was like the General Manager or the Floor Manager and Annie Pocock was the house (lighting) designer, so those very early days.

You say they ran the Rainbow?
Dick and Annie ran the lighting at the Rainbow, not the sound element, there was another group inside doing the sound, so they just ran lighting and staging. For instance, there would be a show coming in, and in those days it would either be a band come in and they would have their own equipment, or a band would come in and just have their own designer. Dick or Annie would tell the designer what equipment was actually at the Rainbow and then change the lighting rig to whatever he requested, or if the band had their own system, take out the Rainbow system or take it away in the fly bars and re-rig their system, whatever it might be. A third option was that a lot of bands didn't have a designer, so in those cases Annie Pocock would do a design for them. You would have to check that out with her, but I am pretty sure that was her job there, being the in-house designer for support bands.

Obviously, in those days it was very unusual for a smaller band to have a lighting designer. Bands used to go to a club or venue and use the lights that were in there, or use the house designer, and that was it. It was really only bands like Yes and Pink Floyd, and any of the psychedelic bands of the time, that actually had some of their own lighting kit with them. Bryan Ferry, he always used theatre lighting; it might be quite interesting to contact Nigel Olliff, who was Bowie's lighting designer, I'll put you in touch with him. He was Bowie, Bryan Ferry and Marc Bolan, T Rex's lighting designer. He was one of the very early lighting designers that were touring lighting rigs, along with Mike Tait of Yes and Graeme Fleming of Pink Floyd. You know all those names?

Yes, absolutely. Mike Tait and Graeme Fleming I know of, but have never worked alongside either of them.
It was quite unusual for a band, in fact it was quite revolutionary I suppose at the time for a band to have their own lighting. Those guys were at the forefront of making their shows look a little bit different.

Now do you recall when Dick and Annie moved into the Rainbow?
After the Rainbow version one closed down, new management came in, three months later at the end of May '72. They had to move equipment back in to start all over again, with the Deep Purple concert being the first in July '72.

What they started doing was running lighting tours from the Rainbow as well. So David Essex would come in and do a show, and Dick would talk to Mel Bush and talk to David's management and say, "Yes, we can do your lighting elsewhere", so from the lighting that was in the Rainbow, they actually started doing tours.

They had a rig in the theatre full-time?
Oh yes, they had loads of equipment there, Hartman had excess equipment, Rainbow Productions had excess equipment as well. In general Hartman owned all the control equipment, all the dimmers, all the mains, he built all of that from scratch. He was the first person to have dimmers, 30-way pack dimmers on wheels. Before that it was all Rank Strand six-packs, which weren't on wheels, they were really built just to go in the theatres, not built to go on the road.

Other lighting companies, touring theatre companies and ESP, which you probably already talked about with Croftie, they had the Rank Strand six-packs which were about that big (John opens up his arms) and toured with those in road cases, whereas Hartman built dimmers to go in road cases so you could wheel them into trucks. He was one of the first people to build them like that. It seems naff now, but at the time it was quite revolutionary. Anyway, they had the touring systems that worked out the back of the Rainbow, and Dick and Annie went out as a pair. Dick called the follow spots and was the crew chief, on I think it was David Essex, Slade and Rick Wakeman. They were the first three major bands that they toured outside of the Rainbow, using the Rainbow as a base.

So how long were you at the Rainbow?
I didn't work there at all. No, I didn't join them until after they had moved out of the theatre in 1975.

Okay, so you moved straight in to Bethnal Green?
Yes, they were already there, Richard; they had already moved out of the Rainbow, well, they didn't move out, the Rainbow closed down.

When they left the Rainbow they moved into an old sack warehouse in Bethnal Green. Dick and Annie had all these acts that they were doing, all popular, and so either they would do their shows, or design it and another designer would go out on tour with it, or they would introduce a new designer, so they were out quite a lot of the time.

Rainbow Productions lighting crew, 1975
© David Clark

L to R: Jim McCarthy, Martin Disney, Willie Porter, Mike Pia, Dick Parkinson, Anne Pocock, Pip Robinson (½ a head), John Coppen, Alan Goldberg, Sandy Imrie, Lee Francom (½ a head), Martin Coppen, Simon Franklyn, Mick Healey, Bill Martin, Martin Gilfeather's friend, Martin Gilfeather, Amanda Stanley, Peter Rayle, Bill Duffield.

Martin, my brother, said to them, "Why don't you try John, he has just been made redundant from Virgin Rags, why don't we get him down here?". Basically I went to see them, they didn't have any options, it wasn't really an interview, Dick just said, "Sit over there and we will give it a try", and essentially that's what we did. Three months later they were off to Australia and I was running it.

But I have to say it would have been impossible without Hartman, absolutely impossible. He was holding my hand and teaching me the ropes, what was what, and how to organise the crews, and really I came in as a general manager, co-ordinator, overseer, just making sure everything was going right. That's how I really got into it, Dick and Annie going away more and more, they got busier and busier, so it really entailed me getting involved more and more and picking it up as I went along, but it was enjoyable, it was always a challenge.

So the Rainbow Theatre closed its door in '79.
I remember going to the Rainbow when I was working at Rainbow Productions. Dave Hill was involved, remember Dave Hill? Well he was involved in the third coming of the Rainbow.

Dave Hill was the stage manager, wasn't he?
Yes, well, lighting designer, stage manager. He was the main man in there.

By then Dick and Annie had gone from the Rainbow, and Rainbow Productions were basically flying. I suppose the two main companies at the time were ESP, which was Brian Croft's company, they predominantly were doing the Deep Purples, The Stones, they were doing all those bands which were the major majors, and Rainbow was really, I don't know, biting at their ankles trying to compete if you like. But luckily some of the bands Rainbow Lights were doing were getting bigger all the time, and of course they did all Mel Bush's work. I think ESP was doing Led Zeppelin, Bad Company as well then.

So there were just two companies then?
Yes, well, ESP was the big company, no two ways about it, any band coming over from the States and ESP were where you went. Rainbow existed in a small way, but when they moved out to Bethnal Green it became a major competitor.

A big reason for that was because all the crew that they had, had been training at the Rainbow. They had been doing the load-ins and the load-outs every day, a new lighting rig every other day, back-to-back shows. There had been a huge number of shows there, so the crew got to be absolutely honed to perfection, really pulling gear in and out and focusing.

Dick, to his credit, trained them. I don't quite know where Dick got his training from, but he trained them to be really particular about lighting cues, how you focused (and this obviously came from Hartman's practical side of safety), and how you used things, about electricity and about electronics. Hartman was teaching them every day, too.

Out of the initial crew that they had, they probably got more of a grounding than the ESP guys. Put that together with revolutionary equipment that was faster than what ESP had, and you have a cracking little lighting company, which was probably, for two or three years, the flavour of the month.

Richard, you would know that as a tour manager, having to make the decision about which sound company, which lighting company you are going to use, you look at who everybody else is using.

Hartman was like making his own cans, he had square PAR 64 cans, which were before their day, and they were lightweight, made out of aluminium. Back then, the only alternative was a steel can made by Altman in New York, which were very, very heavy.

The more weight you are putting up, the more problems it causes, with either Genie towers or trussing or theatre bars, so he invented a simple can. Basically, it was folded aluminium 'round the square of a PAR lamp shape with little holding screws, but they were a quarter or an eighth of the weight of the older PAR 64, which essentially was designed to go into theatres, where they would be put up permanently, and it wouldn't matter how heavy it was. In fact, the heavier it was, the longer it would last, and it would be less maintenance in the future, but he sort of turned that all on its head and invented the square can.

Why square and not circular?
Because it was very easy to make! Richard was always very practical, and essentially it was like a square box, open at one end, electrics coming through at the other. It was very basic but it made their lighting rigs a lot lighter, so they had the edge on lighter PAR lamps to go in the air. He also made some of the first trussing, the first aluminium trussing.

Yes and towers.
And the Hartman towers as well, but the trussing came along before the towers. His first trussing was not welded, it was all bolted together, but again it was very light and ESP used to buy it off Hartman. Mike Tait from Yes used Richard's expertise to build trussing and various bits and pieces for Yes, and Floyd as well. So not only were the Rainbow making equipment for their own road shows, but if the Floyd were on tour Hartman would always be involved, and if Yes were on tour he would be involved as well, so we really had the main brains.

And he is still out there.
He is still out there and doing it, Richard. Essentially, I suppose he was the brains, because of his experience at Fillmore East and Woodstock. I know you are going to interview him, but I am pretty sure he was doing a drama course focused on lighting in New York, and that's where he kind of cut his teeth and learnt the trade. Coupled with all of that, the real thing that I suppose pushed Rainbow forward was the link-up with Showco, the big American sound, lighting and trucking company. Predominantly in the States, they were the major rental and supply company for big bands such as Genesis, Led Zeppelin, Bad Company, James Taylor.

They were all going to Showco, and they needed a British operation, or European operation to look after their tours over here in Europe, and that's where Rainbow really kicked in when that happened. We suddenly found ourselves with American designers doing tours for Genesis, Bowie, Zeppelin, and at that point we were probably edging ESP to being the biggest company, certainly doing the biggest acts at the time. By then Hartman had honed in, he had done a wrap with an aluminium PAR can, because we wanted a better-looking round PAR can, and he modernised that end of it. It was continually moving on, and it was major to get those American acts, it was a major, major step forward.

How did that liaison with Showco come about; was this before or after you joined?
Oh, this was while I was there. I am pretty sure. They did sound, lights and trucking. There was Rusty Brutsche, Jack Calmes and another character, it's a strange tie-up, a British lighting designer he designed the lighting rigs and he lived in

Finsbury Park. He obviously used to get involved with Dick when they were at the Rainbow—Ian Knight, that's his name.

I actually worked with Ian with Cockney Rebel in '75; he designed the show, always had a beard.
Ian lived near the Rainbow in Rock Street (very apt) and later lived in L.A., but anyway he was designing the lighting rigs that Showco used and I think he was instrumental in putting the first one we did together. It was Bad Company, and actually I think the first tie-up we did was with a lighting designer called Delton Bass. The tour started over here, and they needed a lighting company. The sound came over, they flew the sound over, but they didn't fly the lighting over, so it was a major thing that we got Bad Company. Martin went out as crew chief, they had an American lighting designer, it was like a 'major' to do it, to pull it off was like a major coup you know, like the renaissance for Rainbow if you like, it really was a turning point.

Now was your brother a lighting designer?
No, although strangely enough, he was Alex Harvey's lighting designer before Bill Duffield. I think Martin getting on with bands and all the rest of it never really happened somehow, he lacked the political skills that you need (laughs). He was actually a better crew chief than he was a designer probably, but he was crew chief for Bad Company.

Alex Harvey was another band we did for Bill Duffield, that was the first band that he designed after Martin, with tour manager Dick O'Dell, Sensational Alex Harvey Band's tour manager.

Was he really? I remember seeing them at Oxford New Theatre, fantastic band.
Well, it would either have been Annie, Martin or Bill Duffield doing their lighting. Martin did an American tour with them; I think Bill probably did a world tour with them. I only saw them once, strangely enough, in Coventry, and they were tremendous. So yes, Martin was the crew chief on Bad Company, and we pulled it off.

What year now are we?
We are probably '76/'77, 'round that time. I wish I had kept all the wall charts with all the tours that we were doing, then I would be better at dating it! So every Showco act that came over, plus the acts that we were doing ourselves, it was a major time for us and a great deal of investment in equipment.

Ian Knight, I recall, did David Bowie.
Yes, he did Genesis, Bowie, Bad Company, Led Zeppelin. Now the designer would design the show, and what Ian did was design the truss arrangement that he wanted, and where the lights were, and then hand it on to a designer, who would then design the show musically, put the gel in and focus and everything else. It was rather a strange job that he had, but it was deemed to be required.

I think Showco bought some set lighting in for something, and it wasn't successful, different voltage from 110 to 240; it was much easier for them to work with a British partner, and so we just carried on doing all Showco's European tours, which was rebuilding equipment, building trussing. Hartman was, I don't know how, he must have been staying up 24 hours a day building everything, the road cases, finishing off the cans, making dimmers, making desks, quite how we all kept up with it I don't know, but we did.

All under the roof at Bethnal Green?
Yes, well, Hartman, he had a flat in Highgate, then he moved out to Hampshire, Waterlooville in Hampshire, and he had a big garage there where he built the stuff. The Rainbow team could build stage sets, we could make the trussing, and we could make road cases ourselves, so that was all done, and we could do that in-house. What we couldn't do, we would buy off Hartman. He came up with the Hartman towers and all that sort of stuff in the very early days. Before that it was these things that were quite worrying called the Vermettes, remember Vermettes? If you couldn't suspend a rig anywhere you needed the ground to support something.

Weren't they just wires on steel?
Yes, it was like a square steel mast, you wound it up, and it was designed for putting in air-conditioning systems.

And they had, sort of like, extendable telescopic legs.
Yes, that's it, very, very dated, but that's what we used in those days.

One on each end of the truss?
Exactly, and even square box rigs as well, when I think about it now it almost keeps me awake at night (laughs), because you could really hear them creaking, but they were tested and everything.

Really?
Yes, you would have to have them tested. You basically engineered them yourself, and it was common sense, a lot of the early rigging was absolute abject common sense, nothing else.

So, from a safety point of view they were built to withstand twice the weight that got put on them maybe?
They were probably two to one; I wouldn't think a lot more than that. When you think about it, a Genie tower, which is a telescopic CO_2, with 24 cans on it, that's at the edge of their limit as well. But typically, in terms of lighting rigs in those days, it was a truss and two Genies, or a back and front truss with two Genies, or a two or three Genie lighting rig, that was the common thing you were renting out all the time. Probably the Steve Harley initial one was a back truss and two Genie towers with two follow spots out front.

Do you know who invented the Genie tower?
Vermettes and Genie towers are all American lifting equipment, and I think it's used to put in ducting, air conditioning when you are building, it's a building tool, it wasn't made for lighting at all.

And Genie towers?
Exactly the same, it's a piece of building equipment for putting conduits into buildings. Vermettes were available over here, used in the construction industry, Genie towers as well.

Were Vermettes being used when you joined Rainbow?
Yes. Hartman invented the Hartman tower, because he thought Vermettes were unsafe, and essentially all he did was made a steel base with square sets of metal tower with a chain hoist on the top.

Triangular or square?
The tower was a square metal pole inside a base; safe as houses, Richard, you could put a half-ton hoist on it.

With feet?
Yes, with feet but it had a proper chain hoist, lifting chain hoist, like a rigging chain hoist, up the top of it, just like a gantry, like a hangman's gantry, simple. You put it together, nothing could go wrong with it, and we just ditched the Vermettes.

I think it was probably on McCartney in 1976, that's another big act Showco brought to us. He did a show in Venice and it was up on either four or six Vermettes, outdoors, and Martin, my brother, and Hartman were there. The weight of the rig was okay because of the lightness of the trussing and the aluminium round cans, by then it was very light, but when you put in the factor of 'is the stage level?' and the wind factor . . . well, just one little thing going wrong with an outrigger or whatever. You have only got to have a bit of a tilt somewhere, or a bit of stage give way, and it's like a stack of cards isn't it, away it goes. And from that show Hartman came back and invented his Hartman towers. He did the half-ton version, the one-ton version, and later on did a two-ton version simply because of that show.

Are they still used?
No. It's flying, but what happened was, a much quicker and safer Genie super lift and super tower came in, again from the American construction industry, and they took over from the Vermette in the construction industry. You see, the Genie super tower on the back of hire service shops, and they are still used anywhere you can't rig, they are used in the smaller venues, and they are safe.

So the Genie super tower overtook Richard's?
Yes, for the smaller end, yes. I don't think, in reality, people do the venues where you can't rig anymore, because once you go above a certain weight, putting the lighting rigs on towers becomes cumbersome. What I think people do now is, put

a grid in or they hire in specific towers. If they have to do it, it's normally from a grid, like a festival grid, and you put your lighting rig underneath it.

What holds that grid up though?
It would be Tomcat towers; big, heavy-duty, aluminium ones, which are similar to Hartman's old two-ton towers, the two-ton tower is the precursor. They had a hoist at the bottom or a hoist at the top like a big gantry tower, put a grid in, and whatever the band's support grid would be, hung under that.

Most of the major venues now, especially in Northern Europe, have snow loading on the roofs so you can rig them. There are very few major venues where you can't rig. Where you can't it's normally in Southern Europe, where you don't have to have snow-loading roofs. Basically it's governed by snow loading. You can't rig in some Spanish venues because they weren't built to withhold a ton of snow every square metre, but the further North you go, all the roofs are structurally sound, and you can rig off them.

Can I now ask you your recollections on lighting and the outdoors?
The first big outdoor show that I was involved in was Led Zeppelin at Knebworth in 1979, and at exactly the same time we were doing the CBS convention at the Grosvenor House Hotel, which was a massive job. Both of them were with Showco, and I think that the CBS convention was with Mal Ross, because that's where I know him from, the CBS convention at the Roundhouse a couple of years before that, and then the CBS convention at the Grosvenor House. So slap bang together we were doing Zeppelin at Knebworth, which was a massive great lighting rig, and the CBS convention, which was every night with one of the major CBS bands, and it was tension to actually pull it all off. Right in the middle of it, Knebworth, which was supposed to be a one-week hire, sold out the show, and so they decided to do two shows, the second being the following weekend, so the whole system had to stay in there for two weeks. Effectively it was a three-week job. There was the set-up, the first show, leave it in for a week, second show and then the pull-out. It happened with about four weeks' notice, so with all the equipment needed, it was a massive, massive job, and coupled with doing the CBS convention at the same time.

Showco put the bid in to do one show, which was like one week, and then just quoted the next show as another week, all in dollars, and we did the contract in dollars. Right at that time the dollar to the pound rate crashed, and the dollar deal rocked Rainbow. Both those jobs were done as dollar deals, and that's one of the first 'chinks', if you like, in the history of Rainbow Productions. There was quite a lot of sub-hired in equipment that all had to be returned, and it ended up being a three-week hire as opposed to a two-week hire, coupled with the dollar going up and the pound going down. It was a bit of a catastrophic time.

Did you lose money?
We did, lots of money, seriously. Dick was in hospital for a while, it was tense times.

Blimey, so who did the deal then?
Dick Parkinson. You would never have guessed that at the time there was going to be a run on the pound.

You didn't think twice probably.
No, well, it was one show, it wasn't two shows—it was a bit of a double whammy really.

And if you hadn't had so much hired gear?
Well, it still would have been difficult to schedule. It had to stay up, the crew had to stay on site, it was just a nightmare, a catastrophic nightmare, of being absolutely overworked.

They obviously got double the money, but did you get double the money?
Yes, exactly that, it was exactly double but the dollar went down, and instead of it being a week extra it was nigh on three weeks.

You didn't charge them three weeks?
No, couldn't do it, it would have been three times the money.

It cost you three weeks, why didn't you charge three weeks?
The deal had already been done, it was done with Showco, and everybody assumed that you could do the job for two weeks' money. I think that, after riding the success over the years, that was a kind of beginning of the end of Rainbow. We were still doing the large tours, still doing it, but I think Dick had lost his energy with it all, let's put it that way.

I think everybody, and I have thought about it quite a bit since, everybody saw it, maybe Brian would echo it as well, that it was something that wasn't going to last forever. It was, "we are in it, we are enjoying it, it's good fun, it's a challenge, it's always something different, better than working for anybody else, and it's a laugh, and you are with people that you enjoy working with", but I think no one had the foresight, especially Dick and Annie to say, this is going to last 15, 20 years.

Maybe Hartman did, I don't know, but I think very few people did, it was just sort of rolling along, and subsequently every decision that Dick made was very short term. For a while he was ahead of the other lighting companies technically, he had a cabling system that was quite revolutionary, and then the trussing system that was quite revolutionary. Then people were coming in doing welding truss and multicore cables and spending quite a lot of money and re-investing. I think Dick, because of the Zeppelin thing, never saw it as a long-term future, as if it would last, it was always "short-term short-term, let's do that tour and we will last another year".

He never had a long-term business plan, and essentially at the time it was financially ruining the company after the Zeppelin thing. It was the most difficult 18 months, two years of my life. It was shuffling cheques in and out and running a company which had been quite cash rich, not really investing for the future, but quite cash rich. Suddenly it was penny pinching here and there, just to keep things turning over.

Amazing how you can be so busy and make so little?
Yes, I tell you the surprising thing is, Richard, I think something changed in those years. I think new companies came in like Showlites with welded trussing, with better systems than probably Hartman was coming up with at the time, and it's like with everything, they're sniffing at your heels.

Do the Birmingham guys come in here?
No, that's a little bit later, like LSD (Light and Sound Design), they came in a little bit later. Showlites were the first and then the LSD guys. LSD were in manufacturing when they first started out; they were manufacturing their own spun 'round' PAR 64 cans between them and Graham Thomas (James Thomas Engineering), I don't know who really came up with them first, they were probably in each other's pockets, but that's how spun cans came about.

Thomas and Telestage were the people that came up with the welded aluminium trussing, and Thomas came up with round PAR cans spun out of aluminium, whereas Hartman's ones were a folded aluminium circle.

Basically we were becoming more old-fashioned, we couldn't reinvest over a two- or three-year period. It became more and more apparent that Dick and Annie, their hearts weren't in it anymore, or they didn't want to take the next big, big step and say, "Let's go to the bank and borrow £50,000".

I think accountants got hold of the rock-and-roll industry and changed it somewhat, with the element of competition and pricing. You would know more, Richard, about how that was driven from being the tour manager, but it seemed like from '76/'77, other than crew costs the equipment element of it was being squeezed all the time. The crew costs were going up when obviously it was a higher-level tour, and the price would be more expensive because you would have a higher-quality crew or specific people were asked for. Financially it wasn't a growth industry, it was going down in scale rather than up.

Except, we were working with Genesis as well through those years, and that was always, you know, they wanted the best, they wanted everything to work, and they would pay for it. They were a good customer.

Pink Floyd at that point were doing their own lighting, building it all themselves and, rather like Hartman, making their systems, buying Hartman cans and renting them out.

Until Varilite came along, the lighting budgets were very stagnant, very much so, and that didn't help the industry, because you could start off a business but you couldn't reinvest. So ESP and Rainbow started off, they would invest in the equipment, they would get going, but the returns weren't enough to actually go and buy all the latest modern equipment, and it was hampering the growth of the industry.

Subsequently Light and Sound Design and Showlites would come in with brand spanking new equipment, and RDE as well, Richard Dale Enterprises. He went bankrupt a few times; he had a reputation for starting his company owing bits of money and then going belly up and then starting again. There are a few of those in the business, but its indicative of what I was just saying really, the budgets were squeezed and it would only take one project not to pay you, and you were in trouble. It would only take like the Led Zeppelin thing and you were in severe trouble, so on the Rainbow story that was the knockout for them.

So we're in '76/'77 but you were still going in '79.
Yes, I don't know what were you doing then?

'78 I think I had been using you guys, definitely on the Sailor tour of UK and Europe. Then I did The Grateful Dead in front of the Pyramids; that was with Richard Hartman.
Martin, my brother, was on that gig, Grateful Dead and the Pyramids. It might not have been Rainbow, it might have been Hartman's company TTR, and he took Martin as a crew member. I can't remember Martin doing the tour though?

There wasn't a tour, they cancelled it.
And they just did Egypt?

They just did Egypt. I got out to Egypt, and the first thing they said to me was, "Cancel the European tour we can't afford to do it".
Those were the days, eh? (laughs)
  What happened around '79/'80, which is probably why you didn't use Rainbow, is you wanted some of the newer equipment. I know we had a long business association, the young guns, the Showlites, the Light and Sounds . . .

Supermick, and all his PAR cans were chrome.
Yes, Supermick, he was cheap and cheerful, how he did it I don't know.
  At the prices that were going around, you can start but you can't renew, and it's an age-old problem. What happened at Rainbow is that Dick and Annie decided to cease trading, pay off all their debts, they paid off the VAT man and the Inland Revenue and ceased.
  Hartman bought all their equipment off them, he gave them a reasonable amount. He bought all their cases, bought the infrastructure; essentially he bought the company, but not the company's debt. He bought all the equipment, he bought me, he bought Lee, Mick Healey, the people that were running it, the team if you like, and he then went over to Edwin Shirley's yard, and we set up TTR which was his company.

Where was that?
Edwin Shirley's yard down in Crows Road, West Ham. We got a cheap unit in there with a portakabin for an office, so the overheads were not a lot. I think Hartman borrowed some money on the back of doing a Genesis tour. We managed to keep all the Showco acts through all of that period, lost some things though.

Everybody wanted to work with Showco, ESP wanted to work with them, Showlites wanted to work with them, because they just brought big bands in, but Hartman and myself, we just managed to keep it together. Richard then invented the folding truss.

The folding truss?
Yes, for the Genesis tours, and by that point Showco had invented the Varilite for the first Genesis tour, so things had kind of moved on. Suddenly the automated lighting world was happening. We did the first tour over here with Genesis with automated lighting, the very, very first one out of Edwin Shirley's yard over in West Ham.

Again ground-breaking, and with this new trussing that actually folded like a bookend, if you like, took up less space in the trucks. He also invented shorter PAR cans that would pack into smaller areas, thereby saving trucking space. We are talking about '79/'80/'81.

It was the rebirth, if you like, of the business I'd been managing. Hartman managed this re-equipping, holding on to the acts, and that really, from my point of view of running a company, was like a game changer. We re-equipped, we had got the right equipment, we had the acts, we had the contacts, and we had something.

Hartman had a long-term vision, that this was his job, and we were doing it seriously. I think for the first time he got in the black, he ran it properly, he ran it lean and mean but he ran it well, we both ran it as a tight ship. I knew all the mistakes we made at Rainbow and realised, hang on, there are cheaper ways to buy materials, there are cheaper ways to buy bulbs. Being a bit harder about it, not just paying £20 for the bolts or buying the rope from Jack Smith the yacht merchant, but going to yachting manufacturers, the rope manufacturers, and buying it direct and buying 200 metres or a thousand metres at a time. You are going to use it. So we just bought it, put it on the shelf and little things like that. Over a year, between us, we saved a fortune, an absolute fortune. Rather than sending someone down to the hardware shop for 20 bolts, go out and buy 1000 bolts, go and make your own PAR cans, make your own follow spots, just do it, but do it in a better fashion.

We were sharper basically, we learnt how to save money, learnt how to run the business a little bit better. Between the pair of us we got the thing tight, kept the good crew, got some really good technicians in the warehouse maintaining the desks and the dimmers.

And for the first time Hartman is not making his own dimmers. He bought Avo dimmers and bought Avo control desks because that was what everybody wanted. The Hartman desk, the Hartman trussing, his cabling system had all had their day, all gone, you have just got to throw it away, all the bolted truss, gone away. But he was making welded truss in his own design that was again quite a revolutionary thing to do, it was lighter, less space and heavier duty than the competition. Hartman was a catalyst of all that, and from the business and contacts end, I was kind of running the UK operation.

At this point Showco, in the States, dropped their lighting company, and so Richard went and installed himself out there in Dallas and set up the lighting company there as well. We were containerising lighting from the UK and shipping it out to Dallas to service their American and world tours.

We were doing remarkably well, and again we were probably up there with Light and Sound, we were at the cutting edge, it was like the second coming if you like. We had gone up with Rainbow, come down over the Zeppelin thing, and Dick and Annie not seeing the long-term potential of something they had built up, unfortunately didn't get any of the reward.

It's all driven by innovation, somebody coming up with something and then having the right people and the right approach, professional approach in place and that covers rigging skills, building skills, electronic maintenance to keep everything going whilst it's on the road and being inventive, you know.

Genesis wanted moving structures in a rig, and Hartman got the contacts in the States to actually do a computer-controlled moving lighting rig. His old friends from college had come up with the ideas for it. He had all those contacts for being able to do something completely different.

So Richard went off to the States, to Dallas, running the American operation going backwards and forwards and I ran the European side from London, and that went on for three years of real success. Richard was obviously coming up with developments, such as moving trusses for Genesis, and making his own follow spots.

Mainly by then there were set standards. You had trussing towers to support the rig, the rigging was specified and that was when Hartman was instrumental, along with Robin Elias and Unusual Rigging and all those guys in bringing up the standard of what rigging should really be. Hartman was able to encapsulate that and have it made by proper shipping steel guys, who were making the rigging steels and buying the shackles and all of that, and they were all properly tested, the whole thing was becoming more and more professional, so you could look at it as more of a long-term investment, and it would seem a bit more worthwhile.

After Varilite went out on the first Genesis tour, everybody wanted them, everybody wanted moving lights no matter who you were, and Varilites were just shooting snakes in the barrel. They had technicians and people on every tour, whether it was Showlites, ESP, or TTR or TFA (by then ESP had become TFA).

I am wondering who I used with Duran Duran; that was the first tour that was with some moving lights, and I can't remember.
It was us, TTR. It was TTR with Alan Goldberg. A great designer.

What happened after that is that Varilites were on everything, and they could name their price for the lighting systems, and suddenly everybody was getting much more money for the lighting systems. Because of Varilite everybody found the budget to be able to afford this brand-new development, and that was kind of a bit of a sea change as well.

At that point Samuelson's, the film camera and lighting company, bought Varilite UK or Varilite Europe (it was auctioned off, Brian Croft tells the story) and on the back of that they bought Zenith. They then bought Brian Croft's company, TFA Electrosound; they bought Turbosound, Mike Lowe's company; and they were just going through the industry buying companies to facilitate the Varilite.

Because they had bought Varilite they thought that they would be able to take over all the clients that were being serviced by TTR as the lighting contractor, but they couldn't. Genesis and Pink Floyd had closed down their lighting operation at that point, and Samuelson's thought that because they had bought all those companies, that would automatically have an effect on TTR. But the relationship that Hartman and I had with the designers and our crew chiefs was so strong that they couldn't drag them away, so Samuelson's wanted to take over TTR, and of course Richard resisted it for a long, long while. Then it was like, yes, we are going to do it, and then all of a sudden their interest would wane, and they would get one act or we weren't busy or something, and they'd think oh yeah we don't need them, and they would go off the boil and then six months later they would come back again.

But eventually, it went through and Hartman sold out.

Where have we got to here, mid-'80s?
Mid-'80s, yes, probably, '83/'84, so eventually on the back of what happened Richard sold and he went to work for them in Dallas.

Did Samuelson's create their own American division?
Yes, Samuelson's Concert Productions Inc, that Richard ran which was the lighting company for Varilites. Also, Samuelson's Concert Productions in Australia, which took over Eric Robinson's company, Jands.[1] They took over everybody.

I went sailing with Eric about five years ago.
He's all right, isn't he, everybody liked Eric. We have got quite a lot in common actually because he is from a long line of farmers, Australian farmers though (laughs). His reputation goes before him.

So Samuelson's took over and we were covering their world tours wherever they went. Europe could make enough money, Australia did quite well but unfortunately, the States never did, it never took off.

Richard went out there without an infrastructure and had to build it from the remnants of the Showco lighting guys from years ago. Whereas if you take Brian Croft, Chris Adamson, Jon Cadbury, Jimmy Barnett and me, there were five of the top account handlers in the UK all working for Samuelson Concert Productions Ltd. When you think of the acts all those five people had worked with, it's just formidable, and then there was Richard all by himself in the States!

When I first went there it was headed up by Jimmy Barnett, he was running it, the lighting designer from ESP, who also did ABBA and Queen. He was running the company for Samuelson's and subsequently Eagle Trust and Varilite. Essentially Brian, Jon Cadbury, Chris Adamson and myself were all working for him running different tours, as account managers if you like.

It then became more corporate, they had taken over Theatre Projects as well, and the dream was to get everybody out of Cricklewood Broadway and into one big facility, which eventually happened over in Greenford.

It was successful but we weren't making any real kind of money. I think the Varilite end of it was doing well, but if you looked at the concert production side it wasn't really holding its own. It was a fairly big overhead, I don't think the prices were that bad at that time, I just think the overheads were high, it was very hard with the amount of people they were employing. Running a company on a more regimented basis, more like a PLC if you like, and finding it difficult to knit all that together to make a profit, and then to make a profit for your shareholders as well . . . I think in that industry it normally works better with an owner/driver of some description. Varilite, that went from strength to strength, and really they should have sold that at some point, technically allowing other people to come in with competition for moving lights.

Joe Brown created his own moving light at TASCO at some point in the '80s called the Starlite.
I can't remember him having any moving lights. The first time Varilite had a competitor that I remember was the Telescan from France in the early days, which was basically a follow spot with a mirror on the end, and that was Varilite's main competition.

Hartman and myself would try to convince the company that the only way you can have Varilites is as a package with our touring company generic lights included.

But rightly or wrongly, they never saw it that way. They did not use Varilite as a tool, to say to bands, "Okay, you want Varilites you have got to have our package to go with it". Our idea was that Samuelson's Concert Productions and Varilite wouldn't be two companies, it would be one company, one lighting company, which is what it is now.

Samuelson's being a PLC company, went through a very traumatic time when they lost the Panavision Camera end of things, and I think they were buying these companies to protect themselves, diversify, and it all went wrong, and they ended up selling it off to Eagle Trust.

Let me ask you what you did subsequent to Samuelson's. I take it at this stage you would in your spare time be coming up with the Cirro Lite mist?
Yes, I suppose I was. I had surprised myself at how easily I took to working for a large company; I didn't think I would last in there at all, but I did. The States is where it all went wrong. With Hartman, they closed that side of it down, I think Richard was paid off and went to be a consultant for various trussing and rigging companies and working with U2 amongst other things. The UK side had kept going well though.

In the meantime, my brother Martin was a crew chief involved in MTV live shows in L.A. His partner was Marcy Levy, the backup singer for Clapton, he went out to L.A. to live with her. On one of the shoots he was on, he discovered a special effects guy that was using this system of running air through oil. It was a very basic form of it, and had even been used in the silent movies.

The cracked oil system?
Yeah, he saw it and thought, 'that's pretty good', and he bought one of the barrels off the guy. He brought it over here to the UK and gave it to me and said, "Put a compressor on it, put some fluid in it, smoke fluid", and I did some more research,

found out really what you should put in which was pharmaceutical grade white mineral oil, and developed it. I thought, 'this is brilliant!', and the first show we did with it was the Brits, the Brit Awards, the famous one with Mick Fleetwood and Sam Fox and designed by Marc Brickman. It looked very good. The guys at the Albert Hall had said, "You are not going to be able to fill this place with mist", and I said, "You wait and see", and we filled it with a haze that was just beautiful.

So I developed it, honed it down, tried different compressors, different pressures, and different fluids. Subsequently I was getting in at 6 or 7 o'clock in the morning to Samuelson's making the machines, assembling them and putting them out. By the end I had 40 or 50 machines, renting them out to Samuelson's, and they were being used on tours, touring productions, the BBC, films, pop promos, commercials, basically changed the look of Top of the Pops! So that is how the Cirro Lite end of it happened.

Martin also found, out in the States, a company called Kino Flo, a fluorescent lighting company that had just started up in a garage. He said to them, "If you ever want to come to the UK, give me a call, my brother and I would be interested". I met them in Reno with Martin and strangely enough Richard, of all the other offers they'd had, they said, "Yes, we will go with you. We want to go with somebody that can think out of the box, not somebody that's tried and trusted". I was gobsmacked.

So all the money we had made out of the mist machines we plunged into starting Cirro Lite with Kino Flo. Everybody told us we were mad, and here we are 15 years later.

I miss the rock-and-roll industry though; I miss the excitement and the production end of the rock-and-roll industry. Financially this is much easier, the film industry is difficult, nothing is easy, but from the financial side it's a lot more rational.

I have asked everyone this so far: Who was your biggest influence; who do you think has been the biggest influence to the industry; and what do you think has been the biggest influence in the creation of production?
Well, it's got to be Hartman really, it's got to be Richard Hartman. He invented the stuff and guided us along; he trained all those crew, along with Dick Parkinson. Really, for what we are talking about here, it has to be Richard Hartman far and away. What was the next bit of the question?

I was going to say, who do you think has been the biggest influence in the lighting game, but you're going to say Richard again!
Yes, I think he has to be. For everybody he was involved with, if you take it from Floyd to Yes to Genesis, he was the catalyst and is still involved now with U2 making things happen, inventing things. He's like the Leonardo da Vinci of the rock-and-roll world if you like. There are lots of other brilliant people; Annie Pocock was a brilliant conventional lighting designer, and Alan Goldberg was too.

What do you think has been the biggest influence?
I would say between Pink Floyd and Genesis, certainly in Europe and the UK, they have driven the changes, the barriers and the boundaries, in my opinion, of what

you can do and what's possible. Specifically, Pink Floyd in a rock-and-roll show. To this day, (although I haven't seen a lot of shows recently), for production and a total show, The Wall is far, far the best show I have ever seen while working in the industry. My mouth opened up, and I thought 'how did they put this shit together?' I haven't seen anything to touch it, Richard.

I would have to say, they artistically have pushed production and lighting production to the edge and put the money into it as well. Everything is driven unfortunately by money, but the Floyd were willing to say, "There it is".

I saw it. It was probably one of the biggest shows ever talked about with the fewest shows ever performed. I think it was only performed in three cities, wasn't it?
Yes. The control area, where the people controlling the puppets, the animation and the sound were, there must have been 12 people working together, which is just like crazy. Normally you've got a lighting designer or two, you have got the Varilite designer, the lighting designer who is calling spots, and the sound engineer. There must have been at least 12 people in the control area doing different things for The Wall.

And Genesis, they did moving lighting rigs, so you have got to give it to them as well. They were maybe not quite the trailblazers of Pink Floyd, but they developed the Varilite, they put the money in to develop it and they had shares in it until recently.

Finally, over the years is there a story that you have told friends over dinner parties more than once, some sort of funny story or event that stuck in your head.
I guess, because it was so traumatic, it has to be the Zeppelin show at Knebworth, and the CBS convention, although it's not a funny story, it almost drove lots of people bananas. It was a story of success and failure. It's probably the one that's come up more than any other one in conversation, I suppose. Through the years lots of funny things have happened, but that one has been recounted many times, let's put it that way.

Thank you, John, that's great.
Thank you, Richard.

As John explained, he created Cirro Lite and continues to be an active MD with the company providing a comprehensive service for those looking for lighting for film, television, video production and still photography where special lighting effects or a controllable precision light source is required.

Note
1 www.jps.com.au/category/history/

Part 2
Pioneers of Sound

I think most people in the world of sound would agree that Charlie Watkins[1] was the 'Grand Daddy' of UK early live sound. Charlie concentrated on manufacturing and selling his systems, rather than hiring them out for tours. He was very generous though, as the early festivals were always helped along by huge amounts of WEM equipment, at no charge to the promoter, albeit great promotion for Charlie's company.

Weeley Festival stage, 1971
© John Sellick

During the research for this book I came to acknowledge that Jim Marshall[2] was doing a very similar thing to Watkins in the late '60s, and he helped to develop a range of PA systems for rent, first with David Hartstone, who created IES (International Equipment Services), and then a little later with Joe Brown, who created MEH (Marshall Equipment Hire). The reasoning behind these collaborations being, I believe, in the pursuit of marketing their other products to the professional musicians and the general public at large.

David Hartstone, the Kiwi founder of IES, was the first real entrepreneur to begin hiring PA equipment on a large scale, primarily across the UK and then worldwide. Although he quit the business relatively early in the scheme of things, I managed to track him down to his yacht, in the Caribbean, for a great couple of hours of conversation on Skype in 2015.

Bryan Grant, also heralding from New Zealand and currently MD of Britannia Row, tells me of his early days in New York before coming here to the UK to work for David Hartstone at IES. I first remember meeting Bryan in 1973 when I asked him to come to the Speakeasy Club in Margaret Street, London, to help me get to grips with a brand new Soundcraft mixing desk. At the time I was the soundman for Cockney Rebel, and Bryan was trying to teach me the ways of crossovers at a soundcheck.

Tony Andrews built the first PA system (Purple Phase) that Cockney Rebel toured with, and he later went on to co-create Turbosound. I sat with him fascinatingly discussing his early days developing sound systems.

The UK's second large sound rental company was owned and created by Joe Brown. When IES folded, a lot of the staff and equipment moved into his MEH warehouse. The way that Joe got off the ground in the first place though is via the Weeley Festival, at which Joe showed up with a Marshall PA system. This collaboration with Jim Marshall turned into his first rental company, MEH, which subsequently morphed into TASCO.

In the early '70s were no rental companies to speak of, everything we used was bought and paid for by the band. I recall, very clearly, the first thing Cockney Rebel did when they signed a record deal with EMI, was to go out and buy a PA system. They already had backline equipment and instruments and, as a new band, were most often a support group for the headliner, so they played through the main acts' PA systems.

Pretty quickly they got to headline gigs, and a PA became a 'must have' to be able to play anywhere. As the band grew in popularity they played to larger and larger audiences, and soon their original little PA became inadequate, thus in 1975 we hired an IES PA and their crew, recommended by the then sound engineer. When the sound engineer changed for another, quite often the sound company changed as well, and this happened with Cockney Rebel. The next company we used was Kelsey Morris, the brains behind some of the early British sound desks and speaker systems, in conjunction with the Dave Martin[3] bin and horn inventory.

As I progressed as a freelance tour manager, I employed, on behalf of the bands that hired me, a variety of sound companies, including IES, Kelsey Morris, Entec, ML Executives, MEH and TASCO.

Notes
1 www.soundonsound.com/people/charlie-watkins
2 http://jimmarshall.co.uk/
3 https://martin-audio.com/about

7 David Hartstone

Dave Hartstone

June 3, 2015 with Dave Hartstone by Skype

I'd like to start off with, where did you go to school and how did you fall into the music business? Start right at the beginning, if you would.
Whakatane High School in New Zealand, North Island. I never did very good at school. I left at 14, and I went and worked for Ford Motor Company, who allowed me to work under the Labour Department Laws, if I went to school three nights a week. I became an A-grade mechanic by the time I was 19, and worked on big bulldozers and tractors, out in the bush with the Maoris. But I was in a group all the time, in a band. We never made much.

What was the band called?
Oh, shit! I think it was called The Templars. Fucking terrible band, we never did much.

What did you play?
Guitar and sung. I was doing concerts when I was 12, 13 years of age, country and western concerts in Tauranga, and the guy who put me on, Johnny Walker, said as long as I sang for free he'd teach me the guitar. But I helped to form a band, and that was the Four Fours. They had a lot of hits in New Zealand, and then it

was changed to the Human Instinct when we went to England. That was a difficult time.

We worked in a small place in Auckland, a small club called 'The Platter Rack'. We used to play a big hall with 3000 people in it earlier in the evenings and we'd play there till 12. At 12 o'clock we pack up and go down to the Platter Rack. We had two sets of equipment so there we'd play quiet music, cool stuff, you know, club stuff. It was at this time I got into sound, stage sound and had the band use gooseneck microphones. The microphones were to sing our songs through, so we didn't have to hide behind a mike stand, which is what everybody does.

The secret was that we had a little on-off switch on the horn of the guitar, it was the same colour as the guitar, black. The leads to the amps behind us were stereo, to carry the guitar sound and our voices individually. I made up the leads, and they were 60 feet long so we could move around the stage.

I made the little mikes out of a short piece of copper pipe with a fitting on it to attach to the guitar, and it was all no more than six inches long. Since it was black as well, you could hardly see it, and it made it look like we were singing into nothing. Bill (Ward—lead guitarist) didn't like this at first because he couldn't hide behind a mike stand, but in the end he gave in.

I don't know if you're familiar with feedback; it is caused by an open microphone being too close to the speaker system. If you turn the amp up loud enough it'll start to feedback with the open mike, but before it feeds back, your voice is as loud as it's gonna be. So playing music in the Four Fours was simple, people could hear us sing very clearly with the amps behind us, which were very loud, and we learnt to switch it on with our thumb when we were going to sing, plus play our guitars, then turn the switch off so we didn't get feedback. We were very clever with the speed at which we operated the turn-off switch. This was a technique of the Four Fours, and we learnt to do it very well.

From this I learnt a lot about sound systems; we had two stacks of Marshall equipment for each one of us behind us on the stage. One double stack would be just for the voice, another for the instrument being played. This is how I learnt to build sound systems. You must remember that at this time you couldn't hear any singing in rock groups, just a bit of guitar and a lot of drums, but people got used to it and figured that's the way rock groups are, a lot of people don't understand this progression. Nobody had this system, only the Four Fours. The Small Faces wanted me to build it for their system too, but we never got 'round to it. It became obvious to me then, that singing and guitars had to be separate.

When we went to the UK, the BBC did a television show called 'Tomorrow's World', and they did an episode on how we did our sound; they brought a crowd of about a thousand kids for me to demonstrate the sound we had. I would lean forward, switch the switch on the horn off so there was no feedback, and then I would lean back and turn it on, and I would describe what I was doing to the audience. They called it 'the anti-feedback sound system', it was a bit of a con really, but nobody knew. We went and told the BBC what we had done, and they wanted to do another episode showing how rock groups will do anything to get on the air!

So you came over to England when?
In '66 as a band. We didn't even know how to operate the telephones there. It was horrible, London was terrible, but we managed to survive. We had a manager called Maurice King,[1] he used to manage Dusty Springfield, the Walker Brothers and various others. I think he took us on as a joke really, because we were New Zealanders. He just thought New Zealand was a joke; I mean, the first gig he sent us to was a bomb-site. There was nothing there, he sent us there for a joke. We saved our money, just to pay for gas, and it was fucking terrible.

Maurice King. He went to jail in the end, but he got out, and he knew me when I was in the sound business. We had another manager out in Wembley, and he was really good. He pushed us, I mean, the first gig we played at, we got beat up by Mods and Rockers. One guy came on stage and head-butted me straightaway. Put us all in hospital . . . Tipped our van over and smashed up our equipment.

Where was that, in London?
In London, there was a big pub, it held a lot of people, there were three bands playing. There used to be big pubs in those days, I don't know if they still exist. I mean, it had to be 3000 people in there.

No, they don't have them anymore.
No, because of the fire risks, I guess. Anyway we had been in London about two months and we had no equipment. Our manager then took us out to Marshalls, this was where I met Jim, and the manager just said, " Pick what you want". So we picked two stacks of Marshalls each, and then I said, "We've got no van". So he bought us this brand-new van, a dual-wheeled transit, really popular during those days.

This is a nice manager.
He's a good manager, he's an owner of Bingo clubs in England.

What was his name?
Peter Newbold, I think his name was. Yeah. He was a good guy, he had plenty of money, and he said, "The first gig you got to do is back in that place where you got beat up again, that's where you've got to go". Well we said, "We're not going back there", none of us wanted to go back there. He said, "You've got to change your music; you've got to rock. You can't play other people's music; you got to write your own". So we were sat down with the Troggs.

Reggie Presley helped us rehearse the songs we'd made up. He was a big help, Reggie Presley; he was a really nice guy, he helped us through. Then we went back (to where we'd done the first gig), and we went over a storm because people like New Zealanders for some reason, because we were neutral; we weren't Aussies, we were just New Zealanders.

So, what was the name of the band when you were in England?
Human Instinct. Yeah. We had some good records, none of them were any hits in England, but they were played a lot.

Okay. And where were you living? Were you all living together in a flat or something?
Yeah, in West End Lane. It was really hard. We lived on cat food for a long time. Cat food and toast, and we used to steal milk off the front doors of people's houses; you know in England they leave the bread and the milk outside from the dairies. We'd go out and steal a loaf of bread from here and there and live on that. It was terrible.

I don't know how I stood that. Then the band broke up and went back to New Zealand, and left me alone. We had this record of Renaissance Fair (written by Crosby and McGuinn of The Byrds) coming out which Mike Hurst produced, and it's going to be a hit, and it's paid for and everything, you know. It was a good song, it's well done, we had part of the London Symphony Orchestra backing us. It was good, but they'd all left, they didn't stay to get the record out. I've got it; I've got a copy.

Then I'm alone and I think, what am I going to do? There's a girl in Maurice King's office called Angela. Really pretty girl. We used to stay there waiting in his bloody office to see Maurice King, and he'd never see us. I rang Angela, and I was sort of desperate, and she said, "Well, you've got a lot of a equipment". You see, the band, they'd left all the Marshall equipment we had; they were going to try to take it back to New Zealand, but I found out where it was.

Also, I happened to strike up a relationship with a bloody good lawyer, because we played in the Bunny Club for three months, where I got to know the legal advisor to the Playboy group in Britain, Arnold Finer. He helped me get back into our flat; I'd got kicked out of that but I got back in there, and all I had to pay was a shilling a week!

You found the equipment?
I found the equipment at a shipping office. You don't know how many shipping offices there are in England. With the lawyer's connections, we found the gear. I boxed it all up, and took it all back. I had it all in the hallway of this apartment, and it was stacked to the ceiling. There was so much Marshall equipment, you couldn't believe it! Then Angela said, "I've got Hot Chocolate coming up, and they've had all their equipment stolen. I know you've got all that Marshall equipment, why don't you go out and set them up the sound and stuff like that?" Well, Maurice King was the promoter, the very guy I hated!

Oh, hell!
So, I went out and I did it. Apparently I did a good job because Maurice King started spreading the word around everywhere, saying how great it was working with this new sound company! Now, Clive Coulson,[2] who was our UK road manager with Human Instinct, he worked for me. A Captain Beefheart tour came in, so Clive went out and did the sound on that. Well, now there's the two of us out there doing two tours. Then somehow I realised, it's no good me being out on the road, I've got to be in an office. When we started, I mean it just ballooned, and within a year, I had a million dollars turnover. That's how IES started (International Entertainment Services).

With Marshall equipment?
Yeah. Well, that was for the *stage* sound. We had to sort of make up the sound systems, with Marshall towers and that. It was bloody awful, you know, not a real sound company.

And that was what Charlie Watkins and WEM were doing then?
I knew Charlie pretty good; he's a nice guy, but his stuff wasn't right for what I wanted. Jim's wasn't any good for sound, it was good for stage stuff, but not for the sound. Then I ran into a guy called Bill Hough, and he worked for ITT, he was the R&D (research and development) guy. I became friends with him, he was good with Hammond Organs, and I'd started to get into those, and I needed him to repair them.

He would work sort of freelance for us at first, but when he found out I was developing a sound system he became more interested in that than anything. He had other gigs aside from me; he used to look after Jon Lord of Deep Purple and Keith Emerson occasionally to fix their organs.

I met him through Keith Emerson, I needed an organ split. I had to split these bloody great C-3 Hammond Organs. It was the English version of the Hammond Organ, the American version was called the B-3. It looked like something you'd put in a very expensive house, it had round legs . . . it looked like a piece of furniture. They were the same organs, but they just had different woodwork on the outside. Bill was older, a gentleman, had been in the Second World War in Israel. He was just the man I needed, I could tell from the moment I met him. He was also Hammond Organ's top man, but at the present job he had, he was the R&D guy for ITT. He was big-time and well respected. For me to ask him to split an organ (so it could be carried easier) was a paltry job, I thought. Regardless of what I thought, Bill worked in my apartment after hours, after we had wrestled a brand new C-3 into the hall of my apartment, we had to split the organ but we also needed the 90-pin plug.

The lower part of a C-3 is complicated, with many cables from the lower bass section, so Bill cut all the wires and put in this 90-pin plug and could connect the bottom to the top via that plug. He was there to wire all this up, and when he had done this and it had worked, we used it for a Ray Charles concert and it worked perfectly.

The first audio tech I had was Bill Hough, he could listen to what I wanted and figure it out. He was a genius with a photographic memory, and exactly what I needed. We could talk for hours about sound, figuring everything out and how it had to be done. I knew what was needed because I had been in a band playing shows and the sound was crap. It was all a slow progression of time, it just evolved through the late '60s and early '70s.

I gave an open credit card to him, to use for anything that involved sound or made some kind of noise, and for travel, because IES was open 24 hours a day, and our trucks were always coming in to load and unload equipment. I bought him a brand-new V12 Jaguar and renewed it every year, and after that he came to work for me permanently.

He was a really good and loyal employee and friend. He always picked me up from Heathrow, twice a week, from New York, where I had another company with more then 200 employees. We would have meetings in the car on the way back to London, which took an hour.

When we first started to tour, I said to him, "I've got to have a sound system, a big one. One capable of doing half a million people, not 3000 people. Nobody makes any money with that". So he says, "Well I don't know how we're going to do this". It was a terrible scene to get going, slow and horrible.

But we had to have JBL, so we rang JBL, we became friends with them, and we started buying stuff from them left and right. I mean, tons of stuff.

We bought our JBL 'parts' through Dag Fellnor. L.A. thought we were trying to take them over, due to the amount we were buying, so they insisted that we buy our parts through Dag. He was the agent for JBL in London, therefore they wanted us to honour his position as the agent.

Let's not get confused, we were not purchasing a sound system from Dag or JBL, just parts to assemble an IES sound system. Buying the parts was easy, putting it together as a sound system was the tricky part, and that was separate from Dag. Neither Dag nor JBL were involved in the assembly of the equipment, which suited us.

Did you ever use any other components such as Altec or Vitavox? And what did you think about that equipment?
Altec was the original company of JBL, the technicians got together before JBL was formed and called themselves the 'all technicians company' or ALTEC for short. James Lancing designed all the speakers, but he called his speaker the JBL, and he wanted that name, and he wanted the royalties for JBL but the courts said he was working for the Altec company when he designed JBL, therefore the design belonged to Altec, so he left Altec and formed his own company, JBL. Lancing has since died, but his name lives on, although in a more commercial sense. The ones we used were a professional series. Never heard of Vitavox, by the way.

The bins I found first were called Perkins bins. I redesigned them with plywood and asked Jim Marshall to make them special for IES, providing I ordered 100. There were doubles, 5 feet high by 3 feet square, and singles were smaller, half the size. Those were the bass bins.

We had a new thing that happened which greatly improved the bass bins; Jim Marshall rang me and finally agreed to build our bass bins, as he had a special machine for his Marshall equipment that interlocked the corners of the cabinets; it was an expensive machine. The deal was that he wouldn't build less than 100 cabinets at a time, this was kind of an honour coming from Jim Marshall, as he didn't build shit for just anybody; in fact, we were the only ones he ever built anything for because he saw something great happening, and he wanted to be a part of it.

Yeah, they were chunky in those days, weren't they?
Well, they got to be still, unless you're looking for computer sound.

So we were getting trouble from JBL; they said we were buying too much, and it looked to them like we were trying to take them over, so they cancelled us! We said, "That's okay, we'll build our own speakers", so we started to manufacture our own. It was a long deal, we already had speakers but we started manufacturing anyway. Mike Philips was in charge of that. We had to use square wire and stuff like that (for the voice coils), and we'd got the winding machines. The only thing we didn't have really were the drivers. Now, that was going to be complicated, but we were ready to make them. We had plenty of money, money wasn't a problem. Then JBL changed their mind, so we decided there was no point in going on any further and JBL made the horns, and then I put them in boxes that initially Marshall made. A mid-range horn, and one with fins. Then there were high-range horns. It was a three-way system. All made under the name of MAVIS: actually named after an old girlfriend of mine. I taught her guitar.

We needed amplifiers, bass bins, then a good speaker and mixing console (eventually WEM made one, much later). We needed around 15 microphones to run a band, and there needed to be a good mixing console to run it. It was expensive and, at the beginning, costs were out of my league. We also needed really good audio techs, but there wasn't any such thing because there was no sound at the time, and we were creating something totally new!

Okay, wow!
So the sound system took about 18 months just to get the speakers together. We did that in Carnaby Street, because we were in Sharpenhall Street. We had a big place up there, but we wanted to have separate spaces, so we went to Carnaby Street and rented a place. Just to check the speakers out. It was a long process, and I think Bryan (Grant) was with us or started to be with us then.

Did you know him before? He's a Kiwi looking for a gig and you bumped into each other?
No, I think he came for a job, and I employed him. He's a good guy, sensible, very sensible.

So the business when you started was basically a backline business?
Well, it was, yeah.

Renting backline, and then as you were making money from that, you decided to develop speakers and mixing desks?
Well, you know why we were making money, because Vox's owners went bankrupt.

In those days, The Beatles used Vox. Vox gave equipment to them for free to get advertising, but roadies used to sell it off! Because Vox had the contract they therefore had to provide the gear.

Jim Marshall came to me and he said, "I don't want that shit happening to me. I've got Hendrix, Clapton, Bruce". So he said, "I'm taking all my equipment back and I'm giving it to you". So two tractor-trailer loads of his equipment came to my offices.

Can you explain that a little more please?
Jim Marshall supplied systems the same way (as Vox did) to many big names of the day, Hendrix, Clapton, Beck, The Pretty Things, Yes . . . and had the same problem. He called me and said, "Here is a bunch of equipment for you . . . you use it in IES and then others will buy it from me". Hence the huge success of Marshall and IES. Marshall's secret to successful amplifiers was the old 80 valve, which gave the amp its hard mid-range sound.

We were given about eight tons of Marshall equipment from Jim Marshall; it was given to me because he trusted me and was a friend of mine. Renting it out to bands would ensure that the equipment wouldn't be sold off by the roadies, and he felt as though he received as much publicity from giving it out to me as Vox did from giving it out to The Beatles. It took a lot of figuring out and considering how the rock business was at the time. It was all done out of love—the love for music and creating great sound.

That's pretty cool.
Yeah, right, he said, "You can have it, rent it out". Jim Marshall and I had an arrangement you wouldn't believe. He was a very good man to me.

So we were still developing this sound system, and we came up with a big hang-up with the amplifiers to drive the speakers. They were going to cost, I'd say, probably about a million and a half in R&D to develop. It had to be a big amplifier, and it had to be quiet. Then we learnt that you can't use ICs (integrated circuits) and you can't use chips, so we had to go another way.

Bill was doing Judy Collins. Bill used to do these private little gigs on his own, he loved Judy Collins, and he was in America doing her shows. He was walking along a road one night and he saw some guys levelling concrete in a car park. He used to walk a lot, he had a beard, he was one of those guys, you know . . . and, he said to the guys, "How are you levelling this concrete, I don't hear any compressors?" They said, "We can't use compressors because the hotel across the road moans and grizzles, so we've got some power supply". That pricked Bill's ears up. He said, "Power supply? What sort of power supply are you using to do that?" and they said, "Well we'll show you" and they showed him. Bill said, "Can I borrow that afterwards?" and he laid the deposit down, a couple of thousand dollars they wanted, went back to the hotel and did some tests on it, and then he rang me and he said, "We've found the nearest thing to a straight wire, for an amplifier, 800 watts each side. Sound as the day is long". And that was the Crown DC-300.

Oh, Crown amplifiers . . .
Yeah, so I flew over to America immediately, and we went straight up to Elkhart, Indiana, and we ordered 800 of them. Well, I had millions of dollars, so I just ordered them and paid for them. The Crown DC-300, a coupled amplifier, 800 watts at 4 ohms on the speakers. The trouble was with Crown, as soon as it went wrong, it would blow out all your speakers, so we had to develop a safety mechanism on the back of it like a crowbar thing. In other words, if the amplifier developed a fault, as quick as the fault went to the back end of the amplifier, the

crowbar system would come in and it wouldn't blow the speakers. We put a couple of light bulbs on the back of the amp, we caused a fault, and it had to switch off the speaker system before it would blow the light bulb. A fuse was way too slow. This was better than the usual crowbar system and was Bill's design. Genius.

These amplifiers were so good you could put a one-and-a-half-volt torch battery on the back of the power supply, and it developed 160V AC, that's how good they were.

I'm not that technical, but if I'm reading you right, Bill discovered through sonic compression the way to make an amplifier?
We didn't have to make them, we bought them.

Right. But they weren't being used for sound; they were being used in the construction industry?
Yes, that's right, to level concrete. The guys at Elkhart didn't even know they were an amplifier. They thought they were a power supply. The amazing thing about it, I don't know if you know, but Crowns are pretty much used everywhere throughout studios all over the world, or they used to be.

Now, I remember Crown amplifiers at festivals in the '70s, definitely.
Oh yeah, we were the first to ever use them, first to ever discover them.

Now that we had our amplifier, there was only one thing left, the crossover unit. This had to take an amazing amount of power, and it crossed over from bass to mid-range and mid-range to high in an overlapping system, like a big V but crossing over just before it got to the bottom so that there were no gaps in the frequencies.

As the DC-300s were in stereo, so of course did the MAVIS crossover unit have to be in stereo, too. This is where IES turned from just love of the music to a full commercial business. Bill and his techs built the crossover unit, and it was a beautiful piece of equipment. Bill built and designed the crossover, and Phil Bowman would check what Bill had built, and then another guy would check and test everything after him. There was a whole process, which took weeks.

Bill was building two or three things at the time, and we had about 30 girls putting the circuit boards together. Since there was no one competing with us, this was not a race against time. If there had been another company we would have used their equipment, but since there wasn't, we just had to do what was necessary and invent and build it ourselves. We had to build other things ourselves too, like power supplies which would regulate power from 9V, 12V, 24V, 110V and 220V and could have no variation in voltage. This was the MAVIS power supply; this would drive both the mixing consoles and the crossover units. Many of these were sold to universities, as they needed accurate power and saw that the MAVIS was the best. Hawker Siddeley bought many crossover units so that they could test the vibrations in their planes, and altered their designs accordingly. None of it was cheap; for instance, the big 30-channel mixer was £149,000 to buy and had a lifetime guarantee.

Were you still in Carnaby Street when this was all going on?
Well, we took the Crowns down there, to check the speakers; we still didn't have a means of driving the Crowns, for that you need a mixer, a pre-amp as you might call it.

So we had to build a mixer. We knew how to build a mixer, but it was going to take time, took us nearly a year to do it. We built the 30–30, or 15–2, which are 15 channels, we're building the 30–30 at the same time: 30 channels in, 30 channels out. The idea was we could record the bands at the same time we're doing the sound, perfectly flat. We bought a series of Ampegs, 24-track machines, which I sold to The Kinks, I think. Yeah, they were good friends of ours.

And Bill designed the MAVIS desk, didn't he?
Well, he did all the electronics on it; I designed it, the basics of what it had to look like, how much it had to weigh, and what it had to do. He designed the electronics.

Without him, there would be no MAVIS desk. He was a good guy, really clever, he and Mike Philips. And Phil Bowman who went to Russia, he was a university professor, he used to work for us, too. He defected to Russia, and we had the bloody defence department come around wanting to know where he was, it was a strange situation.

Anyway, Bill Hough was the key to my success, there's no doubt about it. I loved old Bill. The mixer was the key, of course, the mixer drove the Crowns. Well, they didn't strictly drive the Crowns, they drove the crossovers. Now, the crossover was

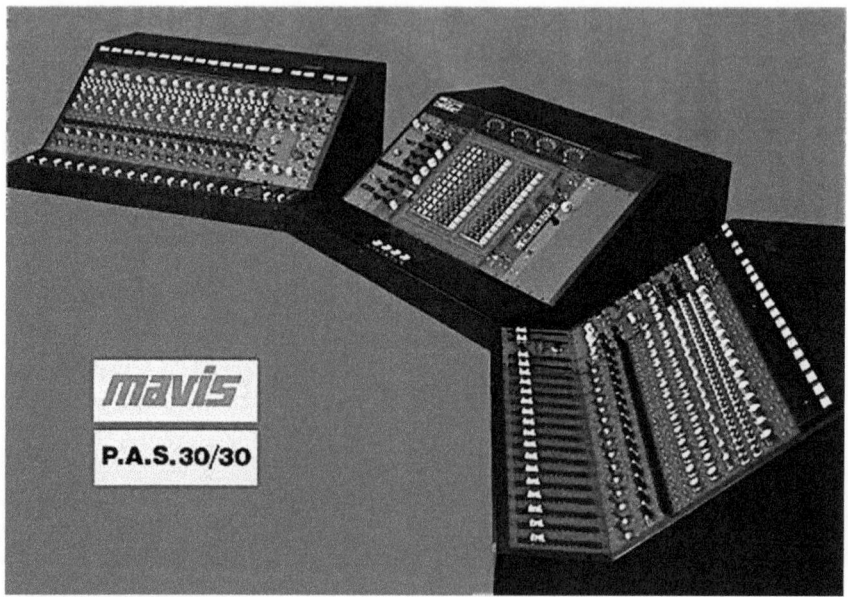

MAVIS mixing desks
© D. Hartstone

a big thing, we had to have something that would go from bass speakers to mid-range horns to high-range horns. All you have is passive, you can't use that with big speakers and big amplifiers because it just blows the passive drivers out. It had to be electronic, and nobody had one except Bill Dilley from Spectra Sonics in America; he had a two-way. Now, Bill Dilley did all the nose cone electronics on the rockets going to the moon, first ones and Spectra Sonics did that.

Bill Hough and Bill Dilley were friends, so we rang him and Bill Dilley said, "I'll send you over a card for a two-way electronic crossover", and we thought this might save us some time, but we could never get it to work, I don't think we ever did.

In the meantime, we were building a three-way electronic crossover and ours worked perfectly, beautifully. We crossed the bass speakers with the mid-range speakers to the high-range speakers and overlapped them perfectly, so we've got no noise.

The key to sound systems is when you get a person say, like, Andy Williams singing, he goes out on the stage and he sings and he says, "Cut", you know? No sound, no music, no nothing. You can't have sound systems hissing and blipping and making noises. You had to have analogue, to have the sound and get the noise down, which makes it hairy. We tried the chips, we tried everything, Westinghouse chips, everything you could lay your hands on, but everything had a noise in it.

So you basically built the business up, and you became the only company in the UK, if not Europe, with a large sound system that could be rented out. Do you remember anyone else at the time that was renting out stuff?
Oh, yeah, Roy Clair, you have Clair Brothers and Tycobrahe Sound, they had sound systems.

No, I mean in England.
Oh, in England, there was nobody. There was nobody, I had it on my own for three years and then, some people who left me tried to start sound companies and did it, but you had to have a lot of money.

I mean, The Beach Boys wanted us to build a sound system for them, but I said, "What's the point, next week, it'll be a different scene, it'll be out of date in a few months, better to rent it". I went out to their place in Denver and I went out to their place in Chicago, we're all friends, I was friends with all these people.

But I said, "It's not worth doing it". I would've liked The Beach Boys to have used my system because we were the first to use quadrophonic sound, I don't know if you know that. We invented something that did all the sound, any sound you wanted so we could choose channels on the mixer to go through the quadrophonic pot, and it would swing that sound around when you linked it together. So it was pretty, it was almost like a computer.

That was on the MAVIS desks?
Yeah, that's the MAVIS desk, and I've heard people say they still have them going, they still go. You see, the MAVIS desk didn't have anything on it, it was flat and that was the secret of it. It was a flat mixer, pure sound.

There were constant improvements, and I was never satisfied, nor were our technicians. Imagine, you use the MAVIS 30–30 mixer and without any extra wiring can plug in two 24-track AMPEX recorders, and the recording would come out better than most recording studios could produce. ELP, The Moody Blues and John Denver were just a few I can remember that did this with our system, because you could do a live recording straight from the mixer, because it's already equalised to the band to come out perfect. This was an unbelievable thing, all because of this mixer.

All of our equipment was up to Dolby standard, which is a type of EQ, used in the UK mostly. The MAVIS had a whole Dolby system built into it on the right-hand side, for a clean recording sound. Not only that but MAVIS had a quadrophonic pot in each channel and two pots on the output, and you could switch any channel you wanted onto those pots.

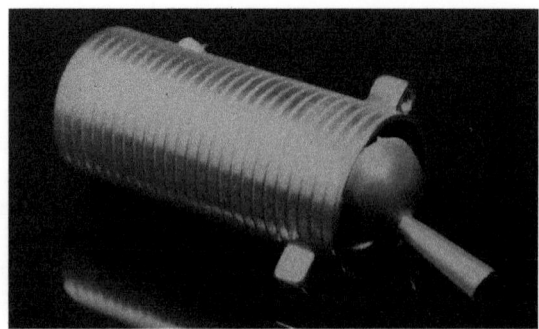

Quad pot for MAVIS desks
© D. Hartstone

It was way above the DIN standard, which you have to have to export. I remember we had to buy test equipment from Norway and Denmark to test the DIN standard with our equipment. I've heard most MAVIS mixers still go beautifully today and using analogue is the only way to go for sound, whereas digital has a hiss and hum. Analogue is far from being forgotten.

We sold a 30–30 PAS system to Eddie Offord, who was the sound engineer for Yes and many other groups. He was the first to buy one of our 30–30 mixers, the cost was around £149,000, which was an unbelievable amount of money at that time. The mixer was under complete warranty, parts and labour. He was working with Yes, with that mixer, and we rang him to see how it was going; his answer was 'bloody beautiful', and we asked him if anything had broken on it, and he said that the teamsters had dropped one of the cases, of which there were three, and on close inspection in the auditorium they found it had broken one of the little switches on one of the modules, but Eddie said it didn't matter because he wasn't using that switch. I flew Bill Hough out immediately to Pennsylvania to replace that switch and check the mixer and the sound system and check all their stage equipment while we were at it.

They were totally surprised to see Bill arrive and do the work. They were very grateful. IES warranty going to work!

Yes, I think the only other desk that was sort of comparable in a way was when The Who's road crew created their own production company called ML Executives, and they had Neve make two road mixers.
Well, Neve are good, and we knew them very well. We bought a couple of mixers off of them. The problem with Neve is the output of a Neve is only half a watt. That's not enough to drive Crowns or a crossover unit, you have to build something on the back of it which creates noise. Our mixers would put out 10 watts per channel, 10 watts, not half a watt. Pure, clean, flat sound and that was the point. Neve knew it, but they were building stuff for studios.

Chris Quayle with two quad MAVIS desks in 1975 at a Crystal Palace Bowl show

Yeah, it was getting bigger and bigger. When we did Emerson, Lake and Palmer, we had eight tractor trailers just for the sound! When we did the Californian Jam in '74, there were half a million people there, we did that in quadraphonic sound, and you could hear it, it was just like a studio.

Yeah, well, you were the hard edge of it all. When an English group went over to America, did you just ship your stuff over there?
Yeah, we used to, not ship, we used to hire planes, we used to hire 747s, have them take the seats out and put it all in there.

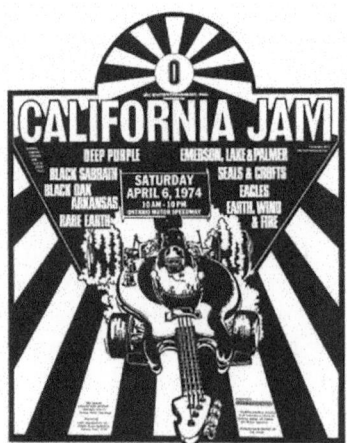

California Jam poster
© Don Branker

Holy moly, didn't even send it by sea?
It was too slow. We couldn't do that. In the end I formed a company in New York; Perry Cooney was in charge of it. I said, to Roy (Clair), of the Clair Brothers, "Come be partners with us, we've got English groups coming over, we are sick of paying the freight over. We want to build the sound systems right there". But he declined, so I went on my own. We were building sound there, and Perry was there, and that saved us a lot of money, and we were able to do the sound a bit cheaper.

What year was that?
That could be, I don't know, '73, '74.

IES trailer in New York
© D. Hartstone

But you got fed up in the end?
It was more than that. Harold Wilson was my key, do you remember Harold Wilson?

Yes.
Right communist he was.

Oh, this is the '99p in the pound' tax times.
Well, yeah, that's right. It wasn't him personally, but all of these fucking people came out, and they just basically told me that my business was their business. I said, "Well, you can stick the business up your arse" and I walked out. I paid my taxes, what I didn't like was the VAT.

And I wouldn't pay it, I never paid a penny to VAT. I went to the Bahamas, I had a boat in Florida, and I went to the Bahamas and I saw the sand, the sunshine, I was like, 'what the hell am I getting up in darkness every day and come home and go back every day with 400 people giving me a load of hassles'. So I left it.

So at the height of your success, really, you just quit.
Yeah.

Did you sell the UK side of the business to TASCO?
No, I never sold the business. I sold it in bits. You know, I couldn't sell up because of the taxes. VAT put a hold on anything like that, so I had equipment everywhere all over the world, big sound systems, and I sold them off piecemeal. Maybe in New Zealand, might have been in Australia, I just don't know, I can't remember who I sold to, we were doing 22 sound systems a day.

Renting them out?
Yeah, big ones too, Perry Como, Frank Sinatra, you name it, Hendrix, the whole lot.

So the business, well, just grew out of all proportion once you'd moved to America?
Well, not only that, America was big before I went to America. America was the airlines, that was the problem, the cost of freight. That's all that was. But the Americans don't like you coming over and sticking your nose in their businesses, even if it's your business, they liked to control it.

Yes, but did you like doing business over in America?
It was okay, we had a nice flash apartment over there, and we had another apartment for our guests, for our crew, and you had to deal with the mob.

Now, as I understand it, before you actually broke up the American side of things, your sons, your children ran the business for a little while, didn't they?
Yeah, it was a disaster. I knew it would be, they were too young, but they said if you don't let us go and run the New York Company, it'll always be said that

you never gave us a shot. But I just wrote it off because it'd probably go in taxes anyways. I just let it go and, in the end, they just gave up. It didn't go bankrupt or anything, I just closed it down.

You just shut it all down.
I shut it down because there was no point anymore.

Did anyone take over the majority of the equipment in America?
Well, Showco had a bit. I don't know, different people in America bought sound systems to rent or use, I forgot what. They were quite expensive and I'd say, "That's the price of it" and they'd say, "Jeez!" I'd say, "I'll tell you what, give me half in cash and it's yours". So I had all this money in paper bags so Harold Wilson wouldn't get a hold of it.

We had a mezzanine floor in London and I said, "Who the fuck are those people, sneaking around in the big sound systems we got down there, sneaking around and taking photographs?" So we rounded them up, four of them from Showco!

I said, "What the hell are you guys doing?", and they said, "Ah! We're sorry", they said, "We had to sneak in and we're just trying to photograph your sound, that's all". I said, "That's fucking nice, isn't it"?

They said, "Well, I'll tell you what, you come out to Showco and you stay with us for a week and we'll put you up, feed you and we'll make it up to you". So I did, I went out there. Got some of the best steaks in the world!

That was in Texas?
It was in Texas, yeah. They were really nice people it turned out, and they really weren't onto my business, they were onto lighting more. Lighting became the thing, and I wasn't going to go into lighting. That's another thing that put me off, when sound became not important anymore. We were already done and everybody said, "Well, the sound's there, now we're going to need the prosceniums and the lighting", and I wouldn't go into that. It basically ended a natural death, and I was quite happy, I ended up with a lot of money. I am quite happy with what happened.

Dave, thanks very much for your time.
All right, cheers!

As David explained, he sold up and bought a boat. To this day he and his wife Miranda sail the Caribbean and offer charter holidays on their yacht Taboo.

Notes
1 www.express.co.uk/entertainment/music/122669/The-monster-who-made-Scott-Walker-a-superstar
2 www.audioculture.co.nz/people/clive-coulson/

8   Bryan Grant

Bryan Grant

February 2, 2007 with Bryan Grant at Britannia Row Productions, Wandsworth, London

Hi, Bryan, let's start at the beginning. Where did you go to school, where were you brought up?
I was born and brought up in New Zealand in Auckland, and in 1970 I left there and went to places like New Guinea and the Middle East as I made my way to London.

Why New Guinea?
I had a brother living there, and it seemed like a good idea at the time. I earned enough there to get the fare to come here (to the UK), because you could earn relatively good money living in the jungle. I worked for an Australian heavy machinery company, nothing to do with entertainment at all.

I had started doing a law degree in New Zealand and got bored with that after a couple of years, and as I say, the world was going on here, things were happening here, and I wanted to see it for myself.

I eventually arrived in London and stayed for a few months until early '71. Then an American guy I was sharing a flat with and I decided to go to New York. It was April, it'd been a shitty winter and we had a choice between paying the rent or getting a cheap flight to New York—a no-brainer.

So I went there and then hooked up with a friend I had grown up with in New Zealand. He was living in New York working for this company called the Megaphone Company that did sound, lights, backline, rehearsal studio, all this exciting sort of stuff. They were based in a loft building in Crosby Street down on the Lower East Side, and my friend called me up one day and said, "Would you like to drive a van and gear down to North Carolina?" and I didn't see any reason why not, like you do, you know when you are a kid; didn't know where North Carolina was, never driven a truck before, never driven on the right-hand side of the road, but it didn't seem as though it would be that difficult, so I did it. I was so terrified by the whole experience, although excited by it, that I got there and back in quite a short period of time, and Whitey Davis,[1] the guy who owned the company, was so impressed that (a) I got there and (b) found the band and (c) got back and probably just as important (d) brought him back the van and the change from the float, that he gave me a job.

It was great in those days, he and I ended up doing tours with Nina Simone and Miles Davis and Dreams, which was Billy Cobham's band, and various one-offs around the upstate New York area. Those shows with Davis were all over the place, and ended with a residency at Shelley's Manhole in Los Angeles. That was the first

Newspaper advert for Gaslight club

band I ever mixed on a Marshall five-channel mixer amp at that club! You couldn't do that now, but there was no one else as Whitey had gone back to New York and left me there on my own to look after the PA. After I got back to New York, Whitey told me that we'd taken over the management of a club on Bleeker St.

Called what?
I think it was called Gaslight Au Go Go.

A jazz club?
Maybe it had been, but we also had people like Hot Tuna and Mason Prophet and Alex Taylor. We also ran rehearsal rooms in Crosby St and had people rehearsing in there like Mahivishnu Orchestra.

Hot Tuna was an offshoot of Jefferson Airplane.
I just remember the Hot Tuna gigs because they had come to play a date in upstate New York, and the date got blown out so we had them in the club for three days. We gave them all the takings, and we just kept the drinks money, and there was this whole bunch of bands that played over that three days, and it was just mad, mad stuff that never let up for that whole weekend. You couldn't drink anything because people were spiking drinks with acid, just fucking crazy, anyway that's another story.

As well as running the club, hiring out equipment and doing one-off gigs, we were getting up in the morning opening the recording studios for Mahavishnu Orchestra, who were early risers, starting about 9.30, 10 o'clock in the morning, working through the day and then at about, I don't know, 10 or 11 going down to the club, hanging out there until that closed about 2.30, and then probably going off to Nobodies or somewhere similar and hanging out until about 5 o'clock.

So I didn't sleep for several months as well as misbehaving badly when I was awake, and so I cracked, came back to London, slept for several weeks and just started doing freelance engineering for people like Stoneground and Roy Harper.

Then I joined IES as a tech engineer, did tours throughout '72 for their customers including Captain Beefheart and Johnny Cash. That was great, because it was a hire company so you did whatever came along, for example, Andy Williams. I remember that conversation with Hartstone; "Oh I don't want to do Andy Williams, my mum likes Andy Williams!", and Hartstone who owned the company said "You will fucking do it", and so I went off and did it, and of course Andy Williams was a complete hooligan, as was his entourage, they were more rock-and-roll than anything I had ever seen.

Tell me a little about IES.
Well, IES was a company owned by a guy called Dave Hartstone, another Kiwi, who had come over to the UK with a band, I think in the late '60s.

Anyway they had struggled here and the rest of the band went back to New Zealand, and Dave stayed on, and he kept the equipment, and he started renting it out.

He had these great guys working for him at the time: Bill Hough, Bill Kelsey, Dave Martin, they were all working with Hartstone and all became legends. Kelsey and Martin had left to form their own companies by the time I joined, but Bill Hough and Hartstone were making some very cool gear. They had just started developing actively crossed-over systems using JBL components and Crown amps, which was quite revolutionary for those times, and they began manufacturing the MAVIS crossovers and consoles, which were really cutting-edge technology in live PAs.

MAVIS consoles.
Yes, I mean this is '72/'73, and this equipment was way ahead of anything else around.

When we weren't touring or doing shows we worked a six-day week in the warehouse for £25 a week. Hartstone sacked me after about nine months for being a shit stirrer because I wanted more money and more time off, so I went off and worked with Zeppelin for a few months as John Paul Jones' roadie, which was a strange experience as I'd never done backline before. Nice guy, John Paul, I have to say, as was Robert Plant, but backline wasn't something I wanted to do, and my girlfriend Catherine, now my wife of 35 years, didn't really want me to go on a world tour, so I quit and did a tour with Genesis as head PA tech for Sound City Hire.

Anything to do with the shop?
Yes, Ivor Arbiter, I think, owned the company, and Sound City Hire used to have this top floor in a warehouse just off Moorgate. Bill Hoade, the manager, talked me into going and working for him full time, and he and I ran Sound City Hire for a bit, and we started doing more tours. We built more PA systems, and I attracted a couple of guys from IES to come and work for us.

After a few months I get a call from Hartstone saying, "I always intended, it was always going to be like this, I let you go so you could get more experience, now come back and run IES". Anyway, I went to work, as general manager at IES, and IES at that time were building pretty sophisticated stuff, including bigger MAVIS consoles, active wedges etc.

Multicores.
Yes, big Belden, individually screened multicores . . .

Stageboxes.
Exactly, transformer isolated splits, way ahead of their time in the live touring business.

Was this the point where—I am trying to get some history on when multicore appeared. Who maybe created it?
Well, multis had been around for a few years by then; various people had them, and various people had different ways of getting a signal to front of house, I remember in those days.

Because most sound was done from the stage until the multicore?
That's right, when we did The Who, The Who wouldn't let Bobby Pridden go out the front, he had to be there so Townsend could see him. When we did The Who first time, Bobby was still mixing from the side of the stage, and when we persuaded them it was a good idea for the console to be out front, Bobby had to stay on stage.

So he switched to monitors.
Yes, and they had an American guy as the front-of-house engineer. We actually did one of their tours in quad. We were exploring things like quad in those days.

You did quad for Cockney Rebel at Crystal Palace Bowl, and we got Alan Parsons to do the mix instead of Bob Gross.
Yes, with the MAVIS 30/30, which was a 30-channel console with 30 outputs. Every channel had its own quad pot, then there was a master quad routing section.

We had an Ampeg 24-track that would sit next to the console so we could record direct onto tape, and in fact we did some quite extraordinary recording with that. The concert I remember most was the Rainbow, Finsbury Park, where we did a Warner Brothers package show with Little Feat, Doobie Brothers, Tower of Power, Grand Central Station and others. This extraordinary package toured around Europe, and every show was recorded in this way, and they did an album from it. Little Feat's album, *Goodbye Columbus*, was partly recorded on that tour.

When you first started working at IES, apart from Sound City were there any other sound hire companies. Was IES the very first one?
No, there were others but we didn't seem to run into them. MEH, Marshall Equipment Hire, headed by Joe Brown, which later became TASCO, was around. I think Rikki Farr was just starting Electrotec then. I quit IES in early 1975 and got everyone jobs at TASCO because I was pissed with Dave.

This is the third time.
No, I was fired once and I quit once, that was it. But in that period '73 to '75 when I was running IES, we did some great stuff, we did ELP through America and Europe in full quad, the first Knebworth festival, and loads of major tours around the world. We sold a desk to Eddy Offord, who was Yes's producer at the time, one of, I think, only three MAVIS 30/30s that ever really got built, because even in those days they were 40, 50 grand to buy. That was a lot of money for 1974.

Built where?
Built at IES, hand-built; Bill Hough was the designer, he was brilliant.

Where did he go after IES?
Hough went to work for TASCO, last I knew he was working at the Palladium, working for Moss Theatres. Now he would be a mine of information.

Do you recall monitors coming into the frame?
Do I ever? In '72 we often used to mix monitors as well as front of the house from the same console.

Roger said, with The Who, that they had got monitors when The Faces were supporting them, and they both had PAs, WEM PAs, and one PA went towards the band and the other towards the audience.
Well at that time, that's how I did Roy Harper; no, in fact he didn't bother with monitors, just as James Brown, the Godfather of Soul, never did. We just turned the PA in a bit.

When I joined IES in '72 we were making our own wedge monitors using JBL components, 2470s, I think, on the high end, and something like a 2220 on a 15-inch speaker. There was a passive crossover network, and I remember the first gig I did with Captain Beefheart. Van Vliet called me backstage after the first gig and he said (fucking imposing guy, you know, he had these really piercing blue eyes that seemed to drill right through you), "I know you are not a sadist man but your monitors suck", and I said, "I am sorry about that", and he said, "Well I will do a deal with you; from henceforth, when I walk on the stage at the beginning of the concert, I will kick your monitors off the stage and we will proceed from there".

I called Hartstone up at about 1 in the morning and said to him, "You got to sort me out, you have got to get me something that works!". Beefheart's vocal power and range (and he had a clarinet as well) was just mashing this passive crossover, so we ended up modifying one of these big active MAVIS crossovers and bridging the mids and highs and changing the crossover points slightly and went from there and it worked, so he was happy with that.

We'd been using active crossed-over systems in front of house for some time, but that's when I started using active wedges, and of course, once we had started that's the way we went with them. Still in those days in '72/'73 most acts' monitors were passive and being mixed from front of house.

You know what, I just had a huge flashback, in 1973 I remember you coming down to the Speakeasy to show me how to set up the crossover in our brand-new Soundcraft desk. I kept blowing the fuses on our H&H amps. To this day I have not got a clue how crossovers work. Before long we hired Bob Gross from IES to mix Cockney Rebel because I was doing a terrible job, and I fortunately became their tour manager.
So I was a good teacher then?!

You had 10 minutes or something. I know you set it up for me so that particular night I didn't blow any fuses, but I blew them up at every other gig.

Where were you based, where was IES then?
We were in Chalk Farm in an old dairy that later became a studio. The loading door was down a mews that had been converted into really expensive flats. There were two ways in; you could either wheel the equipment down an alleyway where

the council flats were, which used to piss the residents off because of the noise, or you could actually manage to back a truck down the mews with expensive flats into the factory, which used to piss the residents off because we kept running over their pot plants.

I think Roger Searle said that's where he got the gig for driving the truck because he was able to actually reverse an artic down your mews.
That sounds about right. It would never happen now. Jesus, the things we got away with!

Before IES, everyone just had their own stuff, right?
Pretty much, I guess. The Who started ML Executive with their own trucks, backline and lights, and the Floyd had all of their own PA and lights. I was on their first tour with Led Zeppelin. Zeppelin had their own PA—it was a Heil system, which wasn't up to much—and they hired The Who's lights from ML, which was quite good fun because The Who's road crew weren't taking it that seriously. They had the feeling that Zeppelin were a bit effete really and not like The Who.

Alan and Heath had built consoles for The Floyd, they had built their quad console.

Alan and Heath?
They are a company that's still around.

Was it studio based, Alan and Heath?
Possibly, many people were. You have got to remember in those days we were borrowing technology from wherever we could borrow it, the cinema, theatre etc.

There were very few specific manufacturers around at that time building equipment for the touring market; it was so small and new. Midas was still very much a garage operation, as was Soundcraft and Avolites and so forth, all kind of garage operations. Bill Kelsey had his operation in Notting Hill building consoles and speaker systems. Dave Martin had split off from IES, and he was just starting Martin Audio. Bill and Dave were always accusing each other of ripping each other's ideas off, it was quite funny; and Tony Andrews of course was just starting around that time as well. So you had all these people building stuff because no one else was, things like consoles; people sitting down thinking "how do I go about this, how do I go about being able to be 100 feet away or 200 feet away from the PA and the audience?" Because that's where you had to be; it made sense to be out there (in the auditorium) because you could hear what the audience were hearing. Trying to mix arenas from the side of the stage was an awful idea; you couldn't hear a bloody thing.

I remember in America the first time I went in '74 it was as assistant tour manager (baggage-man) for Wishbone Ash, and there was ML Procise from Showco on the sound crew and their mixing desk was in a rack.
It was probably an Altec mixer.

Stuck between the seats in the auditorium.
Yes, it was green, right? They were rack mounted, and you had EQ for every five channels. I mean, the Americans were relatively behind the UK, especially with consoles.

Is that why the big UK bands took equipment over there when they toured the US?
Yes, IES were doing that all the time, for bands like Humble Pie, ELP, Jeff Beck etc.

Were you sailing the PA over?
Well, we would airfreight, and sea freight, depending on the itinerary and where we were going. Jeff Beck took a whole PA to Japan because there was nothing there. At that time if you wanted to do Japan that's what you had to do, and when we say the whole PA the actual size and power of theatre PAs then were probably about the equivalent of what you would find in a club these days.

2/3000 watt.
Yes, funny how the level of expectations change, isn't it!
    The first Knebworth, 125,000 people, that was probably using about the same amount of power that would do the Hammersmith Odeon now.

What year was Knebworth?
I think the first one was '74, that was the Allman Brothers, Doobie Brothers etc.

Knebworth stage, 1974
© Barry McCulloch

Were you at the first Shepton Mallet in '70 that was the precursor for Glastonbury?
With Freddy Bannister?

Which was 100,000 people.
I think Mike, my business partner, he might have been there. Charlie Watkins probably did it with a WEM PA because again Charlie Watkins was the kind of father of all this really; he did most of the big UK events in the early days.

Sold WEM systems to everybody.
He not only sold them but he would lend them. I remember going 'round to his factory and picking up a PA when I was touring with Roy Harper. Oh man, yes, what a lovely fellow. He produced the WEM copycat, the Audiomaster; I mean these things were innovations in their time. I read an article in some magazine about him not so long ago, and he was still making stuff. You know Charlie did that Stones concert in Hyde Park in '69.

Yes! Brian Croft recalls all his crew disappearing from the ICA one afternoon; they all went across the road to Hyde Park.
I remember doing Buxton Festival, I think it was the Beach Boys, Humble Pie and so on, so we are doing Humble Pie and they were either headlining or second on the bill, and we brought our own PA, an IES JBL PA, and wouldn't let anyone else use it. We set up our PA next to Charlie's WEM PA, and of course you know this was a load of active JBL stuff, and I remember Charlie coming up to me and saying "Be gentle with me boys", and we were real bastards, we just cranked it up to number eleven and blew the WEM away—at least we thought we did.

Do you remember in those early days actually doing stuff with other sound companies, the big shows?
Collaborating with them?

Actually tying in your gear together. I seem to recall IES probably having the biggest amount of power, and The Who used it.
Yes, up until Charlton football ground. We had to pass on those shows as we had all this other work going on, and MEH did it. Bobby Pridden never forgave me for that.

There was Charlton, Celtic and Swansea, three football stadiums. It was amazing, did you come along to one of them?
I did, I think I went to Charlton.

I was on stage when those lasers went up, awesome.
The Who's lighting designer, Roger Searle, was innovating, he was into some good stuff.

So, 'round about '75 you had gone back to IES?
No, I was running IES, and then I fell out with Hartstone again and quit. By this stage Hartstone was living in America because we had opened up a branch in New York.

So you must have been the first UK production company that opened up in America.
I would think so.

Before Rikki Farr did his thing?
Rikki was just starting in the UK around that time.

So Rikki was Electrotec (formerly Electrosound).
Yes, around '74, '75, you've got TASCO, Electrotec coming on, I think there was a company called Ground Control, IES of course and there were probably several others. Anyway, so Hartstone by this stage had made a bucket of money in '73/'74/'75, he was doing well, he had gone to live in America, I was running it over here, and for whatever reason he and I fell out and I quit in March/April '75.

Hartstone based himself in New York, Perry Cooney was running the company with him there, and then he got paid a pile of money for doing the delays for California Jam, I think. Anyway, he went and bought himself a big yacht, a Goldstar 70 or something, and started getting excited about that. He moved down to the Bahamas because he decided to rent the yachts out instead of PAs, and the last I heard he still is. He sold out all his US equipment, and he folded IES probably late '75/early '76. He had lost interest.

Here in the UK as well?
Yes.

By this time now you are how, where, what?
Well I got a job with Pink Floyd doing their advance work. Mick Kluczynski at that time was running audio for Floyd, and when he heard I had quit from IES he offered me a job. Harvey Goldsmith actually offered me a job as site co-ordinator on The Who shows, though the Floyd tour of America seemed to me a lot more fun than doing that, so I joined their circus. I worked with Floyd for a couple of US tours in '75 ending up at Knebworth.

Had Britannia Row been created by then?
Britannia Row was being set up in mid-'75. Mick and Robbie Williams were running the sound and Graham Fleming was running the lights.

Mick Kluczynski wanted me to join Britannia Row, but I really didn't want to. I'd had enough of audio rental for a bit, so I tried my hand at band management. I started working for a guy called Tony Secunda, who was managing Steeleye Span in those days, and worked for him for a couple of years, and then I was tour managing bands like Wire and Steve Hackett. I met up with Robbie Williams, who was now running the UK side of Britannia Row Audio, on a Roxy Music tour in '79 and, after some discussions with Norman Lawrence, Nick Mason's financial advisor, we decided to set up Britannia Row productions.

Mick Kluczynski, who had been living in America, running Britannia Row's American operation, quit in '78, and Seth Goldman and Perry Cooney took that over. The idea of Britannia Row Productions was to try and do packages of lights, sound and set and have bands record in the studio—the whole service if you like.

Mark Fisher was working for Britannia Row at that point. He has gone on to being one of the most well-known and respected set designers on the planet. He is a fascinating guy, a fascinating history, he is an architect, and he designed those biospheres for the film *Zardos*. He got involved with Pink Floyd in the '70s and designed the set for The Wall, still to this day one of the most extraordinary sets ever. He went on to design for The Rolling Stones, U2 and some of the biggest live events around the world.

So there were all of these great people who came out of those early Britannia Row days. During the early '80s, Robbie and I had been given control of all of the various companies under the blanket of 'Productions' and were doing some very successful tours for artistes such as Stevie Wonder, Bob Dylan, The Cure and Roxy Music, but we had become a political football between the Floyd's various financial advisors, so in '84, having closed down the US operation, Robbie and I bought Britannia Row Productions and the equipment from Nick Mason and David Gilmour.

We sold off the lights and concentrated on audio and carried on doing that until the mid-'80s when the Floyd decided to go touring again. Robbie joined them as production director and went on tour for two years, leaving me to run the company, which was by now very busy as we'd teamed up with an American company, MSI, and were looking after each other's acts on each side of the Atlantic. We'd be doing Tours with Whitney Houston, Neil Young, Gloria Estefan etc. over here, and they'd be doing Pink Floyd, The Cure and Depeche Mode over there. It was a great arrangement which lasted until the mid-'90s when we decided to open up in the US ourselves for a few years.

So, I needed help! Mike Lowe, who had been running Turbosound rentals and then joined Samuelson's when Turbosound rentals got sold to them, wasn't enjoying the experience, so in '87 I invited him to join Britannia Row. Mike joined and in '91 Mike and I bought Robbie out. Mike and I have been partners in Britannia Row ever since.

What is Robbie doing now?
He has got his own very successful production company, RWP. He does a lot of very high-profile events like the Queen's Golden Jubilee Celebrations, The Prince's Trust concert at the Tower of London, Nelson Mandela's 46664 concerts etc. He does a lot of events for the BBC.

Tell me a road story.
I will say that those early days of touring were just mad shit, weren't they! You got to travel the world and you got to misbehave, not too badly, and you got paid for it, and there was great music going on, and it was just wild, wasn't it, and yes at Britannia Row we got a bit of a reputation for doing gigs in strange places; we did Sinatra at the Pyramids, Robbie Williams went out there and again, you know, 'what do you do for power?' He actually got hold of The Dead's office and said,

"You guys have done it, what did you do for power?" and the guy said, "Well what you do is you go to the Great Pyramid and you walk so many paces this way and then turn left and walk so many paces that way and so forth, and you dig down and there's a three-phase supply down there" and Robbie said, "You are shitting me!" and he said "No, it's true!" and Robbie did it and it was there. Unfortunately, apparently every time they turned the PA and lights on, the local village went through a blackout! And then there was India.

Who with?
We did a stadium tour with Boney M, I think it was in '85, Mick Kluczynski was the production manager on that, and a guy called Malcolm Ross was the tour manager.

Malcolm took over from me in '75 working for Wishbone Ash.
So Boney M at that time, '85, apparently were bigger than The Beatles in India, they were just huge, and a guy called Bryan Miller who was their agent hired us to do the sound and lights, bring it all over from here, fly it all over and do a tour of India. It seemed like a good idea so we did that, and I flew over for the first gig in Delhi, and it was a bit of a disaster really.

Were you playing indoors or outdoors?
Outdoors in a stadium and there were these huge posters all over Delhi, 80,000 watts of thunder power for Boney M, or maybe it was 800,000 watts, they were prone to a bit of exaggeration. We set the PA up and during soundcheck the local electrician had turned the power off and turned it back on again under full programme, which blew all the high end, which we weren't aware of unfortunately until the show.

So we had to do the first show without any high end, which was quite a strain, and the next morning I had to go out to the airport and spend all day extricating from customs 60 of these very expensive diaphragms that I'd had to fly in from America, and had to refit them before the next show that night. When we'd done that, I said to the local promoter, "Get me the chief electrician", and he came with his little entourage, immaculately dressed in his white uniform, and I stood nose to nose with him, covered in sweat and grime, and said "If you do that again without warning, I will kill you, okay? It's very simple". Kluczynski is standing there going "He means it! He means it!" The electrician looked me in the eye and said "You are very cruel man" and from then on, I would get this call from him saying, we are turning the power off in half an hour, we are turning the power off in 25 minutes, in 10 minutes we are turning the power off etc. . . . so I used to get this solemn countdown all the way to switch-off; I guess he believed us!

I went back to London and the tour carried on to other cities around India. The band crew and gear trundled around India for about five weeks, the gear travelling in six open trucks, and I finally get a call from Kluczynski saying, "Listen, it's all going very bad, it's all going very wrong".

Most of the shows were being cancelled, the promoter by this time was going broke, neither Malcolm nor Mick trusted each other, and no one trusted the promoter and they couldn't figure a way out, so I went and saw the agent and said, "I need 5,000 dollars in cash, and you aren't going to get a receipt for them, I'm just going to get us out of there", so I flew out to Mumbai and proceeded to try and extricate us from the situation.

We had six trucks full of equipment, the band and the crew to get out, and the promoter was responsible for paying for all this. Mick, Malcolm and I had a meeting with the promoter and his new financial partner, who as far as I could gather had insured this final show in Mumbai *against* happening and was going to make sure it didn't! It was a surreal and chaotic meeting that went on for hours with people shouting, storming out of the room, having whispered conversations in the bathroom etc., and I am negotiating with these promoters. After hours of conversation, I said "Look we have had enough, we have to go", and so the guy said "Yes, you can go if you sign this piece of paper, which says that because of technical difficulties we cannot do the show, otherwise you don't leave". Eventually I agreed on the promise that the band and crew would leave that night and they did, which left me, Malcolm Ross and one of my crew to try to get ourselves and the gear out. All the equipment had been temporarily imported on Malcolm Ross's passport, so if the gear didn't leave, he didn't leave.

We spent the night finding and paying the trucking company and arranged with them for all of our equipment to be at the airport the next morning at 6 am. We had one flight to get it on, it was a Friday afternoon flight, and if we didn't make that flight, things were all going to go very wrong.

Six trucks worth of gear?
Yes, six of those Indian trucks—beautifully painted but open—and the gear had been travelling around India for five weeks, so you can imagine the state of it.

I had found a local freight agent the previous day; he was wonderful. I was recommended to him by an embassy official, who told me he had allegedly managed to steal a western helicopter back off the Indian army, who had it on approval and wouldn't give it back. So I thought, he is my man, explained the situation, gave him a bunch of money and said "I don't care how you spend it, just get us out of here".

So five of the six trucks turn up at 6 o'clock in the morning. We unloaded them and piled them all into the customs shed in Bombay on the premise that if we took up enough space and caused enough chaos, they would want to get rid of us.

The trucking company had held one truck back to try to get more money out of us, and the freight agent spent the rest of the day trying to find it. In the meantime, the customs officials in Bombay airport made us unload every piece of equipment, open every case, and show them to check everything off against the manifest.

Of course, normal life in the Mumbai airport freight warehouses was carrying on, more and more stuff was piling in and things were getting busier and wilder, getting really crazy. I'm stalling the airline people and shmoozing the customs people and just generally trying to keep our place on the 6 o'clock plane. At 5 o'clock this last truck arrived, and we had to unload that and show them the equipment against the manifest, and by this time everyone was panicking.

So the freight's all there; we've got the paperwork filled out, and we're standing there in front of this customs official, and we are just absolutely filthy covered in grime. You remember the opening scene in Midnight Express? I can still see him in his immaculate whites, this customs officer, and we show him the papers and he looks at it and goes, "No, it hasn't got the proper signature on it and stamp on it". I think, "Oh no, this is it!" The freight agent grabs me and whispers, "It's okay, come with me". So we run through all these different buildings and offices, and we walk into this office and it's just chaos. I guess it's Friday evening and the airport's about to close, all the freight offices are about to close, and everyone's heading for home. My man walks in and he looks around, and he opens this drawer and pulls out this stamp. He licks it and he stamps the document and squiggles on it, and then he says, "Come on, this will do" and we run back to the customs official.

I am not quite sure what is going on, but we go up in front of this guy with the white outfit again, presented the paperwork to him, and again it's that Midnight Express moment where he is looking at you and you are looking at him, and he knows it's dodgy, you knows it's dodgy (laughs) . . . he just stamped and signed it and off we went.

Ross and I drank a bottle of Scotch each, I think, that night!

And because that happened, Ross could get out?
Yes, because his passport had to be stamped as well.

By the man in white?
Yes, the stuff had gone, so we were still in India for that night, and then we got out the next night.

Did you get paid for it all?
Yes, we did.

So, Bryan, who was your biggest influence joining the music business?
Different people at different times really, I suppose. Whitey Davies, who was the guy I first worked with in America, who owned the Megaphone company; he was mad and a complete workaholic but he had a passion.

Everyone was like that though in the early '70s, weren't they?
Yes, it was quite eccentric, wasn't it?

It didn't seem like work either.
No, well it did when you were left on your own to do a load-out! Whitey could be a bastard, a complete bastard. We were doing a Miles Davis show in L.A. We've loaded out, I am driving a truck, and he is being a total prick, and he is balling me out over something. I pulled over to the side of the road and I said, "That's it, fuck you, I am out of here". There are only the two of us doing the P.A. We had a truck

full of gear, and we were going to fly the stuff to the next gig in Phoenix the next day because in those days you could blag your entire backline and PA systems onto aeroplanes, as excess baggage by bribing the sky cap (porters).

Anyway, so we had this big fight and I said "Fuck it, I am out of here" and he said, "I am sorry, I was being out of order" and he said, "I tell you what, have you ever seen the sun come up in the desert?" and I said, "No". He said, "It's fantastic, we have got to do it, tell you what, instead of flying to Phoenix, we'll drive!" I didn't know that Phoenix was 500 miles away in the desert. So we drive this fucking 22-foot bobtail truck with no air conditioning, out into the desert. It's 2 o'clock in the morning by this stage, and we are driving into the desert, and sure enough the sun's coming up and there is this extraordinary sight, but once it's daylight it gets bloody hot and very bright, and we are sweating through it with no sleep, and we drive and drive for 500 fucking miles.

We get to Phoenix for the gig at about, I don't know, 6 o'clock in the evening it must have been, just in time to load everything in, wire everything together and do the gig. The gig finishes and Whitey comes up to me and he says, "Well, kid, well done, now here's another lesson in life". I said, "What's that?" and he said "Well it's not fair! I've met a woman, so I'll see you in the morning, you can do the load-out!" and walks off! Well my lip's going at this stage—What! So he has gone and I have got to load out on my own—no stagehands or anything like that in those days. I am pushing this gear out, feeling very sorry for myself, and this girl comes up to me and she says, "Do you want a hand pushing the gear out?" and I said, "That would be really good!" So we start pushing stuff to the truck, and it's all going well until she pulls out a spliff at the back of the truck and she says, "Do you want a hit?" so I take a hit of this dope and it's just ridiculous. I didn't smoke much dope even in those days, and so we get incredibly stoned, and we are trying to push this gear out, and I am tired, I have had about one hour's sleep in the last 48, and I am now stoned and she is too. So the load-out is going slower and slower, I think it took about five hours to load this 22-foot truck! In the end there is the manager of the venue and this huge fat security guy with a belt full of guns, he is helping as well, we are all trying to load this truck, what a circus! Yes, that was an object lesson in life, I have got to say. What was the other question, sorry?

Who do you think has been a big influence in the history of sound production in the UK?
In those early days people like Dave Hartstone, Bill Kelsey, Bill Hough, Dave Martin, Tony Andrews; deep respect for those guys for their technical skills and for kicking it off. Mike Lowe, I have to say, is a guy who still has a passion for what he does and an ability to really focus on any issue, and I am really honoured that he is still my business partner, quite honestly. Mike has done a hell of a lot for this industry.

And *what* do you think has been a big influence, not who.
I would say that sheer economics has been the biggest influence in our business. Like any industry it's grown up and has had to become more efficient, to deliver

more bang for the audiences while delivering more to the artistes and their business managers and promoters and so on. We have more efficient packaging, more cost-effective solutions and manufacturers who have built huge businesses making equipment specifically for the live music business.

Bob Geldof said a thing at the TPI Awards last year, it was after Live8, probably the biggest show worldwide that the world has ever seen. From the conceptual discussions to the actual show was only a matter of weeks, and he said he couldn't think of any other industry that could have mounted such a logistical operation in terms of getting the materials, the people and the logistics of the whole thing including broadcast together in such a relative short space of time. He didn't think the military could do it, certainly the civil service couldn't do it, and definitely governments on their own, given their own devices, couldn't do it, and I think he's fucking right. What it is, this business is just full of people who are passionate about what they do and are willing to try. It's full of people who see a barrier and try and figure out a way through it; they just don't see it as a barrier, they see it as a challenge, and as long as that mentality prevails we'll keep the creativity going. Do you know what I am saying?

Very much so, the show must go on.
Yes, if a show has to happen at 7 o'clock that's when it has to happen, and how do you make that happen? You just do it. I am unemployable probably, I have lived all my working life in this business, so I don't know if the 'real' world works in the same way. It's not to say we are not realistic and we are not business-like, but we don't have the mentality that at 6 o'clock I can go home now whether we've finished or not. I think that's what it is about, it's what attracts people to what we do; it's those lack of rules, lack of boundaries, so what really is the biggest influence, it's still, I guess, that passion, it's still the excitement of being part of an experience which is unique; it's never going to happen again, that concert will never happen again. There will be other concerts with the same performers and all the rest of it, but it will never happen again in quite the same way; it's one unique experience whether you are in a crowd of 100,000 or 200,000 people or 20 people, 100 people or 1000 people there is something unique about it, and when it is right, when it's magic, that's kind of as good as it gets, it don't get any better. You've been there, you have been in concerts, you have been in that magic moment. It doesn't happen all the time, sometimes you wish you were somewhere else, but when it does it's better than any drug, it's better than any other work experience I can think of.

Yes, just being around when there are 100 people or 10,000 people all there to have a good time, and you are getting paid for it, too!
Yes, but you are the poor bastard who will still be there when everyone else is in the pub! (laughs)

Bryan, thanks very much. That's brilliant.

Bryan Grant is currently the managing director of Britannia Row Productions, having joined it in 1979. The company was originally set up by Pink Floyd so their road crew could hire out their equipment, back in 1975. Bryan was awarded the Outstanding Contribution prize in 2015 by the UK's TPi Awards, which is run by the TPi Magazine.

Note

1 http://rockarchaeology101.blogspot.co.uk/2010/04/some-notes-about-whitey-davis.html

9   Tony Andrews

Tony Andrews

November 29, 2010 with Tony Andrews at His Home in Kent

So, Tony, where did you go to school?
Actually all over the bloody country, because my mother was always moving house, but I think the important one was probably Brighton, Hove County Grammar School. I think that's where I spent all my formative years, as far as the relevance to this goes.

That's your last school before you came to London?
Yeah, I got a place at Chelsea College of Science and Technology. I was really into geology as it happens. That's why I went to Chelsea College of Science, which is part of London University.

To study?
To study geology. If you did well enough at A-levels, then that was the natural thing to do, you would go to university, and I was always very interested in geography, and it was my best subject. I didn't do that though because they used to split you into arts and sciences after O-level, and I got streamed into the science area because I was pretty into chemistry, biology and physics.

 I couldn't do fucking geography, and I was really pissed. This, I think, was the beginning of me thinking, hang on a minute. What is this system? Furthermore, why have I just spent five years learning every bloody chemical reaction you can think of and then when organic chemistry started, that was the end of it. When all they had to teach you was valency theory from which you can work out how elements will combine because in simplistic terms they will try to combine to give an outer shell of electrons containing a stable eight or multiple of eight electrons.

You went to college. Did you finish the course?
No. No. I think I lasted about two or three weeks because I was going into the class, and they were giving me all these facts which I had to write down and remember and then regurgitate them all at exam time. I thought "I can't do this anymore".

Too much like school.
It was. I said, "Look, I need a year off". I needed a sabbatical. During that time I started building swimming pools and also getting into the music thing big time that was all going off around me. I can remember seeing the Aynsley Dunbar Retaliation transit van go past us, and that's where I wanted to be.

Were you going to see music at that time?
Oh yeah. Yeah, remember the Bath and West? We were talking about that. When you said you got the poster, is that right? That was a big one.

When did you go to college then? What year was it that you went to college?
'69. Took that year off, did a whole load of stuff, tried to go back the following year. I think I only lasted two weeks that time. I thought "no, that's it", and I went on the road as a drum roadie with Pete Brown and Piblokto.

Pete Brown?
Yeah, you know Songs for a Tailor?

1969 leaflet to a gig at Oxford Town Hall

Well no and yes. I know his name for some reason.
He had a band called The First Real Poetry band. Well his original, his first album was 'A Meal You Could Shake Hands With In The Dark' and then 'Things May Come and Things May Go But The Art School Dance Goes on Forever'.

Great titles.
It was an education.

No acid there, I'm sure.
No, no, no, Pete didn't like dope smoking. When I got busted in Sweden, that didn't go down very well, and it wasn't long after that I kind of left, but I was having enough of it anyway because I was looking for the real thing. At the time bands like Quicksilver Messenger Service, they were singing about the proper stuff. That's what counted for me because I thought the world is going to change. I still think that, by the way.

It's got to change dramatically. This is all wrong. I didn't understand maybe. Over the years I've understood more of the mechanisms, but at the time it was just, this can't be the way it's supposed to be. I suppose what got me really thinking deeper about it was that Bath and West Festival. I was talking to Ann about it because you've got my mind going a bit. Some friends of mine were getting hold of good quantities of acid, and it was really nice stuff.

I have some memories with them, and there was about a dozen of us. We're just all gone on it, sitting there and Zappa's playing, and somebody said, "Is he God?" We all knew what he meant because we were just feeling it because he was taking it skyward, and we were allowing our minds to go where they were going. This fantastic matrix was forming, and I thought God, this is powerful stuff. This is something here. The world needs really strong sound, and the sound system as you know was a Charlie Watkins thing.

Yeah.
He had those big parabolic dishes on top, which actually didn't do anything, and although they were just four 4 by 12s in a cabinet, they were phase coherent because the drivers were on the same plane. It just wasn't strong enough, and they were direct radiators. The WEM column, that's the original line array, right?

Wasn't part of the problem the amplification to his speakers?
Yeah, but amplification at that time was a terrible problem. A hundred-watt amp was something quite serious. The transistor amp hadn't long been with us. They were bloody unreliable in the early days. The H&H's were terrible. I came unstuck with those at the first Glastonbury. Well actually not quite the first because Eavis did the Pilton Pop festival[1] the year before.

What I was trying to get to about Bath and West was that, I think, was where I realised that there was something really powerful. I remember feeling how inadequate the sound was.

Pete Brown, I mean that's all pretty hazy. I know that I was taking LSD, not on a regular basis, maybe, I think I did it about six or seven times, and the trips kept building up and getting more and more serious.

It got to a stage where I got talking to God, as you can. I said, "Okay, so there is a master plan for humanity. What can I do to help?" He said, "You mate, no, you sunshine. You build loudspeakers. That's what you do. You do that". It just started with the discovery of extra bass by pointing a normal speaker backwards into the corner of a room.

You didn't know anything about speakers particularly.
No, I knew I liked them.

The trip told you, that's what you've got to go and do?
Yes, definitely. I'd already built my own a couple of years before that, like you, bought a couple of speakers, put them in a box for my own listening pleasure, as it were. I think I had just begun to really be interested in it, and then that happened. I mean the two kind of went together. So, it became a determination.

Now when I met you at Lillie Road in 1973, I came to pick up a PA system that you'd built for Cockney Rebel, you were working with your brother, weren't you?
Yeah, it was me and my brother Billy and occasionally a guy called Spot, who was Joe Cocker's roadie. So Billy was working with me probably at that point, but he's a very unreliable soul. I don't think any of us had a very straight-forward upbringing, but mine, I had a completely hysterical mad mother and a completely absent father. It wasn't sensible.

All right. How did you get into Lillie Road? Is that where you first started making them?
Because I met Spot, I suppose. That wasn't the first place we were in, by the way.

So, tell me about how did you start making speakers then.
You know, I don't really know how it all happened.

Did you make a PA system to sell the very first time?
I made a PA system, the first serious PA system I guess was for the Pink Fairies. It came after Pete Brown. That's what we determined we were going to do, and we made them and we painted it pink, too. Well pink on the front, and it worked really well. That's where I meet Boss Goodman, who's a lovely bloke. Boss ran Dingwalls for years after the Fairies broke up.

I remember the Cockney Rebel PA system, it wasn't pink. It was mauve.
Yeah, good. That was the purple phase. By the end of the '70s people had a real bad reaction to hippies, and purple went with being a hippie. That's why Turbosound was blue.

Now what do you remember about building speakers? You see, Charlie Watkins didn't rent his equipment out, he didn't have a rental side to his business. He sold stuff and bands basically didn't have any choice in the UK then, and if they wanted a PA, it was just WEM. If they wanted monitors, it was WEM. There just wasn't any other way.
Yeah. Yeah, it's true.

It wasn't until some of the American equipment came over that there was any real alternative apart from you. You were the alternative to WEM and Charlie Watkins.
Yeah. We were, yeah, because we had been discovering horn loading. I got pieces of kit that's all horn loaded from about '71 or '72.

Were you just building and selling just like Charlie Watkins?
Well no, we built it to use. We rented it out and went out with it. We made something, and it was pretty well how we've always behaved in a way. Yeah, we built it to use it.

Who was the first? Pink Fairies in '70?
'71, I reckon, and Hawkwind.

You did Hawkwind?
They were with the Fairies. They were their support band. Later on, they became successful in their own right. There was also the Edgar Broughton band, who I encountered at a demonstration in Hyde Park about something or other, they had a good attitude. They reminded me of Mick Farren the way they looked. You know, dark hair and lots of beard.

Okay.
We got onto the horn loading thing very early. We had a double garage at the back of the house we were living in, and that's where we started to put this stuff together. Do you remember the Graham Bond Organisation?

Yes.
In fact, that's what kicked it off. We're all in a room. We converted this loft at the top of the family home, which was on Kingston Hill right next door to Richmond Park.

Anyway, we're all upstairs, we're getting stoned together and we're playing around. We've got this speaker in a box, and my brother said, "Let's just point it into the corner", which you wouldn't do because you're going to mask all the frequencies apart from the bass. I'm sitting there and I'm thinking "the bass just got loads louder". See, a lot of life is just observing and paying attention as much as possible, it seems to me. Anyway, this bass increase intrigued me. We finished up making the corner of a room into a box as a portable, foldable thing. I thought, "this doesn't look right, we want to have the corner going the other way, we'll reverse it so that it points at the speaker and splits it and then we'll put a lid on the whole thing". Anyway, we made this and Pixie, the bass player from Graham

Bond Organisation, he came down to give it a try because he was a mate of Muz's and absolutely loved it.

Who was Muz?
Muz Murrel, who was the main road manager with Pete Brown, and I left home and followed him. We both left to do this together. Yeah, you know the thing was to get high, get spaced out for a bit and see what's 'there' kind of thing. Anyway, he's still around.

You designed the reverse bass bin so the speakers go backwards bouncing off and coming out in front?
Yeah, and there's a bit more coming back now—my memory! Pixie came down to try this out and plaster fell out of the ceiling. I mean, it was pretty amazing, and that's how we kind of discovered what we do now. We just refined it immensely. We were all knocked out, him especially; I suppose that wasn't a whole PA, but we sold that to him. He just wanted one, so we made it and sold it.

That was your first bass bin?
Yeah. You can honestly say that, and it was to this day, quite clever because it was big enough to do some serious grunt. And yet, it is just like a small cabinet for the speakers, and the rest of it was foldable and it actually worked.

Yeah, so we did that, and then we did some stuff for Mitch Mitchell along similar lines. We then got down to 15s, and we evolved the technique a bit further. By the time we met you in '73 I'd come a reasonable distance with that. It is only probably about two stages before the one that we got a patent on. We applied for the patent in '77. It really did work.

Do you remember who else was hiring PAs at the time when you were doing this?
Yeah, Joe Brown and TASCO.

Do you remember IES?
Yeah, Dave Hartstone, wasn't it? Yes, and not to be mixed up with Richard Hartman. What a lovely man he was.

I know. I've interviewed Richard.
Oh, beautiful. What a fucking trooper that man was.

He's still doing it.
He was wonderful. I remember him welding trusses just before a gig and coming up black. Involved, oh man.

Did you get to know him much?
Not well but from a distance, you know, a man after my own heart. Gets stuck in and makes it happen. Health and safety nowhere to be seen. Fabulous time. Fabulous.

He was very intrinsic to an awful lot of the lighting technology.

What do you remember about IES?
Only that he wasn't on a good vibe with it. He was renting out JBL horns, which absolutely screeched. They were just horrible things but loud as anything.

You know what? In a lot of ways it hasn't got a lot better because they're still using those bloody components. Even today, and that's one of the fundamental differences between us and everybody else, but that's not the subject right now. Although everything we're talking about is what this is evolved from.

Whatever else has gone on, I can honestly say that it's been a pretty true trajectory holding to the ideals of that time, and expressing it, if you like, through the way we make audio equipment for people to use for the purposes of getting closer to God.

Hmm.
Although everyone has forgotten about that now! (laughs)

Tell me, how did you get involved with Glastonbury then?
Well at the time of the Pink Fairies, we were all into whatever, you know in this innocent and probably confused manner but realising, as a lot of people thought, that we had to come together. We recognised it was around music and sound and, I guess, a lot of people in Notting Hill Gate because that's where they all lived at the time. Fairies, Hawkwind, the *International Times* people, Mick Farren, Steve Mann, etc.

We got to hear about this bloke who lived on Strand on the Green called Andrew Kerr,[2] who'd been with Randolph Churchill. He'd been his valet, if you like. Charming, charming man, and he still is. He's great. He's one of the most charming people I've ever met in my life. He's worth meeting on that front alone, and he's 74 now. He got into John Michell[3] and the Ley Lines, and he'd gone down to Glastonbury because that is the spot where a lot of them meet.

He somehow got to hear about the Pilton Pop Festival that Michael Eavis had put on the year before, which was what, 1500, a couple thousand people, with Thomas Crimble, and I think, well, Michael had done it for left-wing reasons, I'd say in a nutshell. This emerges later, but anyway, Andrew had gone down to Pilton, he'd spoken to Michael about doing something serious and getting everybody involved. We figured, *this was it*.

It really did feel like this was it then, because we were feeling this energy. There was, in that generation and the people I was lucky enough to be around at that point, sharing those aspirations with them, there was this spontaneous spirit, and "if we can, let's do it", that kind of thing. I remember going over there with the Pink Fairies, and I think Mick Farren[4] was there, too. We thought yeah, okay, this is good, we should do something with this.

I know I was going to do the sound for it. Oh, we had these wonderful multi-cellular horns, and we really had it sorted out. That's when the H&H amps all blew up. Thank god Traffic came down with their WEM PA. I sat on a soldering iron, in horror, with all these fucking H&H amps blowing up. I've still got the mark on the back of my leg. That was the Glastonbury[5] that Andrew and Arabella (Churchill) put together with Michael Eavis in '71.

And the Pilton one in September 1970, which came after the 1970 Bath and Western Festival, was a completely separate thing?
Yes, absolutely.

They came very close together, didn't they?
They did and there was something. We felt like something was trying to happen. Maybe this was it. An awful lot of attention went onto this because the underground were committed to it. The Pink Fairies were an intrinsic part of that and *IT* was and Mick Farren. They were an intrinsic part of it. It was right on the cutting edge, and Andrew was such a charming gentleman.

Pyramid stage, Glastonbury, 1971
© Mervyn Penrose

The pyramid stage? How did that come about?
We all went down to Glastonbury town at some point, and we were all tripping and going out, going up to the Tor and stuff. One night we went to the Tor, and I think the Pink Fairies were there, and this bloke called Bill Harkin just showed up. He arrived there, the most odd of circumstances. Anyway, he was an architect and, you know, a designer.

He thought, right, well, let's make the stage a pyramid, and he then became an intrinsic part of it. It was when Andrew and Arabella Churchill got involved. I think Andrew put all the money he had at that point, which was 10,000 pounds, into that, and so did Arabella.

The Who at Charlton was when I first realised that the quality of sound, the distortion levels, had a huge effect on the sanity of the crowd. I remember that being a very sad event because I think there was a lot of violence from skinheads around this time. I remember thinking, that's concurrent with the fact that they were using the horns I just talked about earlier?

JBL, right?
Yes, they really put out this atrocious noise, so bad, it is what drove me to do stuff to get the mid-range correct. See, most of the sound out there today, if you were to liken the spectrum of audio to the colours of the rainbow, orange and yellow are practically always missing, and there is too much green and blue.

Changing the subject for a minute, I remember down in Lillie Yard, you guys were building boxes for trees, lighting trees?
Yes, we were just asked to design a box for them, a road case for it. I suppose that's what kicked us off designing road cases, we made a lot of those. It made us some money. We've always made cases actually. A lot of techniques were originated by us, even in road case building. We were just entirely inventive and creative and thought things through. I mean, the team I've got around me today, there's a core of about six people who have been together for 30 years doing this stuff. We refined the art of problem solving. What I can say about life is that whatever people think about the difficulty of what we do with audio and how out there it is, compared with dealing with people and the madness that ensues . . . it's a piece of cake.

'Round about the time with Lillie Road, what do you recall about the Roundhouse, the Rainbow? Did you ever get involved with those guys?
I went to the Roundhouse a few times, but I never actually got involved. It was Jeff Dexter, John Morris was in there.

The Rainbow, did you ever get involved with the Americans and the gear that they brought in?
Only on the fringes when they brought in that Tycobrahe[6] stuff, it took my breathe away. It was all horn loaded and very well organised, and it was beautiful. It really did it for me.

The thing I haven't really put into this interview, Richard, is when I heard sound, I got entirely stimulated and was almost into a whole other landscape of reality. I found it incredibly inspiring, incredibly inspiring.

That drives you, the engine of your thoughts along, and you can finish up in so many good places, no doubt. I thought this is what we've got to do. This is what God has built us for. We've got to go beyond where we are, and these days I could characterise that by saying we've got to discard the politics of selfishness, completely. Otherwise, it is good night, Irene. Our children are not going to have a very great time.

Going back to—
I'm sorry, I keep doing this, don't I!

That's all right. How did Turbosound come about? What is Turbosound?
Turbosound initially was a rental company that built its own equipment. Before that, there'd been Sonic Trucking and before that there'd been Cosmic Boxes and before that there'd been Peace Sounds. Peace Sounds, I guess, was what I did with Muz and the Pink Fairies and all that part of it. Cosmic Boxes was what we were doing when we met in 1973.

We'd had various premises before Lillie Yard, but I don't remember where they were. Somehow Spot set it all off. They had this warehouse, and he wanted to make a PA. That's what it was, and I said, "Well, we haven't got anywhere to make one, so we might as well move our tools in with you, and we'll build a PA and we'll design it together" kind of thing. So we did, and I've got a picture of it. It's fucking great and it's probably something similar that you guys bought. My brother is there as well next to the table saw, which we've still got in the back workshop.

Purple PA
© Tony Andrews

Then it morphed into what? Sonic Trucking?
We'd just got this huge place in the country down near Lingfield called The Red Barn, we set up workshop there.

You and Billy?
Me, Billy, Frank, the whole family. In a way I moved back home, but I had to because I really wanted to continue this audio thing. I wasn't making enough money to support the whole situation.

At the Red Barn you turned into Sonic?
Yeah, that was the Cosmic Boxes place. After that, it turned into Sonic Trucking.

Sonic Trucking. Was who?
Sonic was me and Simon Renshaw.

Simon Renshaw, no way!
Yeah, he went on to be, not quite sure what his capacity was with ZZ Top.

He now manages the Dixie Chicks. He's an old friend.
He was a mad bugger.

He was a completely mad bugger (both laugh).
I'm so glad you've come 'round to see me.

I can't believe hearing his name. Simon Renshaw was the promoter's rep for the 1979 Kate Bush tour. We got to know each other very well because of it. I went to his wedding, and in fact I've still got the laminate that was the pass to get into the wedding reception. How did you meet him, and how did Sonic Trucking start?
Well I think I might have met Simon through an acquaintance of mine, a girl called Faz, it might have been through her that I met him, but see my problem was, I was just completely engineering and logical based. I just didn't do people and politics.

Yeah, and organisation?
That I could do, but not realising all the other agendas that are going on at any given situation.

In 1975 then?
Yeah, we were pretty successful. We were a rental company and we moved stuff around. We did tours. We did our first serious tour with LSD Lights in Birmingham.

Did that go well with Simon?
Yeah it did, because he was great on the phone and he could get business. To be honest, my focus on engineering has cost me dear.

Who did you do on that first tour with LSD and Sonic Trucking?
I think it was the band that Andy Frazier from Free formed. After Sharks he did this funk band. They were bloody brilliant. Trumpet players, girl singers. That's where I met Steve Hall. Oh God, my memory kicks in, he became part of early Turbosound, yeah.

Oh, okay. Sonic Trucking was a sounds rental company as well?
Yeah, it was primarily a rental company, and we did quite a lot of tours. We were good at it, and it was about the time of meeting John (Newsham) in '77. Yeah, that's jumped a bit. I can't believe how much happened in that decade.

It's like four or five companies, and they were all episodes. People came and went and things happened, and then they broke up. Jesus.

Have you got any history written down about what bands Sonic Trucking worked with?
We did, I think we did Stiff Little Fingers.

I wonder if Simon knows? He might recall.
He would. He would probably remember.

I'll email him. Tony, thanks very much for your time, all fascinating stories.
Richard, it's been great, thanks for coming to visit me and good luck with it all.

I did contact Simon Renshaw, and he responded as follows:
Nice to hear from you.

Windsor Free Festival, 1974
© Simon Renshaw

You know, I don't recall how I met Tony, but it was in connection with me putting together a stage for the Windsor Free Festival in 1974. Somehow I hooked up with Tony, and he ended up providing the PA for Stage C that I booked and ran.

While we were doing that, we both got to know a group by the name of Zorch, a fantastic collection of hippy freak synthesiser players with an amazing light show.

I ended up managing the group and living in their commune in Churt. Tony provided sound for the band. We were all broke. We were all very high.

Somehow we ended up agreeing that I'd set up a company and we'd hire out Tony's sound system. Hence Sonic Trucking Services, which I ended up running out of 286 Portobello Road. So I think that would be probably late '74 through mid-'76.

We worked with all kinds of bands, Back Street Crawler, Van der Graaf Generator, Chris Farlowe, Motorhead, and a bunch of others. I actually have my diaries for that period with a lot of the old booking details in. I don't remember how it ended, just sort of faded away.

I was with Tony the very first time that he saw the effects of the lozenge in front of speakers—someone's house in Barnes or Fulham. I had become far more involved in production management by that time and moved into that space. Tony took the lozenge and ended up doing a deal with Rikki Farr, if I remember, to build that first system.

Early Turbosound PA
© Tony Andrews

Tony Andrews' involvement with the Glastonbury Festival led to the development of the Festival System, which consisted of separate bass, mid, and HF cabinets. It already contained the second generation of the "Turbo" phase device, the distinctive, slender, missile nose-cone shaped contrivances mounted directly in front of the mid-range cone drivers and projecting almost the full length of the enclosing horn.

Tony and John Newsham, who were responsible for founding Turbosound in the 1970s, founded Funktion One in 1992. They are highly regarded designers and manufacturers of high-quality point source loudspeaker systems, with numerous patents to their name, are accredited with several generations of revolutionary loudspeaker designs including the TMS-3, Flashlight and Floodlight systems.[7]

Notes
1 www.ukrockfestivals.com/glastonbury-1970-menu.html
2 www.telegraph.co.uk/news/obituaries/11149237/Andrew-Kerr-obituary.html
3 www.john-michell-network.org/
4 www.thequietus.com/articles/11663-mick-farren-interview
5 ukrockfestivals.com/glasmenu.html
6 www.harmonycentral.com/forum/forum/LivePerformanceCategory/acapella-33/385638-/page2
7 www.musictri.be/brand/turbosound/ourstory?active=Our%20Story

Part 3
Pioneers of Stage Design

Genesis 'Lamb Lies Down on Broadway' stage
© Jeffrey Shaw

I think there was a moment in time when someone other than the lighting operator got involved with the 'look' of the show. Bands were persuaded to spend a little money on what was on stage other than them, their instruments and a few lights, thus bands like The Stones, Zeppelin, Pink Floyd, Yes and Genesis brought along something different to the party.

I first met Ian Knight in 1976 when Steve Harley wanted to spice up the visual aspect of the Cockney Rebel live show. Steve had some great ideas, and someone pointed us in the direction of Ian. He sat down with Steve, and between them they came up with a variety of effects that became part of the new show. Ian also designed the lighting truss arrangement, which was set up exactly the same every night around the seven-piece band. It was simple but effective, and that's what we were looking for, so everyone was happy.

When I interviewed Ian, it was during his treatment for cancer, but he was in great form with his memories crystal clear.

10 Ian Knight

Ian Knight

May 13, 2009 in Ian Knight's Apartment, Very Close to Canvey Island

Ian, hi, thanks for seeing me. So, where were you born and where did you go to school?

I was born in the East End of London, during the war, in a Salvation Army Hospital. We moved then to Muswell Hill, and then very quickly because of the war, I moved out to a place called Eastcote, which is near Ruislip on the Uxbridge Road.

I actually went to normal schools from there, and eventually ended up—probably during the 14-plus era of life, not the 11-plus because I was not bright enough for that—I ended up at Harrow School of Art.

That's where I did a fine arts kind of lithography course, you know, and my mother believed that I was going to be a printer, right. I don't know why she believed that, but I liked the idea of going to art school. I got grants because my father died just after the war, and we were not exactly blessed with huge finances.

I stayed there and went on to the intermediate and then the National Diploma, which is equivalent to a degree now. I left there, my sole ambition for the last two years was to go into theatre. This was basically the biggest disaster ever as, in theatres, 90% of people are out of work. I remember my mother saying, "Well, you better go see your uncle, because he is a printer and he'll organise a job. You're not going into the theatre, there's no point, you're not meant to do that". But I did, I went in, I became what was an Arts Council designer, the school kind of acted as your agents, right. Anyway, I got a summer job after finishing my school diploma.

What year would this be?
'59/60. Yeah. I got a summer job as a waiter on the trains from London to Scotland, I conned my way into that, and it was fantastic! It was fantastic money. It was unbelievable! And everything was crooked. They made and sold their own sandwiches, (instead of selling the ready-made ones), it was fucking great, we earned a fortune!

Then the Arts Council phoned me one day and said, "We've got this job for you starting in September in a theatre in Guildford". So I did my bit on the train and laughed my way through it and loved every minute of it, and then I went into theatre. I went into the theatre in 1960, and I then became, probably a year later, like the resident designer just through luck. I mean, nothing in life has to do with talent, it's more to do with luck. Oh, sure, I've had some basic talents, you know, but I mean, luck is a big factor. Then the theatre got burned down.

What theatre was it?
The Guildford Theatre. It had patrons like the Redgrave Family and people like that. Michael was kind of around, and Corin Redgrave acted there. I remember doing *Henry V* with him.

Then, after that, I was in Coventry at the Belgrade Theatre, where Trevor Nunn was a young director. I went to Canterbury, where I did a few things, and then some really bizarre stuff like *Ubu Roi* at the Tower Theatre in Canonbury Tower, Islington, the (Alfred) Jarry play, Theatre of Cruelty and all that.

Were you designing the sets?
Yeah, designing the sets in those days.

Is that the backdrops or the lights as well?
No, just the sets, staging. And then, I went into the Jeanetta Cochrane Theatre, and did *The Changeling* or some weird stuff.

So you're good with your hands, a carpenter?
No, no, I just designed. I could build things, but I drew, I made models and they built them.

At the same time, and because in that business you're a freelancer, you have a lot of downtime, I worked for a studio where you made stuff, you know, like stage props.

It was in Covent Garden, it was Theatre Studios, it was very, very famous, the ultimate prop studios. We did the Royal Opera, the National as it was then and the Royal Shakespeare Company props and props scenery. If it was scenery, for example you want to build a wall and you want to make it concrete textured, that was prop scenery, so not scene paintings, they didn't do that. I worked for some legendary directors; Peter Brook on his productions of the *Marat/Sade*, and *King Lear* with Paul Schofield as Lear. I didn't design them, I worked on them, and it was staggering, I mean, it was staggering. While I was at Theatre Studios, I met a guy who was a sculptor, basically, called Keith Albarn, who's Damon Albarn's father.

Then there was a point the UFO Club came in and technicolour dreams, and I was into drugs but not heavily. I remember in this era going to see Hendrix, I think, not knowing who Hendrix was! I remember going to Powis Terrace to see a band rehearse called Pink Floyd, you know, in Notting Hill, Ladbroke Grove.

I went to the Saville Theatre one Sunday night; I saw this three-piece band—Oh! Christ, they'd got nothing, you know, they got Marshall amplifiers and they looked like yobs, you know it's fantastic, who are these people! Before that, all we had was glitzy kind of rock-and-roll, in the '50s.

Yeah, yeah, yeah.
And it was Cream. I was blown out of the back of the theatre; I thought *this was for me!*

Without remembering exactly the dates when I did things, I ended up working with Keith Albarn, we had a gallery, which was quite famous, in Kingly Street called 26 Kingly Street (Keith Albarn & Partners),[1] which was basically an environmental gallery, all right?

What does that mean, an environmental gallery?
Well, it's a gallery that worked within itself, you built rooms in structural kinetic art, you know, it was kind of mind-bending art. Keith would never say that, but only because he was not a druggie at all, you know! With the old Bag-O-Nails up the road, it was one of the trendy hangouts, and the Speakeasy was just across the road, you know.

Eventually, this gallery was successful, and we ended up in the summer of '67, Keith designed this geodesic dome that we all worked on, and we put it up on the beach in the South of France.

I had met a lot of bands in that era, and I became very friendly with the Soft Machine, who probably, whatever anybody thinks, is one of the greatest bands ever, you know, in terms of musical progression. I think the Floyd would've admitted many years ago that the Soft Machine were really much better than them because they were real musicians, and the Floyd admitted they probably weren't as good musicians in those days.

Well the Soft Machine, they're good friends, and I said to Keith, "We'll put this band in the dome, you'll like them", and when they met him, he loved them. So we put this band in his dome, and Mark Boyle,[2] you know, he did the light show.

Experiment towards Interplay leaflet with Ian Knight
© Hazel Albarn

And where was the dome by now?
South of France. On the beach.

How big is this dome?
It could take about 800 people.

Okay. Inflatable?
No, no, this was hardboard, hardboard.

Blimey!
Yeah, very dangerous. It was a dome, so it was modular. Keith was really good with his hands, and he had a wood shop out in his house in Lancashire.

Discotheque Interplay
© Keith Albarn

How long did it take to put up on the beach?
About a week. It should have been up in a day, but you know, you had to lay a deck, then there was a tower in the middle which the band were on. and Mark Boyle was above and they projected all around the dome. It was fucking unbelievable; this was, you know, the '60s.

It was near St Tropez, between St Tropez and Nice. It was a beer festival and people couldn't believe us; Jean-Jacques Lebel, Brigitte (Bardot) and all the hip people that hung out in Cannes, at least in the summer, everyone came out there to St Tropez. Anyway, like everything else, it went broke. I came back, it was a summer thing, and it was a big success. For whatever reason, I didn't stay with Keith. I was more into my rock-and-roll, the production or the staging side of things, and I, with another friend of mine, Sean Murphy, who I had met at the studios, actually became the managers of The Soft Machine in those days.

I've never been in this business for money, in the sense that, it was a living. I've only kind of almost done things because I enjoy it, at any point in my life. You probably think the same about that. It's not, "How much I'm going to get", it's like, "Oh, yeah, I want to do that. How much are you going to pay me? Okay, well, that's not enough but you know, I want to do it anyway".

I had done The Yardbirds at UFO once because I knew Jimmy Page and we did this 3D kind of thing, it was really good for the people on trips. We had this big white screen, which was a big light box, and we projected cracked oil images, through a friend of mine's equipment, Jeffrey Shaw,[3] we projected these weird slides at the screen. He was an Australian guy, did environmental inflatables, he was into all that stuff in Amsterdam in those early days of the '60s and '70s.

So I built this screen, and inside the screen, which was like tracing paper, we put plastic tubes and we projected slides onto the screens. At a certain point in this Jimmy Page show with The Yardbirds, the whole screen exploded, these tubes came out, filled the whole of the UFO and then, the tubes exploded and a thousand butterflies flew out of them. So if you were on a trip, it was a pretty good experience (laughing).

On a trip or not, I reckon (more laughing).
So I was kind of known in the scene now. I can't remember the sequence, but I got involved with Middle Earth, like a technical director, production manager, which

was the club in Covent Garden. I then did lots of those big events, you know, Ally Pally and those kinds of things. For Middle Earth, I went to the Roundhouse, and Jeff Dexter and Caroline Coon and I ran all those classic shows at the Implosions on Sunday, we were the people that ran it.

So you put on Implosion?
Yeah, we were the people that put on those events, Jeff Dexter and I mainly. Hugh Price a little bit, he came in, but it was mainly Jeff and me and Caroline, and the events were unbelievably successful. I mean, I'm not just saying this, I'm only in retrospective kind of thinking we transformed the whole world of music almost because, I mean, so many fucking bands came out of it.

Poster, 1970

Tell me some of the acts that you put on?
Well, Hendrix, The Who, The Stones, they all played there for eight shillings!

That's what the band got paid?
Yeah, they got paid expenses, you know. Elton John kind of almost became famous there and so many bands, Terry Reid, you know, you name it. It wasn't necessarily done under the Implosion banner, but The Doors and the Airplane played there, and Middle Earth did shows there occasionally, they had some big shows.

What was the difference between Implosion and Middle Earth?
Middle Earth was a club in Covent Garden, they were kind of a competitor to the Roundhouse. It was probably not making enough money and it was too risky, so it kind of fizzled out, and then the market porters of Covent Garden came in and tried to kill everybody and smash them up because there was a rumour that Middle Earth was a club that had sexual rites with children, there were all these stories about what it was. You know, people never understood the reality of it. Then, from the Roundhouse I went with John Morris, as an associate, whom I met when he came over with the Airplane, and worked with him at the Rainbow. I used to live in Rock Street in Finsbury Park in those days. You know, it wasn't such a bad idea for me, but I wasn't sure it was the right venue, you know.

Nice venue.
Oh, fabulous!

What did you do there?
I was associate producer to John. Then I went on to Sundown theatres with John Conlan.

I remember the Sundown theatres. Tell me more, just go back a little about your memories of the Rainbow.
Yeah, my memories of the Rainbow is, I must confess, I questioned why we needed it. I can understand what John wanted to do, John wanted that we make it like the Fillmore, but the Fillmore was not John, it was Bill Graham that kind of made the Fillmore. Bringing the concept and the people over here, it was kind of wasteful. It was good though because Hartman became English you know and stayed here.

Chris Langhart and a lot of the others went back to the US, but there are quite a lot who did stay here. In those days there was WEM PA, there was Dave Martin's Midas Systems, and Dave Hartstone and IES, you know.

There were companies here that did have American-style PAs, and so you didn't really need to go through the grief of creating a Fillmore's PA here. I think it needed an American like John to kind of push this concept through though, because I don't think there were probably any English people that were capable of doing it. Even though the Roundhouse had been incredibly successful, you know.

My first experience other than The Doors and the Airplane was when I did the Holland Pop Festival in Rotterdam, and it was all American bands. I asked John if he would come as a liaison, you know, between the American bands, but strangely enough the American bands, when they came to Europe, did not want Americans, you know what I mean? You understand? It's very strange. They were coming here to get away from that, so it didn't have a great advantage in most instances, although sometimes it gave them comfort.

I saw on the notes you sent me that you were touring with the Incredible String Band (ISB).
Yeah, that's right, the first band I ever toured America with.

With Joe Boyd?
Yeah, yeah. Hugh Price worked at Witchseason, which was Joe Boyd's agency.

I kind of used to hang out there a bit, and I did the String Band, the Roundhouse and then I went to America. I remember being told to take all these amps to see Fred at the TWA airfreight office at the airport, right, and Fred at airfreight said, "Oh, have you got a van?" I said, "It's all in a van", so he says, "Well, don't come to the freight shed, just go straight to the 747". So I drove the van straight to the plane. We took the amplifiers out, they're not in boxes, because we didn't have cases in those days, in this country.

There are no flight cases then?
No, so we just shoved them in the plane. In America we had cases made, you know, because they had cases in those days, they were way ahead of us.

Did you do Woodstock?
No. I did the Holland Pop Festival, which was just after Woodstock, and probably better. It had Floyd, had the Byrds, who were like Gods. They were bigger than The Beatles in Holland, you know. And ISB were on the bill.

Tell me about the Sundowners[4] how did they come about?
John Conlan, he used to work for Top Rank Leisure. He was one of the nicest men I've ever met, and he is still the nicest man, he's been in the entertainment business forever, he used to run Top Rank suites. The guy who was the manager for the old Astoria Finsbury Park, Dudley Perkins, who stayed on as front-of-house manager when the Rainbow opened, was a friend of John Conlan's.

He introduced me to John, and said, "Look, I've got this idea, we got all these stupid old theatres, I want to do something like this". The problem with it even though it was not a bad concept was that it had the capacity to pay American bands a lot of money, which was good and bad.

Yeah, so the four of them; Edmonton, Brixton, Mile End and the Astoria, Charing Cross Road.
Four of them, yeah. Like any corporation, it was a lot of money, there were lots of problems and Rank weren't really in it for the long haul. Then John left, and he went on to run other companies all in the entertainment industry for years. He is a lovely man, lovely man, thoroughly honest. Everybody respects him.

Anyway, after that Peter Grant phoned me up and said, "Hey, you let a load of fucking butterflies out, with all the fucking acid cases!" Then he said, "Fucking great, so you want to come and work with us?" That's how I started with Led Zeppelin.

Ian, do you remember the year of that?
Well, actually that's '73 and '75, the '73 tour and then the '75 tour and then '78 was the end, really. And at the same time I did Genesis, I took Genesis to Showco. Then in '78 I realised all my clients didn't live here anymore, you know, so I moved to L.A.

So you did Genesis with Gabriel?
Yeah. Up to the 'Lamb Lies Down On Broadway' with my friend from Holland, Jeffery Shaw, we did inflatables and slide presentations. I employed him to do that side of Lamb Lies Down On Broadway, which was a superb tour.

Genesis were one of the first very theatrical bands on stage. Whose idea was the head like a flower?
That's Peter, Peter was the genius of that band.

Did you go on the road with them?
A little bit, yeah, a little bit.

And Zeppelin?
Yeah, I went on the road with them, I was like a production manager, well, I was the designer. There was no definition in those days of what anybody did, you know.

Ian on the '75 Led Zeppelin tour

So, did you run the board?
No.

Did you ever run boards?
No, not with rock bands, just in theatre.

So, did you help put the stage together?
 . . . and special effects. We used to have flash pots and dry ice machines, it was a shit load of stuff they had with them. I did run a board then, yeah, I had a board in the States for special effects.

Explosions?
Explosions (laughs)! Showco built me a board because from nothing it was suddenly all incredibly complicated.

When Zeppelin started, the first show we had, in '73, was eight lighting trees and some mirrors and follow spots on stage, which I bought from Strand in

England and brought over to America. It was like, nothing! (at this point Ian whistles). Then Genesis wanted this truss, so I had to bring this fucking aluminium bridge from America from Tom Fields Associates to England to do a tour over here. There was nothing in those days, you know, at all.

So trusses came out of America, really?
Yeah, basically.

Led Zeppelin in Atlanta, 1973
© ledzeppelin.com

This photo sent to me by John Brown,

"if you look hard, the first use of onstage follow spots on top of the access scaffolding towers which I think we shipped from the UK with Rank Strand 765 follow spots".

Richard Hartman kicked in . . .
Yeah. By '75, there were lighting systems, there was Eric Pearce and Showlites, there was The Who and Roger Searle with their equipment that they rented out to people.

These are the ones that started having big lighting systems in England, I can't remember the American ones. Americans, because they played the arenas, just used the follow spots, you know, the hall spots. It would have been bands like Grand Funk Railroad and ZZ Top, those bands, which were Showco bands, were the first American bands. In that era I would have done, in the '70s, a Bowie tour, which used the thin white tube light, the 'white light tour' for Eric Barrett.

I went and saw that at Wembley. Absolutely sensational!
It was, yeah. Well, that was Bowie's idea. Eric Barrett and I kind of did it for him, Eric Barrett was his tour manager. You know Eric Barrett?

I know the name.
Hendrix, he was with Hendrix years ago. Also in that time I did the Wings '75 tour, and I had a relationship with other bands too from the UK like Cockney Rebel and T.Rex. I was asked to go to Showco in the US for a couple of months, and after two months I was like, "What am I still doing here? I've got a wife in England, don't I?" And they said, "You just carry on, carry on". So I carried on for two years, and I thought well, I am an American now. I really hadn't been back to the UK. I did a lot of current tours as like a head of design at Showco in those days. They were mainly a sound company and a very good sound company. Clair Brothers was another good sound company, but you know, different, did different acts. Showco did the heavy ones and Clair's did The Osmonds and Yes and other MOR acts.

Anyway, Showco always had this plan to get into lighting, which began with the Zeppelin. Zeppelin is the first band they did with lighting, and the Zeppelin actually paid for everything but Showco owned it.

Well, they said that they needed to get into lighting. Rusty Brutsche, Jack Calmes and Jack Maxson were the principals, and all Texans. I always said, "You're Texans, you're fucking sound people", but they all totally believed that if you provided somebody with everything, they'd always come to you, because you could control the market, which kind of doesn't really work. It never has and it never will do. People want the best there is, and if it means going to two people, then that's what they will do, so it never really worked. When Jack Calmes left Showco, he stayed in the lighting business; he owns Syncrolites.

They used to have fucking cable this thick (indicates cable 1 inch thick), you know to run the lights because it was welding cable and it was American. People used to get hernias just lifting the fucking stuff, you remember?

Yeah, absolutely.
But as I said, after about '78 I was in America under Showco's wing, and I was really saying, "Hey, you think you're hot shit, what's the next move in lighting?" Then I'd say, "Hold on to your fucking cable, don't be stupid", because everybody is thinking, "I've got to have 5,000 lights this year on my tour because I've got to be bigger than Van Halen and all those bands".

Then I said, "You know maybe there should be lights that moved or changed colour", and I took Rusty some old theatre lights, not just colour wheels, but the ones which had this flag system that used to drop colours in front of light.

At around about that time in 1980 Keny Whitright, who runs Wybron, invented the scrolling Colormax colour changers (he used to work for Showco), and they came on the scene. I don't remember (I was probably drunk a lot in those days anyway, plus I was rude enough to the Texans to hurt their pride), as to when Showco eventually came out with the Varilite. I know on The Bee Gees tour, we had to have a computer system for the disco floor, and that became the basis, I think, of the control system for Varilites.

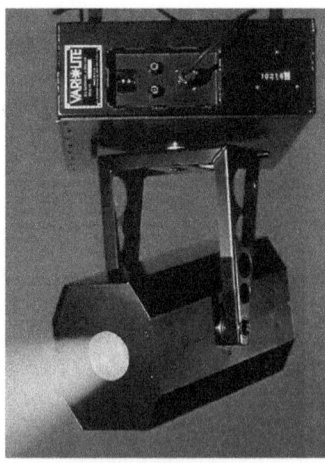

Original Vari*Lite v1

I thought of it as being a light that could move and change colour. When Calmes left Showco he went off and devised a light, the Syncrolite. You got a Leko, right? You put it in a cradle that moved, panned and tilted, and you had a colour changer on it, you know. It used to be fantastic, used to move, then it just jumped up and down, but eventually they did come up with the Varilite.

Genesis was connected to that?
Yeah, Genesis bought into it, they were part owners.

Because the development was so expensive?
Yeah, it was a little bit expensive. I wasn't around then, so I don't really know. I was considered to be something disgusting because I went to L.A. and didn't stay in Texas.

I was more into doing shows, and in L.A., I just did astronomical amounts of shows, you know. I'd do 8 to 20 a year sometimes, one after another.

Just designing the whole thing?
Yeah, designing, doing concepts for lighting, but not physically running the lighting.

Going back to the UK, what are your memories of back here in the early days of lighting design? I recall someone creating a miniature model lighting thing.
Patrick Woodroffe.[5]

Did you ever use that?
No, I was in America by then.

But you knew about it?
Yeah, Patrick Woodroffe created a studio, and the concept of the studio was you put your set in, made a little model of your set and you put it in this stage box, and he had miniature lights. You could programme everything in miniature and obviously record it. And then you could pre-program your whole show, in theory! I never used it, so I don't know how well it worked.

Did that sort of thing happen in America?
No. The original lighting companies in America were Tom Field Associates, TFA, that Rikki Farr bought out. There were quite a few New York– and New Jersey– based companies whose names I can't remember now, and in L.A., there was Obi Lighting, and people like that who were the forerunners.

Anyway, Eric Pearce was ahead of the game because he introduced into the States small cables and the multi-pin Socapex connector you know; the bar of six (bars of six PAR lamps using the Socapex multiconnector); the use of the multicore cable; the Socapex dimming system; the Alderham 60 channel lighting board; and the Alderham 804 lighting console. It was all smaller, more efficient and kind of much better, so a lot of the American companies disliked him for it because he was a limey.

You know that Eric was an innovator, I mean a real, real innovator. And Hartman was an innovator over here, I mean Hartman would build you anything you wanted. You want a fucking light that's got 50 lights and it spins backwards and explodes, "Here, I can do that!" (laughs). Then Varilites came in and all the other versions of Varilite.[6]

Yeah, Joe Brown came out with his version for awhile.
Yeah. Everybody tried. Rikki Farr had one, they all had one. None of them worked, you know, really, because Showco was so fucking big, and it had all of Genesis's money behind it. Anybody who had said they're going to make one, they were sued before they even got off the ground. I mean they covered all the patents you could ever have for moving lights. Even if you haven't thought of it, they got it covered somewhere.

Did it change your concept of design a lot?
Only in the sense that I've only ever thought of moving lights as a supplement to normal lights. You can have lights that have the same quality as a PAR that moved, you see, light is always about the colour and the texture of the lights. If you see shows now, the light is, rather than being yellow, it's more blue, if you know what I mean?

I used to tell Showco, "You're not going to get rid of regular lighting". The only thing that Showco did do, which I pushed them into (and I didn't really work there then), was I said to them they either had to change all their lighting systems, right, which would be millions, millions of pounds or dollars, or they had to do a deal with Eric Pearce or somebody like that who had the basic new standard lighting system, and then add their stuff to give the mixture. Unless you were Genesis, who only wanted moving lights.

That was a show and a half, I saw that one too.
Yeah. Alan Owen was the lighting guy (Genesis's production designer) in those days, he's dead now, God bless. He was the Showco in-house guy basically, which was good for Showco, they controlled the whole package.

You had a fantastic history in theatre before ever setting foot in the music scene, as did Brian Croft.
I think you had to in those days. I know Brian Croft, he was a theatre guy, lovely guy, he worked for the National Youth Theatre before the ICA.

You are lucky, I think, if you came in like that.
Absolutely, I mean, Arthur Max, who did the lighting for the Floyd, I think also came in from the theatre background, you kind of had to because there was no starting point in those days. I mean, eventually there was, for example, a band got lights, a band maybe like Hawkwind or someone like that who would have some acid kid, who would sit down with them and do some lights, but generally speaking, for maybe the first 10 years of rock-and-roll production you had to rely on people in England and America who had had some theatrical background or in shows in some kind of way. People would just come into the business because, suddenly, it was a business.

What did you do on Implosion, what did you do artistically, production-wise?
I was like the stage manager on Implosion, and a kind of production manager when we needed to put in some production (Jon Cadbury was at the Roundhouse then, a close friend, you know). There would be some bands that came in from America who had productions, a band maybe like Sha Na Na, a show band in some ways, and we had to re-create something for them in England in the Roundhouse. Some bands just wanted the backdrop that they were touring with put up, there was always kind of something, you know. When I did the Incredible String Band we had a canopy that was over them.

It must have been interesting for artists in those days, knowing there was a certain amount of equipment they could rent and all wanting to be a bit different.
Yeah.

So when the artists talked with their crews or with a lighting operator and someone says, "You should go and talk to Ian Knight" . . . that's how it happened, wasn't it? Tell me as a designer how did that conversation go with the band?
Yeah, a manager would phone you, and you would sit down with the manager and then with the band, and the band would have some ideas or not have ideas. I'd listen to their music if I didn't know it. You know, in those days you would talk about some very stupid things then suddenly realise that you couldn't afford to do that!

You're thinking of something in particular?
All right, I was thinking like in '78, which was a good example, we had Zeppelin, who were ahead of the game in a lot of ways, they had the double or quadruped Eidophors at Knebworth, right. They had this technology that video or film could give to a live performance, and all Robert wanted to do is have a fucking staircase come out of the stage for 'Stairway to Heaven', which we kind of just ignored. With an old SGB stage, you've got to build the staircase into the stage, and you're talking hundreds of thousands of pounds, even in '78, you know, to do that. There are other ways you can do that, you could have booms to lift Robert up, you can create illusions with video, which you would do now. The business very much works a lot like (maybe not so much now), what somebody else had last week or last year is what we want now, and it was always with the same people, I'm not saying that it was best for the band to have the same people, it's just how it happened.

Yeah, and so that's how you got your work?
Yeah, or friends of friends, it could be anybody. It could even be the carpenter has a relationship with the band, "Hey, the guy you need to help is . . . ". Then there are other people like yourself who came into the business, and you would have had people you employed and were happy to work with. At the end of the day you're going to get the big shit if it doesn't work, you're going to get it every day if it doesn't work, whereas the designer is off, paid or not paid!

Obviously over the years I have actually gone on tours even until two or three years ago. I did Rod Stewart's 'Kind of Christmas' tour where we added the obvious stuff.

Traditional?
Traditional yeah, and quite right, that's his audience. We added all this stuff and Lars (Brogaard) said to me, "It's only 12 shows over two weeks and you want me to bring on another two carpenters and then train them?" Then he said, "Why don't you please just come out and do it, and I'll give you two carpenters, you just do it with them every day", which I did. For my age it was pretty hard going. We went from Earl's Court to Brussels overnight, which is not as easy as it sounds, then to Belfast, as you know there are a lot of ferries involved.

Yeah, they tend to wake you up and tell you to get off the bus!
Yeah, it's not as easy as it sounds, and plus I'm not getting any younger, you know. Lars is pretty organised though, pretty much has control over what he is doing. He just did Lionel Richie and he does Rod, and he is doing sound I think on Michael Jackson. Hopefully he is doing sound on Jeff Beck because I want to see Jeff Beck, you know, he's my favourite. I said to my wife, whatever happens, I'm not coming to Thailand till after Jeff Beck because Jeff Beck could die tomorrow or never play again!

Concentrating on the whole of UK production, what do you think was the most innovative thing in music production?
Well, in lighting I think the Showlites multi-cables system, you know.

Just transformed it...
Yeah, transformed touring. They're even using it in theatres now, you know, so it's not as if it's rubbish and rock-and-roll.

In the building of scenic elements so they could tour and package and fit in the truck, one of the geniuses of that is Michael Tait. He's English but he lives in America now. I remember Michael Tait in the Yes days years ago having made this lighting board, which was very flashy, he had a cocktail shaker on it (laughs a lot). Then Charlie Kale, who used to build sets over here, I forget what they were called. I think going back to people around The Who, I mean, John Wolfe phoned me the other day, he was (lighting designer) around The Who craziness, and lasers eventually.

Have you got a favourite story? One that you've told over and over again?
No, I don't tell stories because I was told by Zeppelin many years ago, whatever you see you don't talk about it (laugh) or else! Funny stories, this is to me my nightmare story. The first tour I did with Genesis, right? Tony Stratton Smith, the great record guy, said to me, "You've got to please help me, come out with this band, they're called Genesis", and I say, "Who the fuck is Genesis?" and he said, "They are a fucking fantastic band". Anyway, I went to see Genesis rehearse, and Peter Gabriel said to me, "We have these screens, they are like figure of eight screens, which are quite nice".

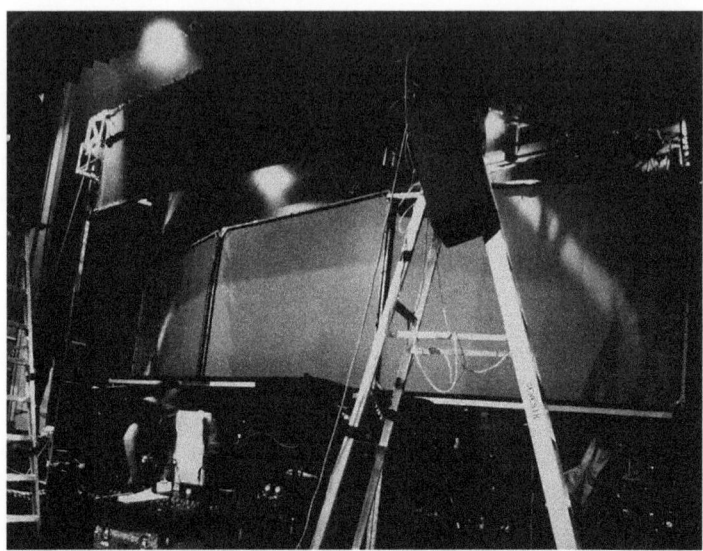

Back of the screens—Genesis
© Jeffrey Shaw

So I think Jeremy Thom had designed these screens and somebody else had made them, and they didn't work. You had to build a figure of eight, and you had to put ribs across them to give them some shape, right, and then you have to stretch the fabric on. Peter said to me, "Do you understand this, Ian? I don't know how it goes together". "Well I don't want you to tell me, Peter, it will all be wrong!" I said. I looked at them in rehearsal, and it was obvious what they do they're all numbered!

Front of the screens with projections—Genesis
© Jeffrey Shaw

So we got to Germany, to some classical concert hall, right, and a friend of mine called Regis Boff, an American guy, he was like their road manager in those days, he said that the sound guy had said, "You'll affect the sound if you start bringing all this stuff in because there's no room, the screens are too early, can you make them outside?" So I said, "Absolutely".

Then I said "Why am I doing this on my own? I'm like a production manager for the production", and he said, "Yeah, but you haven't got stage hands, you haven't got anything here". In those days you didn't have that sort of help, so I had to put the screens together on my own. Well I built all these fucking screens out in the street and about 4 o'clock, Regis said to me, "You can bring them in now because I've got a few guys who can help you bring them in". Of course, they wouldn't go through the fucking door! (laughs).

So as the audience were coming in, I was still rebuilding screens on the stage that I've already built in the street, right, because you never used to think in those days, but they looked all right.

The other story was: we were in a gymnasium in America, and the sound guy never wanted to run his cable, fucking saw the rings, you know the rings you do gymnastics on, hanging from the ceiling in the roof?

Yeah. Who's this with?
Genesis again. I remember Regis saying to him, "Do not use those, right?" Regis said, "Just run the cables, get some mats, run them down the side or, if you want to fly them, fly them down the side". Anyway, he put them all on these fucking rings straight down the centre of the hall, right. Just before the house lights went down, all these rings just lifted up! (laughs)—it was evident they were on a computerised system . . .

Oh my god!
Probably half of the PA fell down, you know! (laughs)

Like the scoreboards they have in hockey arenas, they raise the scoreboards sometimes?
Yeah, it was a gymnasium they were playing. There were always minor disasters with Genesis; they were kind of overambitious for where they were playing, I mean, they would try and take a full set into a club, you know.

Yeah, when you got it, you're going to use it.
Well Genesis were a production band, you know. I mean, the joke of it all is that, Genesis never wanted to be a production band, that was Peter. They thought they were an intense musical band. And then Peter came on stage one day, without telling them, as a fucking flower. They all had a nervous breakdown (laughs). And that was the beginning of the end, you know.

Ian, thanks very much. That's great.

Ian sadly passed away 10 months after this interview in March 2010.
  From his great friend Regis, a fitting tribute:

Genesis and Ian Knight by Regis Boff

Ian Knight died not so long ago, and I miss him. He was a British rock stage designer who I first met when he showed up to design for Genesis around the same time that Peter Gabriel began cutting the rectangular section out of the front of his hairline.

For almost six years we did every show the band did in the US and Europe and Canada. Through most of that time, Ian, who wasn't a very big man, combed his hair upwards from the sides near his ears to the centre of his head then redirected that forward to a point just above the bridge of his nose. He only wore black and covered himself with a body-length, red velvet–lined black cape. He was never without the cape. Peter Gabriel, to his credit, never got jealous. No band at the time had more innovative stage designs than Genesis, and Ian helped much in this.

After Genesis he went on to Led Zeppelin, The Rolling Stones and Rod Stewart. He adored Rod Stewart. Stewart is clearly a sweet man. Between 1966 and 1971, he staged concerts at the Roundhouse for Jimi Hendrix, Country Joe and the Fish, The Doors, Jefferson Airplane, The Rolling Stones and Elton John.

Ian was also involved with the first installation of theatre productions on cruise ships. In 1981 he worked on Ronald Reagan's inaugural ball as US president, at which Ian surreptitiously introduced a troupe of transvestite trapeze artists.

A decade later I hired him to do my country shows and a few in Rhythm and Blues. Ian could be honest with artists. This doesn't get you very far generally, but he was more often than not successful in getting through to them.

He and I travelled to Thailand together and had our lives altered somewhat by this amazing country and its gentle people, him more so than me. Ian wanted to go North towards the Cambodian border, and I was afraid to come, people told me it was safer for the Brits than for Americans. He made many trips there afterward, eventually marrying a Thai woman named Ngeon Khprjunklang and having a son Alistair.

People living or dead are the trees that you have carved your past into with the knives of growing and changing. Ian is tall in my forest.

Notes
1   www.carnabyechoes.com/map/artists-own-gallery.html
2   www.boylefamily.co.uk/boyle/texts/journey2.html
3   www.jeffreyshawcompendium.com/
4   totalproduction.designandgo.net/Chronicle/419292/the_sundown_experiment.html
5   woodroffebassett.com/archive
6   www.vari-lite.com/company/about-vari-lite/

Part 4
Pioneers of Full Production Services

There has never been a full production company that I am aware of in the UK, however, a couple of companies got very close: ML Executives and Entec. ML Executives, a company formed for the crew of The Who, rented backline equipment, PA, trucks, sleeper buses, touring personnel and Super Trouper follow spots.

Original ML Executives business card circa 1975

Roger Searle, an original director of ML, has always been a mine of information, and throughout my career I have always appreciated his help and advice whenever I needed it.

In my capacity as a tour manager, I very often got to choose the production companies that would facilitate different aspects of what was required on a tour, and over a 10-year period, I used ML quite a few times, for example, the 1979 Kate Bush UK and Europe tour for sound, trucking and crew travel; Supertramp and Duran Duran for trucking; and also sound, follow spots and trucks for the Grateful Dead in Egypt.

Entec was a company formed out of the exploits of Harold and Barbara Pendleton when they started their affair with 'Festivals'. Barbara Pendleton, co-founder of Entec, relates to me some fascinating history of how Entec came about.

With Cockney Rebel in 1975 we hired a 40-foot artic truck and trailer and the sound from IES, the lights from Entec, even a little staging and of course our backline equipment. Roger Searle was the driver and even sold the swag for us during the shows, and although he worked for The Who, he was sub-hired by IES to drive their truck, which Roger explains in a little more detail.

11 Roger Searle

Roger Searle

December 7, 2007 at My House in Somerset with Roger Searle

Roger, tell me what happened when you left school?
Left school, got a job with The Who, albeit part time to start with, that's about it really, because me and a mate had a van that worked and theirs didn't.

I thought you worked for Track Records before The Who?
No, well, The Who was Track Records. I got involved with Track Records in going to collect my miserable two pounds a day, which is what I got paid for working for The Who.

And what were you doing for them?
Anything, just driving stuff around.

And what year are we talking about?
'67, they had been going for years, it was pre-*Tommy* so they were still just another band on the road, getting paid £400 a night if they were lucky.

In 1967, that was a lot of money back then.
Like I said, if they were lucky.

I'm after your recollections of production. Were The Who like everyone else and had a WEM system?
No, they had a Marshall system, well, they had an Akuset amplifier, a little Akuset mixer and four Marshall 8-by-12 speaker cabinets. That was the PA system, Akuset was Swedish, it was a reasonably high-quality little system. The Small Faces had a complete Akuset PA system, because they used to produce speakers as well.

So when you did shows, did you have any sort of light show?
No, none at all.

But you ended up being the lighting designer with The Who, didn't you. Were you their first lighting designer?
Erm no, not really, well, yes and no. John Wolfe, who used to drive Keith and John around, was always instrumental in trying to produce a more interesting show than standing on stage underneath half a dozen florescent strips, which is what you used to mostly work under, bearing in mind we were playing theatres and places like Dunstable or Norwich. We weren't doing arenas or stadiums, we were doing Isleworth Technical College, Reading University, Devizes Corn Exchange, for example . . .

We all followed that pattern; they were the same places I was doing with Cockney Rebel.
. . . and so you tended to use what was there, so if there was nothing there, you didn't use it.

At what point did that start to change?
It changed for The Who probably about 1970, 1971, because we came across a company called Electrosonic, or John did I think, and they basically said they could build a lighting control desk that we could move around, unheard of they said. We went ahead and bought this thing, which basically had AC mains going in one side and AC mains coming out the other side to the lamps. It had to go on the side of the stage because no one had invented the multicore, multi-cable, and we used to go to Rank Strand in London and rent whatever we could find, good old-fashioned theatre lamps, some towers, some single stands, some poles and anything we could get our hands on, so we had basically the same sort of show every night.

We talking Fresnel 2Ks?
Yes, 1K things, I can't remember what they are called now, 263s I think and Patt 23s, little theatre lamps that are all still made I think nowadays, but maybe not. We didn't know about anything else because that's what we'd seen in theatres, so we presumed that's all there was available. We used to go hire a couple of dozen of those lamps with some things to hang them on and go make it all work with our little lighting board.

Round about 1970?
Yes.

Had you been to America?
No, not me yet, the band, yes. In 1971 I went to America and bought a load of second-hand Lekos from a lighting company in New York, called Century Lighting.[1]

Were they a music lighting company?
No, there weren't any music lighting companies even then; we took all of our kit from England that we bought from Theatre Projects. We bought miles of cable, we bought the lighting board, we bought four little metal trees to hang the lamps on and we packed that all up and flew it to New York. We then went to Century Lighting, bought loads of Lekos and spent the next three days cutting off hubble connectors and putting on 15-amp plugs in my hotel room and making up series splitters so we could use 100-watt bulbs in the UK if we wanted to, but we found we could convert the lamps to 240 volts, so it wasn't a problem anyway. And that was the start of our touring light show.

We ran 110 volts in America 'cos we had to run 110 bulbs in America, and we had to be careful how we actually operated our desk because of 'half the voltage twice the amperage'. Where we could get away with three times 60 amps in the UK, it was three times 100 amps in America, and our desk wasn't capable of running three times 100 amps, so we had to just tailor it to suit what we could make it work with, without melting it. Then we came back to England and brought all the lamps back with us and had a touring light show, which nobody else did.

PAR cans?
No, never heard of them, never been invented then, well, if they had nobody used them. Because there were no touring lighting companies.

When was the first time you saw a PAR can then, at the Rainbow?
I can't remember what we used at the Rainbow, I think we used our own little touring system there. We used our Lekos for years, and we carted them backwards and forwards to America in the fine wooden boxes complete with a wooden box we had nicked from a theatre somewhere that had Geraldo Orchestra written on it, that we put foam rubber in, and that was home for our set of 110-volt bulbs that used to go everywhere with us. We had a set of bulbs for England and a set of bulbs for America, and we carted a set of 36 Lekos around the world for years, until probably 1975, when we came across PAR cans and a 60-foot truss with so many PAR cans on you couldn't focus it because they were too close together. So the PAR cans on the outside just shone on the back of the PA.

Whose was that?
It was supplied to us, Bill McManus Lighting, Philadelphia. We were lead to believe that they had developed this system of hanging trusses in the air, and bear in mind the truss was what looked like the jib of a crane, it was made of steel, it was incredibly heavy and it had a PAR can every six inches or as many as they could get

on it. We were completely unconvinced by the concept of flying a truss but we went with it, although we still used little pump-up Genie towers at the side of the stage, the gas-filled ones with PAR cans on. We didn't like PAR cans though because you could only point them, you couldn't do very much with them, all the lamps we had you could iris them, you could flood them, they were basically profile spots. The whole of The Who's light show for four years was based on 36 profile spots and three follow spots because we'd bought those as well, which the band didn't know about. Daltry told us we had to send them back because they didn't want them. Then the first time we couldn't use them, it was "Where's the fucking lights?"

We started renting lighting equipment, but that's after we had already then supplied the lights that The Who owned with us, the crew, to the likes of Led Zep, Humble Pie, Eric Clapton, Emerson Lake and Palmer. We did all their tours because nobody else had any lights that toured, so you couldn't have the same show two nights in a row. At this time we got asked to do some work in Europe with Led Zeppelin, we'd had enough of it to be honest by '74/'75.

At what point did you have dimmers?
It was all in the one lighting desk, everything was in the desk, that was what I said, it wasn't like it is now where you have the dimmers at the side of the stage and the board at the front, the whole shit show was in the one board.

And that system, I take it, was with the LD on stage until '74/'75?
Yes.

Were there multicores by now?
No, and there was no one using them because there were no touring lights. You get fixed installation lighting systems as in any theatre, probably installed by Rank Strand, where there'd be this humongous great lighting board, but a lot of theatre lighting was controlled from the side of the stage. If you look in some of the older theatres with great big brass wheels where you used to screw these wheels down onto a big shaft, and then every time you screwed a wheel down, that was another circuit that worked every time you shifted a lever at the end of the shaft. Once they got into electronics there'd be this thumping great lighting board built into a booth somewhere at the back of the theatre, and that would be hard-wired to a bunch of dimmers at the side of the stage, and they'd be hard-wired to a bunch of lamps hanging on pipes, and that was the way it worked. There was no multi as such, it was just hard-wired through a conduit somewhere.

What do you remember about The Rainbow Theatre?
A really dreadful place to load into (laughs), dressing rooms being on the nth floor in the air (more laughs).

Do you recall the production in there that the Americans had brought over from New York?
I remember Joe's Lights being in there with those oil machines, but other than that, no, just another gig.

Whom did you hire from then?
Theatre Projects in London, which was a theatre lighting company in Neal's Yard, until we had bought everything we needed.

So you have created the first touring lighting system, UK, Europe and USA?
No, just UK and Europe.

Then what was the next progression?
The next progression was John Brown and Brian Croft had just started a little lighting company somewhere down in Wandsworth called ESP lighting, and we pointed Led Zeppelin in their direction. We'd had enough and we didn't want to do anymore touring as such, and we said, "They've got some lights, they'll sort you out". I think that was their big break in terms of a decent client to go out and rent their lights to. What they had I don't know, 'cos we just body swerved it and said, "Give John Brown a call, he'll sort you out". And then more and more people started buying lights and more and more American companies started buying lights, and it sort of progressed to where you could just rent things and didn't have to buy things.

So you introduced ESP to Zeppelin and then started using them yourselves?
There was a reasonably large gap at that point in time because in 1976 we toured America, but we used Showco to supply sound and lights. Back then Showco was a sound and lighting company as opposed to just a sound company. We used Showco probably '76, '77, and then we probably start using TASCO from '78 onwards when I came off the road effectively to look after ML Executives down at Shepperton. I dropped Kieran Healy into the piece to light The Who, basically me doing a design and him board operating.

I hired him to direct the live video for Duran Duran in '84.
Well I gave him the chance to move from making tea at Shepperton to being The Who's lighting designer because I didn't want to do it, and he expressed an interest, so there you go!

What was your first gig in the US?
We did stadiums in America with our 36 Lekos; Forest Hills Tennis Stadium, that was the first gig I did in America, Forest Hills Stadium.

How were you putting it up?
On floor-mounted stands, fixed floor-mounted stands that you climbed up a ladder and hung a lamp on.

No truss?
No, truss hadn't been invented; truss was for holding up roofs, not lights.

I probably missed an element of lighting development by being desk-bound but you got the initiation of Avolites[2] with their control equipment and all the other various entities with Thomas making their lamps and their trussing.

Avolites, tell me a bit about Avolites. Where were they out of and who owned it?
Well Avolites were one of the first companies to get to grips with the control side of lamps and dimmers and packaging it so it was easy to use on the road. They built a desk for me for The Who eventually. They're English, in London, they're still going, and their stuff is still used to this day. I can't remember when Avo first started to get involved. They would have been a spin-off from someone who had an interest and seen a niche market and exploited it.

So Avolites came in when?
I'm just trying to think, they would have come in between the 1976 to '82 period, that six-year period because I used an Avo board in '82 that they built for The Who, and the last time I saw it was in a club in Ilford. I think it may now be residing in Light Structures because the guy at LS has got a passion for Avo equipment, and he's got a load of old Avo boards up there, and I'm sure he's got my old desk.

So what happened desk-wise after you had that one made by Electrosonics?
That desk saw us through for quite awhile, limited though it was, it only had 18 channels, but we must have then started using rental equipment in America and probably a desk supplied by Showco, the make and model of which I can't remember now. I'm sure that's what we did, yes, we did, but on the last tour I did with The Who in 1982 I definitely had the Avo QM500 desk and a load of others as well.

What was the first one you rented out with ML?
We never had a lighting system, we never rented lights as such, we had Super Troupers but that was it. We never bothered with lights because it was too much capital investment for not very much return.

Smart move.
The original one was just our own little system. ML had not been thought of or dreamt of yet, it was just The Who road crew. The Who paid us 365 days a year, and these other people paid The Who for our time.

So that's why ML Executives was set up?
It was set up initially because The Who weren't working very much, and they had a crew of seven or eight people on full-time payroll, and because of the nature of the business as with all the other bands you used to buy everything, you didn't rent it. We had a ton of equipment that had been bought over the years, two or three different PA systems, more backline than you could shake a stick at, stupid bits and pieces of stage set that they decided to use that they bought and never used, odds and ends of lighting equipment. It all came about very quickly around '74, '75 when they decided there was not going to be very much touring, so they needed to try and soften the blow a bit and decided to set the road crew up as their own entity as a wholly owned subsidiary of The Who Group or what became The Who Group. To sink or swim on their own merit by doing whatever they could

do to generate income. And then, as I say, we started off just using The Who's equipment they actually owned. We would maintain it and rent it out to pay for the maintenance so there were no overheads.

How many Super Troupers were there?
We ended up with 14, we bought six to start with in 1976, and then we bought another two and then another six. The first job they did was Wings at the Piazza San Marco in Venice, with one of our brand-new trucks taking them down there, and that was 1976. I just had a truck with six Super Troupers in it, for shows in Venice, Zagreb, Munich and I think Brussels, and that was the first job that the truck had done and the first job the Super Troupers had done.

The buying of the trucks was a direct result of the band asking. When we did the three football grounds, 'The Who Put the Boot In' concerts in 1976, Edwin Shirley did the trucking on that. They were a little remiss in the way they handled getting the equipment up to Glasgow, so everything was a bit late, and then the band got wind of it, and I was asked how many trucks we needed if we go on tour. At the time it was three, so they said, "Fuck it, go buy three trucks", so I did. I marched into Dawson Freight in Leighton Buzzard and said, "I'll have three of those please". We had already started ML by that time 'cos that was incorporated in '75 and those trucks we bought in '76.

Brand-new trucks?
Yeah, and they thought I was taking the piss. Then the accountants got wind of it and decided that because it was 100% capital write-off in the first year at that point

ML trucks on their way to Cairo 1978 for Grateful Dead concerts
© John Rivett

ML Nightrider

ML Nightrider interior, back

in time if we could use three, could we use six? I said, "Well, yes, I could use as many as you want", so they said, "Well go buy another three". So I went back to Dawson Freight and said, "Can I have another three, please?"

ML turned into the very first multi-purpose rental company in the UK?
Yes, we had the first ever purpose-built sleeper bus.

Tell me about that, please.
John Bates, a boat builder at Chertsey, he used to look after all of Pete Townsend's boats. Okay, we decided we needed to get a lot of stuff into a small space, who's good at doing that? A boat builder. . . "Batesy, build this bus, here's the shell".

So you bought a new shell?
No, second-hand. We found it at a bus company in Liverpool who'd gone bust, a six-month-old Volvo bus. Drove it down from Liverpool with no PSV license or anything like that, but (laugh) who cares . . . we took all the seats out and let Bates get on with it.
And that's because Roger (Daltry) wanted a sleeper bus. Roger was instrumental in that. He said, "Could we have one of them buses like we had in America?".

Okay, and you ended up with two of them?
Yeah, two and then everyone else jumped on the bandwagon.

So, Len Wright, he was a driver, wasn't he? He drove ordinary buses, didn't he?
Yes, he drove for National Express, as they were called back then, because he drove us, The Who crew, on a trip to France once in a part-way decent bus that National Express owned.

What year is the bus?
'76/'77.

Up to 1976 crews were using day buses, not sleeper buses?
That's right; sleeper buses, there was no such thing.

Didn't you tell me about an American bus that came over here?
Elvis Costello brought a bus over that he had parked down at Shepperton that we used to look after. He could use it as it was classified as a motor home, but I had a Department of Transport examiner look at it, and he said never in a million years would this get certified as a public service vehicle; it just would never comply with English construction and use regulations, and they still don't. You couldn't bring over a Prevo or an Eagle and run it here, the rules in America are so different.

ML Nightrider interior, front

What was the sequence of the purchases, trucks first...
Trucks first and then, well, it was three trucks, a bus, another three trucks and another bus, more sound equipment, 12 Super Troupers, another two Super Troupers.

Did you actually get given a date on which to spend the money?
No. The band had received a large sum of money. Probably from *Tommy*, which would have been taxable, and so within that tax year it was prudent to buy as much equipment as they would need to use, and that was the whole basis of everything. It was only bought equipment that The Who could or would use for their own ends. It just so happened that they didn't get to use most of it because it was always out on hire to somebody else. But that was the whole theory behind it. So it was really, buy what we need, and if you could rent it out, then all well and good. The trouble was we were too good at renting it out, so they never actually got to use it. When The Who went on tour they had to rent trucks and buses from somebody else . . . and sound and lighting equipment!

So was Edwin Shirley the only one that was doing it back then?
More or less, Keedwells were doing a bit, from Wiltshire, Mike Keedwell. What would tend to happen was, if people wanted to use a tractor trailer, which was almost unheard of, you'd do what you did in America, which was you'd go to Hertz and rent a tractor unit and then go to TIP and rent a stepped-frame trailer and self-drive.

Okay, so remind me what you did when we first met, which was probably in '74, I think. It was Cockney Rebel, and it was the first time we had a show big enough to fill an artic, and you were driving it and doing the merchandising.
Well that was Dave Hartstone's truck from IES because IES was self-contained and had their own trucks. They had one stupid tractor trailer but they didn't have anybody to drive it, and they used to borrow me from The Who to drive it. I was one of the few people in the music industry that had a class one license.

That IES truck with Cockney Rebel in 1974

How did you come across that license, because you liked driving?
Yeah well, how it came about was in '71 or '72 when we toured America we toured using just regular trucks, bobtails. Then halfway through one tour somebody turned up and said, "You could put all this in one tractor trailer", and we

went, "Oh, could we?" and so along came this tractor trailer with somebody called Johnny driving it, I can't remember his second name but he was a complete idiot . . .

Why was he a complete idiot?
Because he would get drunk more often than not, and I'd end up driving the truck back to the hotel having never ever driven a tractor trailer in my life but deciding it wasn't that hard. Then at that point in time in England—I can't remember the exact date when it came in but—you didn't have to take an HGV test once you were 21, you could drive pretty much anything. Then whenever it was that it came in, if you could prove that you had experience in driving a heavy goods vehicle for a six-month period prior to this date, you could claim 'grandfather rights' and not have to take the test. I fell into this grey area.

Grandfather rights?
Yeah, that's what they call it; in other words, I fell into the grey area because I was 21 in September, so this must have been 1969 or '68. You had to prove you'd been driving trucks for a six-month period prior to the following February, and if you count September to February it's only four months so I couldn't have been 21 long enough to claim the exemption from having to take the test, so they allowed me to drive any truck I liked on my old red driving license until this date in February. Then I had to take a test for whatever one I wanted to drive. So I quite happily drove around in a truck The Who had bought because by this time we had bought two trucks, we had a 16-ton Bedford and a 13-ton Ford that we owned. They were basically just to carry our equipment around in, and I used to drive that around quite happily, I never had any lessons, I just turned up at the test centre and did my Class 3 test in that truck. Once I got my Class 3 license I said, "Oh, sod it, I'll take my Class 1", so I went on a three-day course somewhere in Willesden and just learnt to drive a tractor trailer.

In 1972 we toured Europe and we didn't have a tractor trailer, so in '74 we did a tour in Europe, just France maybe, where we used a tractor trailer. It was a rented one, we rented it from Hertz, a Volvo F88 and a stupid trailer from TIP, and just me and Mick Double drove it I think or I drove it, I can't remember now. Moving on from that, Dave Hartstone at IES bought a tractor trailer, which was the completely wrong design for the job, but bought it because it was a straight frame, single axle with a tail lift, which was a complete pain in the arse, but he had no one to drive it and I used to get phoned up.

Are you saying it was wrong because it was too high?
It's just a tail lift is a pain in the neck, and any trailer with a single axle right at the back end is not designed to go 'round tight corners. Your partner in crime the late-great Dave Wernham[3] was continually amazed at how I managed to get it into places (laughs) like Preston Guildhall, for example, when you had to reverse in there and Cardiff Town Hall.

David Wernham
© R Ames

You had to reverse into everywhere!
Yeah, but that's how I got to drive that truck because nobody else could in the business; you could have gone and got an agency driver, but that wouldn't have worked because that would have been a nightmare, having someone completely alien to the business trying to drive a truck!

Were any of your crew truck drivers?
No.

Can you remember all your ML truck drivers?
Most of them. The original six were John Rivett, Micky Curtis, Ian 'Cus' Martin, Gywn Lawrence, Ian Penny, Des Seal and Pete Burke, and then there were Dave Taylor, Phil

Egypt, 1978
© John Rivett

'Phlop' Greenfield, Tom O'Connolly, Dave aka Little Legs, Howard Howes (DJ), now a very good keyboard programmer, Trevor Spooner and Eugene McGreesh.

Now, you had some special sound desks built too, didn't you?
Yeah, we had two Neve desks built.

Now, Neve desks have always been used in studios only, why Neves?
Well, they were top quality but they were probably too good for the purpose they were being used for. Once again it was that whole thing of, 'if you can, then do it', which isn't necessarily the right way to go when you're out on the road, but it's like, 'what works best' rather than necessarily 'what is best'.

I don't think you had too many problems with them though?
No, but it doesn't matter how good the board is, it's if you've got 32 channels of crap going in then you've got 32 channels of crap coming out because it's just too loud. It doesn't matter how good the board is. I remember Glyn Johns, the studio engineer, they coerced him into mixing a show one night at the NEC in Birmingham, me standing behind him at the lighting board. The band walked out on stage and he'd miked everything up as he would do in a studio, and the band started and he turned 'round to me on the first number and threw his arms in the air, mouthing "And what the fuck do I do now?" He hadn't expected anything like the level that was coming off the stage, which was consequently going down every single mike, which were always open, and so it was 32 channels of noise.

He was just on a Neve desk recording?
No, he was doing the house, he wasn't doing the recording, they had decided it would be a good idea to bring Glyn in to do live sound! It wasn't his fault, I mean, it's like they were so loud on stage it was very difficult to control anything front of house because it would just spiral. John would be too loud, so Roger would want the monitors turned up, the monitors would be turned up so Roger would be too loud, so then because of the way they were always set up with Keith right behind Roger, every single one of Keith's drum beats would be going straight down Roger's vocal mike because there were no screens or anything like that. Entwhistle's bass cabinets would be going down Entwhistle's bass mike, Pete's cabinets would be going down Pete's vocal mike, and you had the drums going down Roger's vocal mike, so all of the three vocal mikes had instruments going down them as well as the instruments being miked up!

I suppose this is before DIs?
No, well, there were DIs available but everything was miked, period, very little you could actually DI, it wouldn't have mattered anyway because you still had the level on stage, it was so loud every mike picked everything up.

I've asked everyone this who I have interviewed: who was your biggest influence?
There wasn't anyone because there was nobody to aspire to!

Who do you think has been the biggest influence in lighting production?
Well, The Who and the Floyd were both very innovative in both sound and lights. We both experimented with quadrophonic sound, Pink Floyd took their light and peripheral shows to the nth degree and The Who, we just went out and tried to create the 'wow factor' with people in the audience just going "Wow, fucking hell". Now, if that meant standing—as we did at all shows after the first show at the Isle of Wight—half a dozen film 10Ks on poles behind the band and whacking them all on in one fell swoop, then that's what we'd do, but we decided to just do that. I don't know whether John Wolfe had seen it done somewhere else, but that's how that came about, there wasn't anybody or anything to, like I say, to aspire to, you just winged it, you know. It's like the thing with aircraft lights, I'm driving around the perimeter track of Heathrow airport thinking, "Fuck me, that's a bright light on the front of that aircraft that I've come face to face with" . . . "Showco, what sort of bulb is on a wing of a 747?" "Oh, it's an aircraft landing light" . . . "Well can we use them? How big are they?" Enter the ACL, now industry standard or it was for awhile, not so much now maybe. But so far as production elements are concerned, it's just evolved, no one person I think has been responsible. A lot of bands have been responsible for stuff by either one having the money or the foolhardiness to go ahead. I mean, the ELO spaceship, for example, they didn't need that, it wasn't relevant to the music, but it was a damn good effect. It's like The Stones could go out, and The Stones could do as much business without 50-odd trucks full of equipment, but they choose to take everything with them and go for a stupidly large production—the same as U2 do. And good luck to them we say, because 'they can'.

What has been a big influence in production?
Aluminium.

Okay (laughs), that's a good answer. Please tell me why?
Well because everything used to be made of steel, you imagine the first lighting truss we used with The Who, it was 60 foot long and it was a steel truss three foot by three foot, imagine how much that weighed!

Annie Pocock remembers being the truss monkey while it took 12 people to just lift the truss onto the Genie super lifts, and then she had to couple them up while standing on top of it all.
It wasn't until people started to use aluminium to manufacture trusses and aluminium for PAR cans that you were able to start hanging more stuff in the air because you could only put up what the motor would take. You can't go flying four-ton motors in the roof because the roofs won't take it. It's all too heavy, but as soon as you started using aluminium, light shows got lighter, stage sets got lighter, and everything got lighter. Consequently, you could go throw up loads of stuff quite quickly because two people could carry it instead of six. And it doesn't go rusty while you wait for it to go up.

Did you used to have PAR cans with The Who eventually?
Yeah, oh yeah, the whole lighting rig was PAR cans, loads of 'em.

I was going to ask you one last thing. Please tell me a short story.
Every day's a story because it ain't real. We'll all have to go get proper jobs one day!

Okay, let me put it a different way, tell me a close shave.
A close shave! I suppose a high-profile close shave was when we took Elton John to Moscow.

ML Executives?
Yeah, which was the first time that that had ever been done, a western act in Russia, and this is late '70s, '77-ish, '78, and we just had two trucks. I think it was Rivett and Curtis doing it. It was all very new then, as it was with everybody, we used to use a travel agent then, and in this case it was Trinifold Travel, and they were also in their infancy. They used to do all our bookings for us, for trucks and stuff, for ferries, and for some reason they'd been given a completely bum steer on the time. The way that we were driving to Moscow was to not go through mainland Europe and then through Poland or wherever because you couldn't do it then, it was real hard work. So I said to Harvey Goldsmith, "We'll get there, it's going to be expensive because it's going to be the long way around, but we'll get to the Finnish/Russian border before we have a problem. We're going to go from Harwich to Denmark or wherever, and then we're going to drive across Sweden, and then we're going to drive across Finland, and then we'll be in Russia, so by the time we get to the Russian/Finnish border that's the one border crossing that's going to be potentially iffy but we're on the doorstep".

Well, the first thing that happened is that we got given some wrong information. I can't remember exactly if we were given the wrong departure time, but the cross North Sea ferries used to alternate between Felixstowe and Harwich

ML truck in Moscow circa 1977
© John Rivett

depending on the day of the week and all the rest of it, and we were given a departure time from Felixstowe and it was the departure time from Harwich, and the two trucks actually missed the ferry.

Now, there isn't a next one, the next one is like two days later, so I'm thinking, "Jesus this is a really high-profile trip". So we stuck a pin in the Yellow Pages, and I found a company, I think they were called Norfolk Line that sailed out of Great Yarmouth, and they had a ship when there was a full moon and an R in the month, and they had a ferry going and had space, and we had time enough to get to them to get the trucks back on track.

I think that ferry went to Germany somewhere, so they had a little bit of extra driving to do rather than go to Esbjerg. I think we were supposed to be going to Esbjerg, but we ended up going to Hamburg or Bremerhaven or somewhere like that, but we got away with that but no one ever knew about that. It was like rule number one is, sort it out before anybody finds out, you know!

That, of course, was in the days before double drivers?
Oh yes, it's even worse now. I'm just doing some stuff for next year, an itinerary for this Canadian girl I work with, and now you've got to have 'relief drivers to do this', you've got to have '45 hours off here', and it's going to be murder. If it all goes ahead with digital tachographs, then major tours that have got six or seven trucks, if they can't park up the trucks and leave them, they are going to have to hire locally a tractor unit and a driver who's sole purpose in life is to spend all day shunting trailers around because the trucks are going to have to drop their trailers and stay still!

Roger, great story, thank you very much.

After leaving ML Executives, Roger began a very successful career working as a freelance tour manager and production manager at the highest level and continues today for the likes of David Gilmour and Sir Cliff Richard.

Notes
1 www.strandlighting.com/clientuploads/century_flipbook/A_Strand_Century/index.htm
2 www.avolites.com/company/avolites-history
3 www.billboard.com/biz/articles/news/global/1322883/artist-manager-wernham-dies

12 Barbara Pendleton

Barbara and Harold Pendleton circa 1974

November 30, 2010 with Barbara Pendleton at Entec Headquarters

Barbara, I did quite a few tours with Entec in my early days with Cockney Rebel, and we had a great LD from your company called Mike Hounslow.
He went on to look after Eddie Grant. He was Eddie Grant's man with his business in the Caribbean, back and forth. Then something went wrong there with Eddie, stopped touring or whatever. Mike disappeared for a while then he reappeared. Then Ed was back doing something, but I haven't seen him now for two or three years.

Was Mike one of the first lighting guys?
Yeah, he was one of the original crew. He was there quite early on. He worked at the Lyceum as a bouncer, and he was an electrician so he was able to do instals and things. That was John Peels' Midnight Court, was it not? Well we used to promote that, but yes, Mike was one of the originals. The first though was Pat Chapman, the whole thing started with Pat.

That's good. Can I go back before that?
Before Pat?

Yeah, before you actually started Entec.
It started in the basement of the Marquee, where they used to make those psychedelic slides, you know, before Pink Floyd or around that time, and Kenny Sutherland used to sit down and make all these funny effects.

I remember Kenny Sutherland, projectionist.
Amongst other things. We were at the Marquee in Oxford Street from 1958 to '64.

How big was it?
It was quite big.

Was it bigger than Wardour Street was?
Uh, probably similar. Different shape.

Okay, but not where the 100 Club is now?
No. It was under what was the Academy Cinema. There's a Marks & Spencer there now. If you want to identify it, it's the Marks & Spencer at Tottenham Court Road end of Oxford Street, and next door was the Academy Cinemas One and Two. It was an 'art cinema', and the guy that owned the cinemas built this ballroom in the basement, and it was very, very smart. It was designed by Angus McBean, a very famous designer; nothing was too good for the Austrian who owned it, Mr Hoellering. He couldn't open another cinema in the basement because the entrance wasn't wide enough. He had this space, so he turned it into a ballroom, but nobody wanted to hire it as a ballroom although it was beautiful. Carpet on the floor, carpet on the walls, it was designed like a circus theme, hence the Marquee.

A friend of Harold's, another jazz person called Peter Burman, was running Saturday and Sunday nights with Dill Jones on the piano, very good pianist, and it was dying on its feet. So, he stopped running it, and he came and said to Harold, "I've got this wonderful venue but nobody goes". So Harold had a look and talked to the owner, and we took it on for Saturdays and Sundays. Modern Jazz in 1958.

Had you been promoting or had Harold been promoting elsewhere?
Oh yes. Festival Hall, tours all over the country, the Modern Jazz Quartet, Dave Brubeck, real big tours, Woody Herman . . .

How did he get into that then? Were you married at that point?
We got married in 1960, but I was working for him. It was all to do with jazz. He is a jazz fan.

Yes, but also a trained accountant, wasn't he?
Oh yes, which was how we managed to stay in business. He was a trained accountant in Southport in Lancashire and was a jazz fan, a traditional jazz fan, and was desperate to come to London because that's where the jazz was. He applied for a job in the city as an accountant, came down for the interview, and they said, "Oh, you've got the job but you're going to work at our Bristol branch". He was absolutely furious, I understand! So he had to go to Bristol. Meanwhile, I think, whatever this company was, they had lots of branches, and there were internal posts advertised, and something came up in London, so he put himself up for it to get to London, which he did. Funnily enough, the story is, he came down to London from Southport in 1948 and he got off the train, onto a bus and said to them, "Can you take me to the centre?" The bus conductor was the first person he spoke to. He got off the bus in Charing Cross Road, and there was a very famous, dirty, dingy record shop called 'Dobell's', which was very famous in the jazz field because nobody had jazz records. He saw an old '78 cardboard cover in the window that said 'Jazz Records', so he went in and there was a shabby-looking fellow going through a box, you know, of records. So Harold taps this bloke on the shoulder and said, "Can you tell me where the London Jazz Centre is?" because there was a club at 100 Oxford Street at the time. It turned out that the fellow he was talking to was Chris Barber. He had a terrible stutter at that time, didn't play trombone at the time, and was just a record collector. So Chris said to him, "I'm going there tonight. Come with me". Well they went along, and they've been friends ever since. Harold managed his band at one stage.

So, he had a tandem sort of career for a while, in the daytime as an accountant, then in the evening. . .?
Yes, yes. He opened a few jazz clubs in dingy basements in Soho, a shilling on the door, because he always knew a band that needed to play or you were in a band and wanted somewhere to play. There weren't any clubs or—well there were a few but, not that many, it was just sort of taking off. He obviously got to the stage when he was able to give up the accountancy.

Had he done that by the time he started those Saturday nights in the Marquee?
I know he had several businesses then, a printing business and he was running the National Jazz Federation. He had these small clubs, and he had a club in Greek Street, which was before my time. I think it was quite successful. That got closed down, the police raided it because after he finished at 11 o'clock it was taken over by somebody called the 'Latin Quarter Club' with all dodgy things, and the police raided that. He couldn't get back in the building. He always hankered for a club, but there was a long period of time when he didn't

have one when we were promoting concerts, so when this opportunity came up (with the Marquee), he said "Oh, we'll give it a go" because the rent wasn't very high.

We had Saturdays and Sundays. Then we added the Wednesday with Chris Barber, a very successful Trad band at the time. Meanwhile, Harold was running a newspaper called *Jazz News* because he didn't like the *Melody Maker*.

It didn't do jazz really.
Didn't do very much of that at all. It cost a lot of money, but we were running this newspaper, and we had a columnist called Manfred Lubowitz.

On a Wednesday, Chris started doing an R&B session in the interval instead of skiffle, originally it was skiffle. Meanwhile, Harold and the band had toured America and been down to Chicago and Smitty's Corner and met Muddy Waters and all those people. They liked this (style of music), and so they started doing sessions on a Wednesday, which got so successful Chris couldn't get on stage. Anyway, then we decided to start an R&B night on Thursdays with Alexis Korner, Long John Baldry because there weren't many people playing R&B then. That was jolly good. Then Manfred came in one day and said, "I can do that", because he was a very good pianist, and he used to play at Butlins to make money. So Harold said, "Oh, can you?". He said, "Yeah!" So he gave him the Monday nights and he put together the Mann Hugg Blues Brothers, which turned into Manfred Mann, which as you know became a huge hit.

Now, what year are you in here? It's the mid-'60s by now?
No, this is early, I'd say '59, '60, by which time we had Johnny Dankworth on Sundays and then Modern Jazz on Saturdays, our resident man was Joe Harriet, and there was Tubby Hayes and Don Rendall. Johnny Dankworth then went on to another club somewhere else, so we were dying on Sundays. We put somebody called Los Chalchaleros, and it became a different crowd on Sunday night.

Latin then?
Yeah. Cha Cha had just come in; we didn't like doing that, but it paid the rent.

Okay, now, was it a sit-down venue? Was it tables and chairs around the edge and the whole ballroom was dancing?
It was dancing but we did have some chairs for those that wanted to sit. It was banquet seating around the edges, and the coffee bar.

Ah, there were a lot of people standing and dancing?
Oh yes. Then eventually the landlord came and said, "I've got good news for me and bad news for you, Harold. I've just bought the jewellers next door to the entrance, which means I can knock it down and make it wider and put a cinema in the basement". So we said, "Oh", and he said, "But I'll give you six months to

find somewhere else", which was a drama in itself. Eventually after many, many months we found Wardour Street.

Now you didn't have the whole building, did you?
No. We had the basement. Well, mainly the ground floor. There was a small basement that we used as a beer store, and later we got the first floor for the offices. I think that was a different landlord but I've forgotten now. Whoever was there left, and we took over the first floor as well.

Was there a second floor?
Oh, there were several floors but they were all different businesses. I mean, our first floor we sub-rented half of it to Charisma Records, Tony Stratton Smith and then later to Billy Gaff for Rod Stewart's management.

It's a funny coincidence because my wife's father worked as a master engraver in a studio on the top floor.
Oh yes? What was his name?

Syd Fletcher.
Yes, I know Syd Fletcher, Syd and his partner Stan. Oh, we used to talk to them all the time. We used to have letter headings and things engraved, you know, people had all their business cards made then, didn't they?

Yes. Did Syd do yours?
I'm sure he did, yes! Well isn't that a coincidence. Gosh, I can remember. I can see him. I can see the other fellow, too.

He's got a brother called Steve. They were both very big jazz fans.
Yeah, they used to come down to the club. Yeah. But gradually by the time we moved to Wardour Street, the R&B was getting more successful.

So did you move jazz to Wardour Street initially?
Oh yes. But it wasn't as popular. I liked the R&B; we only put on what I considered to be really blues-based, you know, jazz-oriented R&B. We hit it at the right time.

So when was that? When was the move?
March 13, 1964. I know because it was a Friday. Friday the 13th. Maybe that was it. We closed Oxford Street on the Sunday before, and meanwhile we had arranged with the owner of the old Marquee who didn't need all the stuff that was in there, and I asked, "Can I have the piano and the banquet seating"?

Oh okay, some of the furnishings.
Yeah, to make Wardour Street look the same, we tried to make it look similar. It obviously worked because people came in. I don't know if you remember there

was a narrow front door and you went down a narrow passage, and then you turned and it opened up. People were coming in and going, "My God, how did you do this?" because it looked the same except it wasn't. The initial feeling was the same. We spent a week shifting everything, and the builders were still there like half an hour before we opened.

Who opened it?
The Yardbirds, I think, were the first ones. We managed The Yardbirds at the time. That's another story. Giorgio Gomelsky was their manager at the time, he found them after he lost The Rolling Stones to Andrew Loog Oldham. He worked out of our office as Marquee Artists Management. Simon Napier Bell came later because he found a record deal. He found that record 'For Your Love'. We've see him on and off over the years, but he lives in Thailand now.

In the ballroom in Oxford Street, what did the bands play through? Did you have a PA?
I think there was a puny PA, but once we got into R&B stuff they brought it in with them.

And what about when you moved to the Marquee in Wardour Street?
It was the same, but then obviously we put equipment in. In fact, Charlie Watkins of WEM, we got on well with Charlie, he provided the PA for our festival, originally at Richmond then Windsor, Kempton Park, Reading . . .

Hang on; let's go one at a time then. So, you were running Wardour Street and someone came up with the idea to do a festival?
Yeah, Harold did. He always wanted to do a festival.

This was his first festival?
In Richmond in 1961. Richmond Athletic Ground. It was a jazz festival. We opened the gates, and there were a few people outside, and we knew every one of them. It's such a small business, ah, but it was fun! We had two sets of tickets, afternoon and evening. Two shillings and six pence in the afternoon and five shillings evening, or the whole weekend for another price, I think it was. Two days.

Fantastic. And Charlie brought his WEM gear along?
Not originally. Hmm, I've forgotten when Charlie came in. I remember Charlie got involved at Windsor, there were complaints about his 1,000 watts of PA!

How many people showed up for that first one? The weekend.
A few hundred. I should think maybe 500.

That's not bad. Do you remember it? Did you stay up all night?
I remember it because I was cleaning toilets, making the sandwiches; my father was running the bar. You know, everybody chipped in. Yeah, we had a gang of

friends and everybody just wanted it to happen, so they all came along. But then of course it grew, and we added the R&B stuff and it sort of outgrew that ground.

You did it more than once at Richmond?
Well yeah. We did five at Richmond: '61, '62, '63, '64, '65, then we got chucked out. In '66 we were in Windsor . . . in '66 and '67. I know it was '66 because it was World Cup, and we had a TV in the artist's bar and nobody wanted to go on stage because they were watching the football, and also in '66 Cream was formed, and they didn't have a name when we put them on. No, they just decided to get together, and it was billed as Ginger Baker, Jack Bruce, and Eric Clapton.

That was at Windsor. Where in Windsor?
Windsor Racecourse.

So you obviously had some horsey friends?
Somebody suggested it because racecourses were on their knees, desperate to earn some money.

And they've a certain amount of infrastructure.
Yeah. But to get the license was a nightmare because Windsor Council were a bit toffee nosed as you can imagine.

I'm amazed they even had the licenses in those days.
Oh yes, you always had to have an entertainment license, and they turned us down. Harold appealed and the person that we got to represent us at the magistrate's court, because it was only magistrates, was Quentin Hogg!

(Laughs) . . . Fantastic!
And he wipes the floor with them.

Legally, no reason why he can't stop it from happening.
Anyway, they got their way after '67. We lost the support of the owner of the racecourse.

How many people coming by then?
Oh, a few thousand.

Okay, so it's starting to build up?
Oh yeah. Yes, we're building a big infrastructure. It was at Windsor that Ian Knight and Keith Albarn built this geodesic stage. They came up to us and said, "We've got this great, great idea". We saw the plans and they built it, and then it collapsed during the show.

With people on it?
Well it started sinking. We had to rush in, stick it up with scaffold and God knows what. We had to leave Windsor because the Council or somebody got at the owner

because he had put in planning permission to get a marina there or something, and I think he was told it might get commissioned if he got rid of that . . .

. . . That festival, those hippies!
Yeah, so he recommended us to Plumpton racecourse because they all know each other. So we went to Plumpton, near Lewes in Sussex, and again the owner there was a very nice chap, and I think he wanted to do it to stop all the people who complained about the races. Again another load of toffee-nosed people. . . " Uh! We'll be raped in our beds!"

We managed two years there, so it must have been 60—no, no, no in '68 we went to Kempton Park Racecourse, that's right. That's it. '68, Kempton Park Racecourse and then '69 to '70 was Plumpton.

Plumpton. And what was the festival called?
National Jazz Festival, or at least it was by then, and one was Jazz, Blues, and Rock Festival. That was at Reading. That was '71.

Pat Chapman on ladder
© Entec

Is it because of these festivals that Entec came into existence?
In a way it was. In 1968 when we were putting the show in Kempton Park Racecourse a chap called Pat Chapman came to see us who, as a hobby, was involved with a theatre club. I think it was Questors, and he was very interested

in lighting. He came and said, "Your festival has got rotten lighting, which should be able to support trip lights". He was running something called Crab Nebula Lights at the time, of which the logo was one of those squirrely things that were fashionable at the time. Anyway, he came and persuaded us to let him play with the lights.

What did he do? Did he put up a screen at the back and project?
He just put up some lights. That's what he did in 1968. It was a bit of a disaster that Festival, I don't know if you have ever been to Kempton Park Racecourse, the reason it's quite a good venue is there's a railway station in the racecourse. They weren't brave enough to let us have the festival on the racecourse, we had to have it in the car park, and the walkway from the railway station to the racecourse had a corrugated roof, and a lot of the local kids climbed up on this roof. Yeah, and they all started going 'dum dum dum' in time . . . Don't laugh. It was awful. Ambulances and people with broken legs and poor old Arthur Brown, he was on stage at this time. It wasn't designed for people to sit on, we kept telling people to get down. Anyway, so that was at the end. But that's how we met Pat, who then started doing things in the basement of the Marquee.

Did he start doing work in the Marquee in the club itself on R&B nights?
Probably, I can't remember. Then they started a proper business called Entertainments Technicians, which is where Entec[1] comes from, and were servicing lighting installations, not necessarily music stuff. They took an office in a warehouse, it was a small unit in Wembley, which is where Mike Hounslow was hanging out—never quite sure what they did. I used to go up there to do the wages, but that's when Mike went with them full time.

So you moved out of the basement to Wembley?
Wembley proper. It was Pat Chapman's thing, and I think Harold was funding him. Then we discovered Shepperton Studios, which was on its knees because the film business wasn't doing well. They extended their franchise from just film to entertainment, and we got the unit there.

Was that at the same time as The Who got their units or were they already there?
Maybe around the same time, but that's when the staff grew and music business work grew, you know, more things happening. That's when the sound got added, which was an accident because Entec was originally a lighting company. Do you remember The Sweet?

Yes, I do.
Pat did The Sweet. The deal we did was when The Sweet weren't working, they stored their sound equipment at Shepperton, then Pat said, "Why don't we hire this out when you're not using it?" So it was The Sweet rig that started the sound side of the business.

Was it 1970 by now or '71?
Probably '70, '71, yeah. Windsor was '66, '67, which I have to say when we started our little jazz festival we never dreamt that it would get to that size. Now I believe that Glastonbury and Reading have 100,000-plus people. I mean, it's just mind-boggling, something we thought would only last a couple of years.

Harold gave up the accountancy for a short period. He thought there was no way anybody made a living doing anything in jazz. In fact, his office was a café, a Greek café in Old Compton Street with a public phone where you pressed button A and B. All the bands were amateur. No one ever dreamt that you could make a living at it. I'm still gobsmacked, you know.

By the time you got to Reading, was all the equipment your own?
No, no. I think WEM was still doing the sound. Charlie was on manufacturing. Oh, he was a sweetheart. We hired stuff from him.

He sort of did festival stuff really, didn't he?
Yes, and I suppose he thought it was good PR for the company. All those fans got to see the equipment. Yeah, everybody was just making it up. David Hartstone was one of the first.

Did you ever use his equipment, do you remember?
I don't know? Then Dave Martin[2] was the next loony to make a success of it actually. Entec had all Martin's stuff at one stage; that was our preferred system, we became the Martin house.

An Entec Martin PA stack
© Entec

Yes. So who introduced you to him?
It was the sound guys, and Dave was a good blagger. Probably came 'round and did a deal or whatever. Yeah . . .

To this day Entec is not only the first independent production company this country's ever produced, but is still alive and kicking!
I suppose so. Considering that was at a time when (which didn't do the business any good, only the manufacturers) all the successful artists were told by their accountants that if they bought equipment, they got all the tax breaks on it. That's how Britannia Row started; Pink Floyd bought all that stuff. People like us are going "Bloody hell!" Yeah, The Who did the same.

Joe Brown did it as well with other bands.
Yeah, I mean . . . Iron Maiden have still got a load of gear. I don't think they know what they've got, it's all locked in some warehouse somewhere. I'm just saying that through some outside force, business changed.

Bloody accountants, Barbara!
No, it was the government who gave out some tax relief, and the manufacturers were having a lovely time, weren't they? All the bands go, "Oh, I want one of these and one of these". I know all about the costs of desks, and we had some loony who said, "I'll build a desk. You don't have to go out and buy one". I learnt a lesson there.

An analogue desk?
Oh at that time, yeah. Took forever and then didn't work. In fact, we started building our own speaker cabinets, and I'm going, "Oh no, don't do that". Then there was a time when Pat Chapman was going to build our own lights, and he spent hours designing them. Well it doesn't work these days, there are too many manufacturers out there doing it, that's what they do.

Now there is, but back then there weren't.
No, there probably wasn't. Strand Electric was the only lighting company, weren't they? Being very cost effective is how Entec has managed to survive; we do not spend money where you don't have to, because we have to fund ourselves.

Hmm, good story.
I could go on for hours. You can remember the days. Do you remember Bruno Wayte, our sound manager at one time at Shepperton? Somebody came 'round to interview him and asked how he tested amplifiers. He said, "See this amp here, when I drop it, if it works after that, then it's okay!" We were all making it up as we went along.

Who were the original sound people then? At Entec.
Tony Self, Terry Price—

Oh, Terry came out to Entec?
Well, Terry joined us and left after a couple of weeks. I was really cross then!

Terry was the road manager for The Sweet.
Yeah, he was, Terry and somebody else came, and we said, "Oh, that's good, that's the sound sorted". Anyway, Terry only stayed for a little while.

He went on to TASCO, didn't he?
Yeah, we were not very pleased. I liked him.

What were you doing at that period? Did you and Harold have children?
Yes. We've just got one boy. As well as running the Marquee and the festival, I was going to Entec to do the wages and check up on things. We had managers and accountants, and we'd go there once a week or whatever, a bit busy it was. Noreen was at Shepperton then. Noreen is a Director now; she's a shareholder in Entec. She's been with us forever, she knows everybody. People ask why have we got sound *and* lights, they are two different disciplines? Well in a way it's like we've got two companies. They are different, but there are lots of shared costs. Yeah, the warehouse, the office staff and so on. We always found that if one is doing well, the other one isn't, so one supports the other, and vice versa.

How often nowadays do you manage to do the sound and lights?
Quite often, but we don't push it. Obviously, we supply lighting to shows that have other sound companies, and we don't want anybody to feel we're muscling in there, which we don't. Totally separate and vice versa, we'll do the sound on one show and some other company's doing the lighting. We're very honest about that, we wouldn't ever try and blag, because it's not worth it. The type of shows that we do both sound and lighting on are productions like fashion shows or religious ceremonies, sort of things you don't know anything about. They don't have any loyalty.

Yeah, yeah. Out on the road you've got the two powers that are the lighting designer and the sound engineer as well, which is the other consideration, I suppose.

Going back to the beginning when the two sides of Entec were set up. Was it Dave Martin's equipment from day one, do you recall?
Well what was The Sweet's stuff? I can't remember. If that was Martin, that was probably the way we went, we would have added to it all. We were big dealers with JBL too, we bought a lot of JBL.

That's from America.
Yeah. Because they set up here, didn't they? I don't think they manufactured here, but they had offices.

But you bought, you didn't build internally?
Oh, I think we did a bit. At one stage we were talking to Steven Court, do you remember Court Acoustics? They set up the manufacturing side, but we didn't go forward with them.

On the festival side of things then, was there ever a time when Entec did the whole festival?
Oh, yes, yes. Entec used to do the festival every year, we used to do both the sound and the lighting.

And who booked the festival?
Harold in the jazz days. Later it was Jack Barrie, who was the manager at the Marquee, the festival was like the Marquee's garden party.

So we could see how bands were progressing or whatever. You had your ear to the ground, it was a committee sort of thing, but Jack put it together, so it worked.

Do you have a favourite one? Does one stick out in your memory?
Well, no, because I can't look at them musically because I didn't hear everything, we were so busy. The ones that I have memories of are probably for reasons that nobody else has. You know, funny things that happened with staff or problems with the toilets.

Is that the worst thing organising a festival? Toilets?
Toilets are the main topic of conversation. I mean, seriously, when we started there were no mobile toilets. Harold discovered that when they were building the M1 they had all these Irish navies, and they built mobile flushing toilets, to put on the side of the road for the workers. There was a company called Cliff Plant that built them. We tracked them down and we got them for our festival, but we were the first people to have proper toilets.

From road construction companies?
Yeah. I'm gobsmacked now the number of companies that supply toilets to outdoor events.

I'm amazed you didn't start one yourself!
We nearly did, we nearly did, our plumber for the festival, our site manager Ron, we discussed it and then Ron said, "Ah, I can't be bothered".

Okay. I was going to say from your perspective is there something that sticks out over the years?
I cannot believe with festivals how it has created so much ingenuity and businesses and things. I mean, as you say, as well as hiring showers, you can hire your own wigwam, you can have butler service (which was never the point of the whole thing) and crèches now and, I mean it's just phenomenal. It is huge.

Have you been to one of the big Glastonburys?
Oh, yes. Not for a long time. My legs won't do that now, but I've been a few times over the years.

Marco, our health and safety advisor, said, "If you think of all the things over the years in our industry that's done the most good, it's sleeper buses that saved more lives than anything else". I think he's right.

Well, Brian Croft said the same thing to me.
Oh really? Isn't that interesting.

Yeah. A sleeper bus was probably the best invention for the road that he could think of.
See, what Len Wright thinks. This is a guy who has been in all sections of the industry, and he's decided that is probably the one thing that has improved people's lots. I think he's probably right. You can tell Len he did a good thing.

Well, thank you very much, Barbara, that was very enlightening.
You're welcome.

Barbara continues to this day to work as Finance Director at Entec Sound and Lights.

Notes
1 www.enteclive.com/1960s/
2 https://martin-audio.com/downloads/edge/MartinAudio_TheEdge_40yrs.pdf

Part 5
Pioneers of Rigging

Without seeming trivial and pointing out the obvious, it wasn't until equipment was required to be suspended in the air that rigging actually existed anywhere in the music business. At the outset there were no stage lights, just house lights. Then there were lights on stands, lights on theatre bars and eventually through a series of different mechanical lifting devices there were lights on trusses. It was at this point that some 'brave nutter' used to climb up a rope ladder or steel ladder and balance themselves on top of the truss to have instructions shouted up to them by a lighting director, as to where to point and focus each lamp, laboriously one by one. I remember always being slightly in awe of that daring person, clambering along the truss, especially when the whole truss was ground supported, rather like those madmen that built the skyscrapers in New York in the 1920s and '30s. Truss started off ground supported more often than not, and then when aluminium truss came along, it became a lot easier to hang truss from cables above.

One of the first riggers was Jon 'Happy' Bray. In 1979 we worked together on the Kate Bush tour, and he was part of the lighting crew, going on to become one of the original 'pioneer' riggers.

13 Jon Bray

Jon Bray
© Jade Dearling

August 7, 2013 at Jon Bray's House in Kent

I always knew you as Happy Bray.
Yes. It's a nickname I just inherited from my brother at school, and it stuck with me. I think he got it and he was a bit miserable, I think. Still, there's a few people that call me that, but I dropped that when I was, you know, running a business.

Where did you go to school, when did you leave, and did you go to college? Just a bit of your early background.
Yeah, my dad worked for the foreign office, so he lived in Japan, Taiwan, Brazil, all over the place. I went to a secondary school in Cranbrook about three miles from here, which was a grammar school that took boarding kids. There were a lot of the foreign office and army people had their kids there because it wasn't huge amounts of money.

You were a boarder?
Yeah, I went to see my parents a couple of times a year.

Wow. From how old?
A pretty weird upbringing, about eight. Yeah, I left in '73 and had no idea what I wanted to do. I was a typical hippie dropout.

Had you done A-levels?
Yeah, I've got A-levels. I did apply to go to university and study archaeology for my sins. That was a couple of years further on. Actually, that was interrupted by finding out that Debbie was pregnant. Initially, I had no idea what I wanted to do, but coming from here in Kent, at that time there was lots of casual farm work locally. Like strawberry picking, cherry picking, fruit tree pruning, hop training, hop picking and all that sort of stuff. Debbie was still living at home for a while with her parents then, most of that time we were together and just travelling around. We had a van, picking fruit, lived up in our little house in Wales in the winter and then in the spring start picking. We did two years of that. We had an old J4 van and lived in it some of the time, travelled around and went to Windsor free festivals and things like that and lived the life really. Until shock horror, Debbie is pregnant in '74. Becky was born in the Spring of '75. Suddenly, we realised we needed to get jobs and things like that. I worked locally for a company in Hawkhurst that you can see from here, building hand weaving looms for carpeting. I was trained properly as an apprentice carpenter for a while.

Then a friend of mine, a friend from school who was always clever at electronics, rang up out of the blue and said, "I'm working with somebody in London who is developing all this new lighting equipment and needs some people". I put plugs on bits of wire initially. I think I went there for a couple of weeks, they had a rush on making some of these new-fangled multicores that were coming out.

Oh. Who is this?
Showlites. Based in Herne Hill, South London. I did a couple of weeks up there putting plugs on, and that was all fascinating stuff, those bands they were working for. In a very short period of time we moved there, we squatted in various houses locally.

You could then!
You could very easily, yeah, yeah. No problem at all with a few friends and a young baby. Yeah, and work there was really fascinating. I had a reasonable electrical knowledge, I suppose, from doing physics and stuff at A-level. At the time most of the lighting equipment came out in the theatre, so there was Strand dimmer racks and Strand lamps. There was almost nothing bespoke for touring work.

So Showlites, Eric Pearce ran it. He had another company, it was all the same thing really, called Alderham, who were developing lighting desks and new dimmer racks. Instead of a typical dimmer rack, which up to then was really huge with very few channels in it, he was doing them with much, much smaller packaging and much larger capacity. I was quite involved with, not the circuitry design, but having to mount it in the boxes and making panels and really old-fashioned stuff. Fly-pressing holes in bits of metal for sockets and quite involved in physically building a lot of that kit.

Was that kit just for the company?
At that time, yes, I think it was. RDE (Richard Dale Enterprises) were over the road, oh no, that was a little later on actually. There was Rainbow Lights, and I think Zenith were probably around as well, yes, they were. There weren't many players in that market, and they were trying to really get ahead with this new-fangled stuff. From that role working in the warehouse manufacturing I started going out and doing the odd shows, working alongside others. I remember Little Feat at the Rainbow was a famous show where I think there was a punch-up between the drummer and Lowell George in the wings.

Do you know what that was about?
No, too much cocaine, I should think.

His own drummer?
Yes, oh yeah, internal band fight. I remember the Diana Ross show in the Palladium. A variety of bits and pieces I did locally in London while working in the warehouse. Got to know the kit a little bit. In a very short time after being there, they got a job doing lights for Alan Price, who was doing quite a long tour of German clubs. I was appointed as the lighting man. I was the operator, designer, I'm not quite sure if I was the designer, I think there was a guy called Streaky, he was the sound engineer and senior roadie, I suppose. Maybe he was the designer, I don't really remember, but I was travelling with them, I think it was probably three of us. Sound, monitors and lighting.

What lighting equipment did you have?
We had, it was four or maybe six Strand stands. Just a little telescopic pole to the T-bar at the top with four lights on it. You just poked it up in the air and put a sandbag on the legs.

A precursor to trees.
Yeah. A really simple little thing, a theatre stand. I was operating it, putting it up and driving the truck. It was a little truck around Germany, a real eye opener and definitely in at the deep end. Didn't think much about it really, but had a great time.

Now, in those days, and you're talking clubs, did they have electricians at the clubs?
There must have been, I think it was down to size. With four lamps, you're almost coming off a couple of domestic sockets. We were having to make that up every day. Streaky was the guy who had been doing it a while, I think he was probably

acting as an electrician or knew a bit more. Silly things like disconnecting the dishwasher from the kitchen's power and wiring the lighting system. That kind of level of stuff . . . that was definitely my first time on the road.

Okay, then, what year are we talking about?
'76 probably. I joined Showlites in April '76, so it was probably that year. I'm sure it's within a few months of starting with them. I was doing equipment specs and even costing things for Eric Pearce, and I was doing quite a lot in the warehouse. It was a very small outfit with some real boffin types doing the development to the desks and stuff who certainly weren't up for quoting jobs or anything as normal and down to earth as that (laughs), and in the meantime doing, certainly doing London shows. There wasn't really a rigging side to it. They were either using floor stands or the other bit of kit that was commonly used were the Genie, telescopic Genie towers, which you probably remember.

Trees.
Massive boxes because they usually have the frames with lights on. I think six each side of the mast, maybe bolt on a couple more on the way up. The box was sort of the size of a three-seater sofa and really tall. A typical rig would be four of them. Two a side at the front and two a side at the back, maybe a front truss on a couple of super lifts and two of those at the back or something. I'm sure that in the mid-'70s, around '76, '77, there really wasn't any rigging that I remember going on. Super lifts were used which still are, not very much (laughs). Vermettes were another machine, they were a variation on a theme. They were a lift that you had lots of fixed telescopic pieces to, and you flipped it up and again wound up a hook to raise a truss. Frightening to think how overloaded those things were, I think even then.

And then focus from ladders.
It was a very brave man who climbed a truss on super lifts. I think it was done.

You ever do it?
(Laughs) I think I did it, yes, very carefully. Very gingerly and moving slowly, but yeah, it's surprising there weren't more accidents than there were. I do remember one particular incident I was working on, I think it was Steeleye Span, but it may have been Fairport Convention. Folk rock, anyway, it was one of the two, in Hammersmith Odeon, probably '77. The front truss was on a pair of super lifts right behind the iron curtain, and it was as far down as it could be to get a good lighting position, and the curtain in those days always came in prior to a performance, check it, so it was down when the audience came in. The band came onstage all ready to go, curtain goes up. On landing it had moved upstage an inch, let's say, and a couple of nuts on the back of the curtain caught an outrigger on the super lift, and opening of the show . . . curtain goes up, front truss falls over backwards.

Focusing on a ladder—perfect example here!
© Richard Ames

Backwards, thank goodness!
Yeah, yeah. But it was the only thing that saved the day. It must have been a big special of some sort, I suspect, rather than just a day-to-day show, and they built a gantry of scaffolding. We got SGB or somebody in who built this big gantry of scaffolding with some back follow spots, two or three back follow spots, and this thing just fell over onto it and lodged itself at an angle. The band scattered.

Nothing fell on anyone?
Nobody was hurt, I don't think. One of the crew, I think, had something fall off the cowling and hit him on the head, but nothing serious, and the whole thing was stopped and restarted sometime later. They did do a show. I'm sure there were lots of silly incidents with things prior to proper rigging being used.

Well that was the ultimate risk of those free-standing towers, wasn't it?
Yeah. They're made for erecting ducting on building sites and things like that. They're not really made for . . .

. . . Hanging a couple of tons!
Well it wasn't weight as much as stability. They're very high with a truss between them. Yeah, I was doing those London shows, that's just one that sticks in my mind quite vividly.

So at this point in '76, '77-ish, for the most part, nothing is being hung?
I don't remember it. There weren't arena venues really, except Wembley, which I do have some memory of doing a show in there, but most touring shows were theatre shows. They were doing the theatres or they were sports halls, something like that. In the theatres you would always hang the lights and then have tab tracks and cloths off of the bars, so a lot of the so-called rigging, getting the equipment up in the air, was done that way with probably carrying some super lifts and truss for the venues where you couldn't do that.

I remember certain venues like Newcastle City Hall.
Yeah. That was always a challenge.

Is that after chain hoists came? Or were you dead hanging on chains?
Yes. It was after chain hoists came. From the early days a lot of theatres, I think even when I started, certainly late '70s, there would be a couple of holes for a front truss or something, and then you would maybe carry two manual chain blocks, so instead of using your super lifts, where you could you'd hang your truss further forward to get a good lighting position.

So the chain blocks were there because of the existence of the Vermettes?
I think it rapidly changed by 1979, '80, manual chain hoists were used quite a bit and electric chain hoists were starting to be used. It spread over from the States where using truss and hoists had started, probably sometime early '80s.

So was the initial original equipment hand-driven chain hoists, were they American made?
No, no. They were a very standard piece of kit used in all sorts of industrial applications, whether you were lifting girders in a factory or engines out of cars. They've been around for decades, and they were adopted by our business, and for a straight stick of truss on two of them, that's fine. The problem with them is that you can't get them to run in sync. Two blokes pulling and they'll pull at different speeds. But on a single bit of truss it doesn't really matter. It will be level at the end.

So how did you level it?
You'd usually get somebody to stand out in the stalls or up in the balcony and just eyeball it against the back wall or something. That doesn't work very well with bigger grids, it's very difficult to keep control of where the loads are and how level it is, and that's where electric chain hoists rapidly started to be used. Some of the notes I've got in '79, April '79, when we were both on the Kate Bush tour, looking at that equipment we had six electric hoists up there, and I think that was quite a big one.

Really?
Yeah, yeah, it was around that time, that was very early days of using them.

That would have probably been a front truss and maybe a box truss?
It was a little grid, I think. A little box and front truss, though it seemed like a huge rig at the time, I think, because a normal rig would have been a front and a back truss. So it was around that time, I would say '79, when probably it had been going on for a while in the States.

Yes. Now were those held up by electric chain hoists?
Yes.

American?
No, they weren't. The hoist, the commonly adopted hoist at that time, was something called a Verlinde. French made. A couple of years later, when I worked at TASCO, they had some Lodestars, some American Lodestars, they had a few in amongst quite a lot of the Verlinde hoists, and they're still used today. Lodestars are the standard.

Were they just a better quality?
Yes, much more robust. Heavy duty.

Okay, more heavy duty? Americans do things heavy duty.
Easier to repair. Yeah, yeah, they do. They have an agricultural approach to some of their engineering, but in that case it's a good thing, I think. Yeah, so I can remember, it wasn't an established way of doing things. We were all just making it up, and one I particularly remember, Eric of Showlites got a job to put in lighting for Guru Maharaji, I wouldn't say a show, but a large meeting.

A convention.
Yeah, at Wembley Arena. This was quite an astounding thing for me. I had never seen anything like it.

Lots of hippies?
Oh yes. Lots of flowers. He had a fibreglass throne, it was just enormous, built on the top of this scaffolding pyramid lined with flowers. It was a very, very elaborate show.

They got a scaffolding company in to build a tiered stage. If you imagine a Mayan pyramid sort of look, a lot of steps getting smaller and smaller and each step was faced with boxes full of flowers so it was a pyramid with this massive fibreglass curved throne thing for him and his family on the top.

On the top?
Yeah, yeah, yeah. It was quite something.

How did he get to the top?
I think there was a staircase on the back face of the pyramid that no one could see, I suspect. This, I think, is the first time I remember seeing a full lighting rig and

sound flown, and looking back on it, it's one of the scariest things I think I've ever seen and done. I was just a very naive kid, and Eric Pearce was driving this along, and we put in a long front truss, 50 feet long or something, and—

Would that have been aluminium? Sections?
Yes. Eric used to make his own truss out of angle iron with a pillar drill, or we used to, but I think at that time Chris Cronin was starting to build out of aluminium at LSD.

Welded in Birmingham?
Yeah, whether it was one of his or homemade, I mean we had homemade angle aluminium, angle being the section but it was actually aluminium. This truss was literally hung from right at the top of Wembley, dead in the centre, there was a sort of little gap where the concrete edges come up, and literally there were two steel scaff pipes wedged into this gap on a lip and two manual chain hoists hung going out at extreme angles . . . well the angle increased and increased as the truss went up. It was only electric cables coming off one end, I think, that stopped the whole thing sort of spinning around. Looking back, I mean, absolutely ridiculous thing to have done.

Should they have gone up on four straights?
Well I don't think there were the fixing points at that time, you know. Later on they drilled holes in the beams, I think, to establish fixing points further out, so rigging as we then later knew, could go on, but I don't think any of that was there. Even more bizarre with the PA I believe; I don't know why I remember this, but Stephens and Carter, who were an established scaffolding company that did film work, I think, they were brought in and they built a little box of scaffold platform with high edges and handrails around it. PA piled into it, and I think they just took four turfers, being a manual sort of rope-climbing device, fixed them to places in the building, all over the place with—I don't know God knows what—and literally just sort of cranked this thing up in the air.

Holy Moses!
That must have been not long after I joined Showlites. I'd say within a year or two at the very most. I joined Showlites in 1976. It was very basic stuff, and there wasn't an established system of doing things. But things changed really, really rapidly, very rapidly.

Who made the decisions on two very seriously important levels. Where to hang stuff from when it had never been done before, and how much weight could you hang off these points? How did those decisions get made?
That was very different to nowadays. It was simple in the theatres because the fly bars were already in, it was a very established system, the fly bars had a cradle that could take so much weight, and they were designed for so much weight.

It was pre-determined then?
Yes. A fly bar would be designed and built to take 200 kilos or whatever it was. So that was fairly straightforward. The super lifts that we used had a rating which was usually . . . (Jon raised his eyebrows and laughs).

Well, it was almost impossible to winch them up if they're overloaded anyway. But yeah, when it came to rigging in new buildings, the buildings didn't know much and the riggers didn't know much. It was a judgement.

Well, I mean those holes in the ceiling in Newcastle City Hall obviously went up into the roof where persons like your good self braved entry up into, to tie off. Did someone tell you what to tie on to?
In many cases, no. I guess it helped I'd done physics to A-level. I had that sort of background and mind, and I had some pretty basic ideas of the stress and strain of simple structures, so you could identify what were the stronger fixing points. I don't know where I picked it up from.

Were you aware in those early days of what weight points you were tying to?
Yes. Definitely aware, yeah, definitely aware.

You knew the weight of the truss and—
Ooohhhh yes. I think fairly aware. The good thing was that the lights . . . you know, everything was so small. Nowadays I've been heavily involved in the O2, rigging at the O2 recently and, you know, 50, 60 or even 70-ton shows are not uncommon. I think if we were hanging two tons of equipment in those days over a number of points . . . so it was a different league, and in general I'm not saying you couldn't go far wrong, but the buildings, they've got snow loading, they've got capacity. Funnily enough, the older buildings, the Victorian buildings, they didn't have modelling software, they were just built really strong.

They didn't know the limits of their materials well enough so they overdesigned, and it was when you went into the modern buildings you found there were problems. Deeside Leisure Centre and such places, which were just like being in a warehouse, they had been designed with modern engineering software, and they designed it down to a limit without much spare capacity. One of the things—actually I haven't thought about this until now—I don't know when the year was, you can probably research it, but AC/DC did a tour of theatres with a bell.[1] This was quite famous, the bell, quite a big bell, metal. Now I would have thought you'd get a fibreglass bell and use your sound effects to make a nice noise, but oh no, they had to have a real bell! I wasn't involved in the tour at all, I will hastily add!

This is a proper church bell?
This was a big, bloody bell. Compared to a truss of PAR cans, heavy . . . VERY heavy. You kept going into venues with people saying, "That's the beam they bent with the bell". While a PAR can is so light, trusses with PAR cans on, it was a bit difficult to go far wrong, I think. It sort of self-regulated itself. Obviously you

were fairly set to the unit. You weren't hanging it on the most flimsy of things, but as soon as things got heavier, yeah, you've got a real problem.

What comes to mind for me talking about the bell is the fact that if you've got a structure that is so heavy, that it's potentially dangerous hanging off of one point, you start going to two or three points to spread the load, don't you?
If they're available, every venue is different.

When did that start? That's what I was going to ask. Originally, you're ground based and then all of a sudden with chain hoists, you're air based.
Yes, bridling to points to get things in the right position, I suppose from day one. You have to attempt it. That was mainly in theatres where that was what they would have done, they would have found a big beam. Let's say you wanted to put a lighting truss right down in front of the apron for some front light. Some of the theatres had identified a structural beam that was in front of the proscenium and made a hole directly under it. There are lots of places like that. So your rigging consisted of putting a strap around that beam and adding the chain hoist onto it.

Yeah. And the sound systems?
And some of the sound systems. Although, I think in the late '70s, it was still on the ground, piling it up. Martin systems (laughs). For me it was soon after the Kate Bush tour in '79 that I got involved with Bill Harkin, a guy called Tim Davies and a guy called Jade Dearling, who worked with them doing roofing work. He worked with Bill, with that orange roof[2] that The Stones had at Knebworth in '76. Bill was involved with design of tensile structures, and we both worked on the Zeppelin show at Knebworth in '79.[3]

Okay. That was a Dick Parkinson show.
It was, it was the Rainbow Lights show. I remember, because I'd been working with multicore systems and the little compact dimmers, and I went and had a good look at their system. They were using something called an Electrosonic dimmer rack. That was probably two feet by 18 inches by 18 inches square, and I think it had four channels in it. That was, for the day, a massive lighting rig. They were on either side of the stage up in the PA wings, walls of these things, I mean, just massive walls of them and knee deep in cable and all single. Each lamp had a cable that came from the lamp to the dimmer, no multicores. Bill Harkin by that stage had designed this inflatable roof, it was a white roof and it looked like a pillow, basically a huge pillow, and it attached to scaffolding walls. We had a back wall and two side walls and an apron in front. This thing attached to them and blew and hence shed the water without any central structure.

Like a lilo?
Yeah. There was no roof structure as such.

What was the gas in it, air?
Yeah, air like a bouncy castle now, fan motor to keep it blown up.

Right. Okay. One of the first roofs then?
Yeah, quite early. The lighting must have been hung on the side walls. I don't really remember.

It would have been Hartman Towers.
There would have been Hartman Towers at that time. Yeah, but Jay, who worked with Bill and was working for Tim Davies and who I think was actually the roofing contractor, he was really my mentor. It was when I really learnt all about rigging. He was good at rope access and using harnesses and prusiks[4] and all the other sorts of stuff and techniques that I hadn't known. I'd worked on my own up until then, working it out as I went along.

From that meeting, that was a great time we had, we had a really good time there because there was quite a long build period, a lovely summer, and there were two shows with about 10 days apart so you lived up on the site. We were living in a caravan, which we scaffolded into the stage base so they couldn't move us off to a remote area of the site! I remember we had a laser on the show. To cool the laser, they had a tanker full of water to sort of recirculate through the machinery.

Did they have video?
They probably did Eidophor[5] I should think because Tim Davies, again who was doing the roofing, was involved in the early Eidophor, which were these huge machines that came over from Switzerland. They were the first, they literally would weigh two tons or something.

I think it's the first show in England with live video.
Yeah. Could've been because I went on to do Eidophor tower building for Tim up at Donnington for Monsters of Rock.

Safety harnesses, then?
We did have something very, very basic. Yeah, yeah. We did have and, again, Jay knew these rope access techniques that we were learning.

Was it from rock climbing?
Yes. It was from rock climbing, but nowadays it's very separate. Work equipment has different regulations and ratings and everything else to measure equipment. Back then I don't think there was such a thing, as far as I'm aware. Jay then, he was an established roof erector rigger, and it was a sort of blurred line there. They weren't different specialties. He asked me to join him on a tour of America, so I went to America with him in 1980 with The Who. We toured the length and breadth of the States, as the 'two limey riggers', as we were known. The local unions were a bit suspicious of these foreign riggers, (laughs) and that was where I really cut my teeth. I saw how it was done in the States and, for the first time, it was big arenas, Maple Leaf Gardens and places like Atlanta Omni Dome, really big places. Suddenly we were seeing, in a couple months, ten dozen venues who all worked out systems of doing things, and it was the ultimate education, really, working with Jay and working those venues.

That would have been Roger Searle on lighting. Mick Double the drum roadie?
Yeah, yeah, yeah, look at this.

Jon is showing me the original—what is it, Jon? The original itinerary? Alan Rogan, John Hubbard, James Dann, Roger Oakley, oh yeah.
Yeah, that was actually the European tour. That was a really big show, 16 points, I think. That's one of my drawings.

Jon is now showing me the original lighting drawing. It's A2 and it's a drawing of the trussing.
Yes, it's one of my drawings. That's 14 points, and that was a huge show at the time compared to what, 140 points biggest show now? A typical big show coming to the O2 nowadays.

Wow. This is dated the 17th of August, 1982, 14 points. So you're going into an American arena, what was your responsibility? You obviously knew where you wanted to hang points, but were you allowed to go up in the roof and tie it off yourself?
It varied, depending on the union approach. I don't think I ever came across anywhere that wouldn't allow us to go and check what had been done. Some would allow active participation and others wouldn't. It was certainly the first time we found venues that actually knew how to deal with the rigging in their building and, to be fair, most of them were very good at it. They've done it, they worked out a system. As long as you played along with their system, eventually it wouldn't always be the most efficient or quickest or most labour saving. I mean, you had some weird things like different unions putting in sound points and lighting points in Madison Square Garden. I don't know if that still goes on.

Actually, this drawing was from when we got back in the NEC. It was that tour that I came back knowing so much more about rigging, how to do things, how to do things safely. It was from there, I think, I was suddenly an expert. Well, I was. I had that knowledge that not many people in the UK had.

And at that point you became a rigger?
Yes. Just right place, right time, right break.

Who else was doing it at the time you were doing it?
When I went to work at TASCO—The American Service Company, I think is actually what it stood for—this was about '81, probably '82, there were a core of three of us. Myself, Pete Ryall and Eric Porter. We were never short of work. It was unbelievable the acts we worked with. We were talking about Joe Brown earlier on, he was selling his services in America heavily, and what he specialised in was picking up the European legs of American artists. Typically, it was four- to six-week jobs.

Paul Newman, Joe Brown and Terry Price in
L.A. TASCO office circa 1980
© Paul Newman

I was living in London with a young child, and it was perfect. You'd go off and you almost had a choice of work. They had Elton, Diana Ross, Police, you know the list was long, and just those three kept us working enough.

Jumping backwards, there was Robin Elias. Did you cross paths with Robin?
Yes. Quite a bit over the years. Robin Elias,[6] very early days, ran a company called Zenavail Rigging in Brixton, and I'm sure it would be the first dedicated rigging company rather than someone like TASCO, who was a lighting and sound company with some rigging. Robin, he's director of Unusual Rigging, he was bought out by Unusual some years ago and very much still in the game.

I'll tell you actually why. While I'm looking at this drawing, these things, these improbable-looking things are PA cradles, and he cornered the market. PAs were big, bands were getting big and wanted lots of PA. At Zenavail, Robin had developed these PA cradles, which were big platforms. They had nets covering over it all for safety reasons because they were still putting all those little horns on top of the bass bins and stuff.

They look like something that you would cradle onto a boat?
No. They had a solid flat platform so scaffolding sides and, yes, they did have nets, they had a framework around the top and then a net all the way around them because there were so many loose elements and then the whole thing was hung on

a single point. You had to, sort of, shuffle the boxes around to get it balanced, it had quite a high centre of gravity. They did the job, but there was a certain period when they were the only thing around, and Zenavail I think was supplying those cradles. Robin probably designed those actually.

Did he come from a lighting company, do you know?
He must have, yes. He must have done, I think him and Charlie Boxall was another very early guy in the rigging game.

Backwards a little bit from that period. When I got back from the States, having done that Who tour with Jay in '80, I went back to working for Tim Davies. After the Led Zeppelin show at Knebworth in '79, the promoter Freddy Bannister, he didn't get paid, there were lots of arguments, and he didn't pay Tim Davies for the roof. So they took it, they literally took it, put it down, folded it up, put it in a truck and took it away because they knew they weren't going to get paid. Tim started a company called Hijack Roofing (lots of laughs), it actually is a registered company. Hijack then turned into, I don't know the ins and outs of the business decisions, but turned into Tecta Staging. It was Tecta that Tim Davies and his brother Hoagy ran, that went to talk to a company in Glenrothes north of Edinburgh, who were building orbit structures as temporary buildings. That was the beginning of the whole orbit stage, which still continues in a serious way. This was a commercially available building, I think the military used them as temporary buildings, anyone who wanted a quick, temporary building.

Jon is showing me another drawing.
The 23rd of February '81. Bill Harkin again. Festival Stage System number three, from Tecta Staging, 17A, Kensington Park Road, W11. Yep, and all this is standard building. They developed this sort of front cantilever and the stage system and a way of fixing it down. I was their sort of foreman, if you like, I was their head stage bloke for one season. Again, we were learning as we went along, we were making up, sounds the wrong words, but we were having to solve our own problems. We built water ballast bags and all sorts of things for ways of holding these things down in the wind.

Yeah. Wind being the major factor.
I could have spent the rest of my career, I think, doing staging, but I walked away from it at the end of the first year because all the guys who were working on it, most of them were based at Worthy Farm, Glastonbury, and there was just so much dope smoking; nothing was getting done, it was really difficult.

It was very hard trying to run the team when most of the team wanted to lie down most of the time. I'm friends with Tim to this day, but I had to tell him I couldn't stay, and I think there was just so much other rigging work coming up, I went off in another direction.

So you went to America and learned quite quickly from the Americans on this Who tour, and you were learning, also, from outdoor festival

staging. So riggers, per se, in the UK, were just evolving, really, around about '79, '80?
Yes, usually members of the lighting crew who had a head for heights, it's probably as simple as that in many cases.

As a freelance rigger from around about 1980, did you just float from company to company depending on basically who called you, or did you align yourself with a particular busy production company? How did that work?
Yeah, that's it. Certainly in the early '80s it was TASCO I did a lot of work for, I can't remember if I was PAYE. I know they got into a lot of trouble with, you mentioned, the Starlites they were building. They got into a lot of trouble. They were paying all their people, billing them as self-employed, and the revenue had them. Actually I would have been self-employed but as I said earlier, they had so many accounts, there was always something coming up, there was plenty of work.

When you say work as a rigger, work, from a UK production point of view, would be based in the UK? If it was UK arenas, would it be a day job and you'd go off on a daily basis wherever you were needed or would you specifically be on a particular tour?
There was a bit of both, but mostly it was on a tour joining UK and European legs of American tours, because quite often TASCO worked for many American artists.

What I'm trying to establish is the difference between the evolvement of a rigger within a lighting team, and it becoming its own main job, because nowadays it is very much that.
Definitely. Oh, I think that happened quite early on, certainly by early '80s, '81, '82, when I was working at TASCO, it was its own job. Yeah.

You might have been the rigger and helped with the focus or something of that sort. I do remember, I did an awful lot of work with Barclay James Harvest where I was rigger and lighting crew boss, and possibly dimmer bloke, but it depended on the size of the production. In those early days, the productions were so much smaller that you could probably get the rigging done in an hour in the morning, or an hour and a half, whereas, as the years went by and pretty rapidly, things got so much bigger that it was a full-time job.

When did it split between the groundsman and the roof rigger? When did the groundsman come in?
Well, there was always a need for that, even from day one.

Originally it was stage manager, wasn't it, probably?
Yeah, yeah, probably was. Again, it was an hour's work in the morning.

Can you just explain how that works for the uninitiated?
Well, to fix equipment to the roof, be it the bridles, the steel cable, the chains, the lifting chains, themselves, you obviously need to get that equipment up to the

roof. Occasionally, you have a venue where you could load that into a lift or something and take it up to a catwalk and work pretty much on your own until you're needed to pull the chains up, but more often, you'd need to preassemble cable and shackles and slings in a certain configuration and then pull it into the roof. So you need people in the roof and people on the ground, yeah.

But in the early days, let's say in a theatre, you're putting four points in for a front truss and two PA points or something, as I said earlier, you'd probably have a beam straight above it, maybe even with a sling already in there, that the theatre owned and kept there and literally you're lowering a rope in and someone's clipping a hook to it that you'd pull up and attach. In those cases, it's quite a quick job.

So strangely enough, the actual art of getting the chain hoist up into the roof was as time consuming as it was to actually get the structures up in the air themselves anyway.
It could be, depending on the venue. Some were very simple and some weren't.

Yeah, and bridling, especially, I would think.
If you're putting a lighting rig up or a backline or a PA, you're pretty much doing the same thing every day, and the rigger has always got a different building and different challenges. I think it's probably that that attracted me to it.

Sometimes getting production managers to understand that is quite difficult. "Why is it taking you four hours today?" "Well, it's not the same as yesterday!" (laughs)

Yes, and of course, nowadays, production managers and the head rigger go hand in hand, very much.
Yes, quite often. Yeah, there's quite a lot of riggers who've become production managers. Maybe they have a better overview of everything that's going on.

And possibly a sense of calm and patience. It's no good being an impatient rigger.
It doesn't help (laughs). No, but it did change. I mean, going back to the whole evolution of the thing, it changed very rapidly from when I started in '76, where there was virtually nothing flown and it was all ground supported in one way or another to quite early '80s, you know only four, five years later, it was the norm and even though, on quite a small scale, it has just grown exponentially from early beginnings to quite crazy size now.

Going back then to these early days that we're talking about, can you recall, was there someone that created a rigging company for live music production in the UK before you founded Summit Steel?
The earliest I remember is Zenavail run by Robin Elias. Along the way, there have been others, but the music side of the event business has relied on supply companies providing rigging and using self-employed riggers, so the likes of TASCO

would have their own rigging and use self-employed riggers or riggers would come over from the States with the tools. It wasn't that big a market for rigging companies to get into until trade shows and corporate events started needing quite a lot of rigging, and then other companies came up like Vertigo and the company I started, Summit Steel, because there was more of an opening.

I spent all of my life in the company trying to get into rock-and-roll, but it's pretty self-contained with the likes of PRG or whoever wanting to supply the rigging they own, although that is changing slowly.

In those early days was there someone special in your career as you started off that sticks in your head? You mentioned someone actually.
Yeah, there's two. It's got to be Eric Pearce, who ran Showlites, who's still in the business in the States, and who gave me that break and shoved me out on tours, even though I didn't know what in the hell I was doing, sink or swim, which I managed to swim. The other is Jade Dearling, who was working with those early roofing systems, the tensile roofs with Bill Harkin, who was a real mentor and took me to the States with The Who. I think those two, definitely.

A piece of equipment or technology that changed things in your business?
It's got to be the electric chain hoist, I think very basic but still around and pretty much unchanged. There are also some clever computer systems that do things now that we couldn't have dreamt of, but I think the electric chain hoist and probably, more importantly, the control system, that allowed you to run them together. Chain hoists have been around in factories for decades, but that's no good to you, you need it to be able to cable back so you could select the ones you wanted. So I think the electric chain hoist and controllers.

Finally, a funny story. Is there a funny story you still tell over the dinner table?
There have been a few (laughs), I'm trying to think. Certainly, some of the times we had at Knebworth in '79 when we had the long gap between the first and second show. The site was sort of empty, apart from a handful of us living there with our caravan. We had some very interesting times there.

It sticks in my mind one night where things got rather out of hand and we ended up with a gladiatorial match between a forklift truck and an old car. Somebody tried to do donuts with the car on stage, and I don't know how nobody got killed actually, we must have been fairly out of it. The car ended up being absolutely destroyed!

Yeah, I was going to say, I thought the forklift would probably win.
Yeah, the car was more nimble but the forklift had more destruction power. I do remember stopping proceedings while the throttle was redone with two bits of string through the opposite windows. It ended up the car went in a hole and got buried, with nobody in it (laughs). Yeah, that's definitely one that sticks in my mind.

Jon, thank you so much for your time. That was lovely.
That's alright.

After nearly 20 years working as a freelance rigger in all sectors of the professional entertainment and leisure industries, Jon founded Summit Steel in 1994 to provide well-presented and maintained premium-quality equipment and technology, backed up by extensive knowledge and experience. Company high points included winning the contract to supply and look after all the in-house rigging requirements at London's busy O2 Arena.

In 2008 Jon won the Total Production International Award for Rigger of the Year. Later that year he sold up Summit Steel to PRG and to this day remains a freelance rigging consultant extraordinaire.

Notes
1. www.acdc.com/us/acdc-today/acdcs-hells-bell
2. www.ukrockfestivals.com/76-Knebworth-festival.html
3. www.ukrockfestivals.com/79-Knebworth-festival.html
4. www.netknots.com/rope_knots/prusik-knot/
5. www.earlytelevision.org/eidophor.html
6. www.unusual.co.uk/contact/our-people/robin-elias/

Part 6
Pioneers of Trucking/ Outdoor Staging

Cockney Rebel trucking 35cwt and 3 tonner circa 1974
© Richard Ames

I have always thought that the truck driver is *the* most crucial person on tour. Reason: if they don't get to the venue, we don't have a show! Fact. Once they are there, more skills appear as the truck driver becomes the follow spot operator, or in the case of one special person, Roy Lamb, who not only co-founded Edwin Shirley Trucking but also doubled as stage manager and later on as stage builder. I never had the honour of working alongside Roy, but I got to sit down with him to listen to his adventures over the years.

When I first started as a roadie, there were no trucking companies specific to the music business here in the UK. It was very much a self-drive situation, whether connected to a big or small band. I used to hire self-drive 35cwt Ford vans, with tail lifts on the back, from a firm in Rickmansworth called Hav-a-Van. The bigger bands hired from Hertz, Avis or some other reputable transport rental company, but most of their vehicles had trailers that were very difficult to load without a tail lift. It wasn't until 1973 that IES bought their own trailer, but they didn't own a tractor. Edwin Shirley and Roy Lamb saw a gap in the market, and went for it, followed not too much later by four or five other firms, including ML Executives.

Del Roll became an integral part of Edwin Shirley Trucking, with a superb memory of the pre- and post-creation of Edwin Shirley Trucking and Edwin Shirley Staging.

14 Roy Lamb

Roy Lamb

Early April 2015 at Roy Lamb's House near Maidenhead

Thank you for inviting me, Roy. Where were you born and where did you go to school?
South London, Balham. Gateway to the South, and was born right opposite Clapham South tube station, and I lived in South London until I grew up. I went to one of the first Comprehensive's, Spencer Park, which was a big conservative big deal, you know, a thousand kids, and I enjoyed it.

Did you take the eleven-plus?
Yeah, I passed the eleven-plus, and I was a bookworm up until my seven O-levels, and then I lost interest. Got my O-levels and went on to A-levels.

What did you do at A-level?
I did economics, British constitution and maths.

Okay, what was British constitution? Like a precursor to law?
Yeah. I'd just go to the Old Bailey and to the Commons, and I did it because I should've done Sciences but I had to choose at 14, I went to Arts instead of Sciences, but hey, that's another story. Left school in '69. I had not a clue what I was going to do. At school, as a skive, I ran the technical side of the sixth form drama society.

I used to build scenery and light the shows, and we did three or four productions a year, which was just a way of avoiding lessons. When I left school, I got a place at Lancaster to read economics because I didn't have a clue what to do. In the summer of '69, the head of English at school told me about a volunteer organisation called the National Youth Theatre. I thought it only trained actors, but they'd also, as we know, train technicians. So I joined the Youth Theatre for the summer. That's how I met Brian (Croft), who was one of the professionals, and after that summer, he offered me a job with him at the ICA originally.

Where was the Youth Theatre based at that time?
Those days Victoria was where the offices were; it didn't have a home theatre.

We did Macbeth at the Roundhouse, and in the season I was there they had a couple of different shows. I don't remember them because they just took theatres for the summer wherever they could find them. After that Brian offered me a job at the ICA where he was a technical director. On the side there was a small company called ESP Lighting, which started off doing parties, deb balls, liquid light shows, that kind of stuff, that's how we started.

With John Brown.
With John Brown, indeed.

And ESP was already up and running?
It wasn't really going then, I don't know when it really started, probably '70.

When they moved to Wandsworth?
Wandsworth, yeah, I don't know when the company was actually formed but, you know, Brian and John were working together, and I did all the work (laughs).

Alistair Robertson was the one who was always deep in dye up to his elbows, and I started going with him. We started taking a couple of theatre lights in those days, 123s and 23s and stuff like that for the tents, built a little scaffold tower and put the projectors on top and had a laugh, that's how we started. Then, December '69, The Stones arrived and Chip Monck was doing the production, so he met Chip who, to my mind, is still the grandfather of rock-and-roll production in, you know, encouraging bands to actually spend money on their shows.

On production?
Right, in a proper environment instead of just turning up with a guitar.

In December '69, The Stones played Shaftesbury Theatre and the Lyceum Ballroom, that was my first show; my first real rock show was The Stones,

December '69. They sent some lights over from the States that got caught up, snaggled by customs for some irregularities, and they didn't make it into the country. So, at the last minute, we got contacted by Chip to put a rig together for those shows. Resistance dimmers, Strand Electric, I don't know if you're familiar with the old lights, 243s, 223s, 264s, you know, the old theatre-type lights, which was all that was available in those days.

We did the two shows in '69, then the 1970 European tour with The Stones (at this point Roy is pointing to the wall), that's the oldest poster I have. I finally got the American rig in, and we carried a bunch of Altman 2K Fresnels and some Lekos that came from America, and some dimmer racks that were actually built by Electrosonic, the first SCR racks that anyone had ever used, custom-built by the company.

They're only four 4k dimmers in each rack, and basically I had to hard-wire them every day, this big stack of dimmers for The Stones tour, and we carried two Gladiators, the prototype Gladiators, the carbon ones.

Which were bigger than the Supers?
Yeah, they're a Super lens housing with a 35mm projector power source, light source. There's a front half of the Super, standard Super, that everyone knows, the back half is a 35mm projector with a massive carbon twice the output of the Super, a much bigger reflector. In two wooden boxes, we carried them around Europe, and we craned them up onto the top of football stadium roofs, and carried, what was in those days the first kind of mobile rig, as we played outdoor stadiums in Scandinavia, indoor shows in most of the rest of Europe.

So were you still at the ICA for that?
Kind of, yeah, I mean, I was on and off at the ICA, never 100%, not permanently on the staff there. By that time, I was pretty well working for ESP full time.

Now, I mentioned previously I found out from Annie Pocock that you guys shared a flat. Would that be about the same time?
Yeah, it was around then, '70, '71. I moved out of the family home when I started working in London all the time. I moved out and had various flats and probably Buckland Crescent, Swiss Cottage, where I lived with them was my first flat in town. Jimmy Barnett, he stayed there with us. He did Queen's lights for a long time, he also came through the ICA originally, he was the projectionist at the ICA, so he joined the camp.

After the 1970 tour, we did the UK in '71, did the Greens Playhouse with The Stones. I did every show with The Stones—not every, almost every show—in various capacities from December '69 to December '89.

I started off as an electrician, lighting dude and finished up as a production stage manager, and '89 was the last tour I did with them. I went through the whole thing from nothing to 50 trucks with The Stones.

Going back to 1970, the equipment that Chip had brought over, the 2K Altman Fresnels and the Lekos and stuff, we used it all the time. I don't know what deal

Brian did with that. This is before PAR cans. Then suddenly PAR cans started appearing, Altmans cans, all only 110 volts, all the PAR 64s were 110 volts because they didn't make 240-volt versions.

Until Richard Hartman created them?
Yes, that was all series; you had to series them up into pairs all the time. Richard made the biscuit tins.

Yes, the square ones.
Yeah, when we opened the Rainbow version number one, with John Morris, it was '71 or '72, I can't remember, I worked there for the whole installation period where we put in a massive rig of biscuit tins.

How did you get involved then with the Rainbow?
I don't know how the connection came up. I knew Michael Ahern then and John Morris who got involved in it, and I suppose it was a connection through ESP again through lighting. I'm not quite sure.

It seems to me from talking to a number of people that there was a lot of building of equipment going on at the Rainbow at that point.
Well, the original deal was just to build the rig that lived there; we did it as you went along on the fly. The opening night of the Rainbow was The Who, and they are my current band, I'm still working for them. I spent the whole show on stage right, no, downstage left corner of the auditorium where we've got a massive battery of half of Richard's biscuit tins. Half of them weren't working, so I'm still wiring up the control circuits while the show was going on (laughs), the first show. And you know, of course, there were no computers, there were no presets, two presets was as far as you got. Then in, I might be skipping around here, on the '75 Stones tour, we had Tom Fields build a desk in America, which was still just a preset desk, but had a telephone jack, it had a patch panel on the top so we could re-patch circuits, right on the desk, instead of having to do it manually as you went along during the show, that was '75.

So going back to the history again, '71 was The Stones, '73 we did another European tour.

So you're just with The Stones now solely?
No, with ESP. I did lots of other stuffs in between. Moody Blues, Curved Air.

Moody Blues, John Brown tells me it was one of the first tours for ESP.
One of the first ones, yeah, it must have been 1970.

And then he told me that you didn't at that point own anything, you used to rent everything just for the tour.
They were basically theatre lights, because there was no rigging or anything like that, and we used to hang up stuff, you know, sometimes just putting the lights up where we could, we didn't carry any kind of rig.

The first bunch of tours that we did, I used to get up in the roof and dead hang a pipe. We just had a long ladder and we'd have to focus the lamps one-by-one on it.

We did, I think it was Jethro Tull, and we were still using these Century 2Ks, because the standard configuration in those days, à la Chip Monck again, is heavy backlight, a bunch of backlights and cross lights from Lekos and a couple of follow spots. We did the pre-standard rig and Terry, the manager of Jethro, turned up at the first gig, and he completely freaked out and said, "Why are you lighting the backs of all my band?" and made us move and put the lights at the front of the stage, which made them completely washed out and looked horrible. He didn't want his boys lit from the back, you only had to light them all from the front.

This is Jethro Tull's manager?
Yeah, Terry Ellis, I remember him making us move the whole back truss, back pipe as it was, in those days, made us take it all down and move it to the front of the stage, so all the light was coming from the front. It just looked like nothing, because he thought that's how it should be.

That would have been '70, '71. Backtracking to '70 for one second, when we did the European tour with The Stones, we carried two 40-foot containers of production. The band gear was in transit and Ian Stewart, God bless him, put up all the backline on his own. He was the only roadie.

The PA was in a little 30 hundred-weight Commer van with a bunch of WEM cabinets in it. We had two 40-foot containers, we had the All Access, the old aluminium scaffold towers, we built a whole bridge out of that, on their sides, which was, thinking back on it, probably deadly dangerous because they were only engineered to go vertical, with a tower on each side and a walkway across the back. It went up on turfers or something because there were no motors then, with a bunch of lights and curtains and stuff on it, and we carried that all with us in 1970, and no one had ever done anything like that before.

Chip Monck's ideas?
Yeah, that was Chip. Chip ran that whole show, that was his plan.

And a bunch of Americans wasn't it, on the crew?
It was, there were a bunch of Americans. They were a weird, ramshackled bunch.

So, on that Stones tour then, what was this PA? WEM?
WEM columns, yeah, and they built these subs, horns for the subs that were about eight feet wide, they were like a WEM column at the back with a huge, hook-based flare on the front of it, and you couldn't hear diddly shit. Not a hope in hell. I went to see—I digress again—I saw The Beatles twice, never did a show with them, but I saw them twice in London in the '60s, couldn't hear one note, not one word because of all the screaming kids and no PA.

Was that all WEM?
Yeah, it was, yeah, it was like two or three WEM columns a side, you know, couldn't overcome, not even close, the screaming teenage kids.

I can't say I hate music, it has given me a very good living, but it's the technical, you know, the logistic sides that interest me much more than the music, so I've been very lucky that the bands I work for are real musicians. You get what you get, you know, which is pretty good! Warts and all, which is what it should be for a live show. You want to listen to perfect music, buy a record and go and put your headphones on at home.

Edwin Shirley

So, did you become part of ESP? Were you an employee or self-employed?
No, I was always self-employed, a self-employed lampie in the ESP days, which went on until The Stones tour in '73. I can't remember what happened in '72, just did whatever ESP was doing, all their tours. In '73 we did The Stones in Europe again, arenas, all arenas this time. I was, I suppose, head lighting guy on that. I ran the board as well with some of the tours.

On the '73 tour, Edwin Shirley was the special effects man, because we had all kinds of bombs and flashes and smokes, and he was a dangerous character. I met Edwin at the Youth Theatre because he was an actor.

He was an actor?
Yeah, he was a spear chucker in *Coriolanus*, he turned his hand to anything, yeah. Somewhere along the line he got a couple of old transit vans, before we started the trucking company, he would carry for ESP; he carried the lights around sometimes for us in the second-hand vehicles that he bought. He had two 3-ton things. One was called the Edwin Bread Bin, which was an old baker's van, and the other one was a Lotus Sea Gas, which is an old sea gas vehicle they found from somewhere. So that's how I met him. He was on the periphery of us all, he never really came out on the road as a lighting guy until, as I was saying, '73, he was special effects man on The Stones.

He joined ESP through Brian?
Yeah. He did quite a few tours, you know, on and off with us, but he never really became a lighting guy, more of a driver, you know. He never had an HGV, he could only drive little vans around.

Del tells me he did special effects on Judas Priest as well.
Yeah, I think he did, that's right. So he comes out with us in '73, and he had a couple of, I think they were Avis, a couple of Avis trucks. The backline and the sound were completely separate, and the sound, that was Tycobrahe.

At the end of the tour, the bloody trucks were turning up late all the time, and the drivers were miserable to work with, and we actually sat down and said, "we can do this better, because I've always wanted to be a truck driver". I got my HGV license as soon as I was old enough; I just missed grandfather rights for an HGV by about six months. As soon as I was 21 I did it in 1971.

But in '73, they wouldn't let me drive the trucks on the tour because I was too young. The drivers of the trucks were late all the time; they had the completely wrong attitude. Edwin had an operator's license, in those days you had to have an operator's license, and you still do I think, for his little trucks. So, we were determined that we would start a trucking company to do it better, he and I, you know. And then, between '71 and '73, we did. ESP did pretty well every American act that came to Europe because we were the only people who had the faintest clue.

So we met loads of Americans, and Steve, God rest his soul, Steve Kahn, who was the production manager for Santana, we said to him, "Look, you're coming, we want to do the trucking, and we can do it better". We were a partnership then, we weren't limited. Kahn, he was a production manager, he worked for Bill Graham, and in his later years he turned into a chopper pilot. He's the guy who was piloting Bill Graham when they flew into the power line.[1]

He said, "Yeah, we will use you guys" but we didn't have any credit, we didn't have anything. So he arranged to pay for the truck rentals, and I drove one and we found somebody else to drive the other because we couldn't get an artic in those days, no one would rent an artic, so we had two 10-tonners. October /November '73 was the first tour that Edwin Shirley did, with the two 10-tonners with the ESP lighting rig and a Marshall, MEH PA. Santana and Journey, double bill, that was the first job we did as a partnership.

Were the trucks painted in the Edwin Shirley colours?
No, they were Avis, bog standard rentals, Avis yeah, so we were based at Edwin's mum's farm, where the operator's license was based. The office in London was at Bell Street, which was just north of Marylebone Road. Once we started expanding, there were trucks parked all over the place around there, it was before parking restrictions. We used to park artics all 'round the back streets!

Del tells me that they had a sort of a map, a street map, and they used to mark on it where each truck was.
Yeah, because you had to go find it. I believe Del came in '75, something like that. The first couple of years we were on our own. We had trucks scattered around the

place up there, a lot of three-tonners in those days, not too many artics. We bought a trailer, we bought a 34-foot trailer, which was standard height, single axle trailer, put a tail lift on it because there were no step frames in those days, it was a higher trailer. I parked it outside of my house in Balham and we rebuilt it, Edwin and I pretty much on our own, we redid the whole roof, which was leaking all over the place, repaired the sides, put a tail lift on it, and the first tour we did as a limited company was Genesis in January, February '74. One truck, we were a one-truck company.

Did you drive it?
Yeah, I drove it. We rented a tractor from Godfrey Davis, hooked it up to our own trailer, and I did the tour and everything was in one truck. It was ESP lights again. Ian Knight was on the tour, if you remember Ian. Peter Gabriel was still there. The Lamb Lies Down On Broadway. Yeah, we did that European tour, and that was the first job we did as a limited company.

Moving straight on from that we did the Floyd, Dark Side of the Moon, which was four artics, all rented from TIP and ELP, all in about the same period, I can't remember which one was first. Pretty well '74, January to May, that period.

There's a famous picture of three trucks with Emerson Lake and Palmer written on the trailer roofs.
That was in America, you know, not in Europe. We had four in Europe actually, and that really got us off the ground. There wasn't anybody else until ML showed up with their four trucks. When we first started the trucking company, we were all pissed off because Edwin and I had worked our nuts off trying to establish the trucking company, and then ML Executives turned up with all the spare Who cash and bought four brand-new trucks! The exact same trucks that we just struggled to put together.

There was no Transam (Trucking) or anybody else in those days, no one else had a clue about how to do it, and we found drivers who had a kind of rock-and-roll background, which is, you know, what no one else could do, normal truck drivers just didn't get it.

Something I was thinking about with trucking that I haven't asked anyone; it's interesting you point out that you had put a tail lift on the back of them, when did ramps come in?
When we built them, we started renting trailers, step frames from TIP, that would have been '74, and we got ramps from America because they already existed in the States.

We brought some ramps from the States to get started, and we started building them ourselves. In '74 you couldn't use a ramp with a standard-height trailer because it was too steep, so when we started using step frame trailers with the little donut wheels, then it made sense to use a ramp.

In the music business, everything had wheels so you used ramps, but in general warehousing they never had anything on wheels.
No one else had ever heard of them, it never occurred to anyone else.

I guess forklifts were used to get merchandise off of most trucks, or loading bays were built so that trailers just drew up.
For us you got to have a ramp, otherwise it takes too long. I mean, we do load some trucks with forklifts these days because some of the stuff is getting so heavy, but it's easier to load it with a forklift even though the trailers, now they've gone from the old step frame trailers to megas, which have even smaller wheels, but their bed height is a little higher than the step frame. The ramps are quite steep, ramps get longer and longer to make the angle. So the ramp can only fit in underneath between the landing gear and the axles, so there's a finite length for the ramp.

Now, in the early '70s do you remember Hav-a-Van?
Yeah. They were three-tonners, they all had tail lifts. We put a tail lift on our first trailer because it was just too high, you can't push that steep you know, otherwise the ramp will take up half of the truck. Then we started buying ramps from America, because they had started using them in the States by then, in the mid-'70s. We actually bought because there was a dearth of step frame trailers, we couldn't find enough because we wanted maximum cube from the trailers. We bought some trailers from America, second-hand trailers from Clark Transfer[2] that used to do rock-and-roll trucking in the States. In fact, we had a bit of a deal going on at one stage.

I remember Clark Transfer.
Matt Molitch was the boss of Clark Transfer. He came in with us; he invested a bit of money when we bought our first bus, when we decided we were also going to become a bussing company, which was a disaster, frankly. We spent a lot of money, I'm jumping ahead here, that was '76, or something like that, when we bought our first bus.

Okay, wasn't it a horse—
Horsebox, yeah, yeah. No one had one in this country, you know, crews had finally started to actually go and travel around in buses, but they all had to sleep on the floor or sleep in the seats or whatever. We heard about these buses in America, because the country and western acts already had a lot of them as mobile homes. So, we, in our wisdom, said "okay, we're going to build a bus from scratch".

Was it your idea or Edwin's?
More Edwin's frankly, yeah. He was a loony. I went to Scotland to pick up a B58 chassis from Volvo, Biggles hat on and a wooden plank for the steering wheel, just a complete empty chassis.

Oh, and drove it back down the motorway?
I drove it back down from Scotland.

Oh my God! It's like being on a motorbike.
Yeah, to Lambourne, and the company that we commissioned to build the body was called Lambourne Coach Works, who built horseboxes. They had some experience, because they used to build horseboxes for the Middle East, where the front

half of it was living accommodation and the back half was the stall for the horses. So, they had done something similar.

Roger Searle went down a different route, and he used a canal boat builder.
To do the internal, yeah, but they didn't start from the floor. They took an existing body, ripped out the inside and then refitted the interior. We decided against that and we literally started from scratch. Our second one we did do that, we only ever had two buses. Second one was a Plaxton Viewmaster that we bought as a shell. Okay, it was brand new, but it did have a shell on it. We just drove it down and fitted the interior from scratch, by the same people actually, Lambourne they did the interior for us. The first one they did the whole thing, top to bottom. We never really made the bussing work, I don't know why. I mean, we were short of financial advice, a business plan, it was just seat of our pants really.

Who were the drivers? Do you remember the bus drivers?
The first was 'Bitchin Bob' Bob Collins drove the horsebox after me. I drove it for the first year and six months.

Do you remember the first tour?
I remember running it in because we went on a journey, me and Jenny, we got stuck on bridges because we just drove it around the country to put some miles on it and tried to make it go wrong. I think it might have been Bruce Springsteen, actually, the first one.

The crew or the band?
Crew, crew, we never took a band, it was never designed for bands. It was always a crew bus. I think it was Springsteen, the very first one we did. It didn't do enough work, and no one was willing to pay, you know, it costs a lot, it cost a fortune to make, and no management were interested in paying us what it was worth.

So, it sat around quite a lot?
Yeah, quite a lot, and it needed to work 45 weeks a year. It worked about 35 or 30, and very expensive upkeep on it, things broke down all the time. You know, trying to make it something special. It was a bit of a pain to keep it up to scratch, so we abandoned them and sold them both in the end.

Did they get painted the Edwin Shirley colours?
Oh yeah, they were both purple and yellow.

What year was that?
The first one we got, I'm pretty sure it was about '75, something like that.
    Second one I think we got in '77, and we threw the towel in in about 1980, '81.

Meanwhile, the trucks are going from strength to strength?
Well, the trucking company just kept expanding, I mean, we had a huge fleet of self-drive three-tonners at one stage, Saviems, which were huge bodies.

EST Saviem Truck
© Entec

Little sleeper cabs that no one else had and tail lifts. I mean, you know, the perfect rock-and-roll truck. Yeah, we had, I think maybe 15 at one point, and they were constantly working but they got trashed. Kids didn't know how to drive them, frankly the boxes were too big, so massive they'd hang at the back, and the customers kept smacking into things at the back end. They were always overweight, and the rear axle was probably halfway along the box. They were only 7.5-ton gross, you know, with the tail lift, the tail lift six to eight feet beyond the back axle.

And they probably weighed half a ton.
Yeah, they will. We did very well out of them, there was a fella called Clint Jones who ran the self-drive fleet for a while, and they were always working. But they cost a lot of money to keep on the road, and insurance was outrageous, because they were all rock-and-roll hippies driving them. To get the insurance cover was a nightmare.

Did Willy Robertson get involved with that?
No, no, he never did any insurance for us. God bless, Willy, no. Never involved.
  So, where are we?

So you've got your license, you're driving now, have you parted company with John Brown and Brian Croft now for Edwin Shirley Trucking?
Yeah, pretty well, pretty well, yeah. From January '74 I concentrated on the trucking company. I took a couple of breaks from that. I was, yes, divorced from ESP in those days. In '75 I went back to The Stones tour of America, Brian must have been there, I was on a bit of a break from the trucking company.

As stage manager?
No, as lighting, I was running the lighting board actually, well, not really, Jules Fisher designed it, but I ran it as lighting director as opposed to lighting designer, how's that.

Did you do the petal one?
Yeah, yeah. We had two of those stages; you didn't see the wooden one in England?

I saw it at Earl's Court.
We had two of those; one was the metal one with the rams, so when the show started, it was closed up, if you remember Mick appearing on the top of it at the start of the show, and then it unfolded. And we had this other version, which was wood, it looked the same, but it didn't fold up. In America, we didn't use the metal one much, we only ever used the folding one in Los Angeles, Chicago, New York and Earl's Court—it only ever did four venues.

Because?
Because it took about six containers, it was huge, massive, big lumps of metal, too expensive. They brought it over from the States, they'd built it in the States, and they brought it over. Then we stored it down at Edwin's farm . . . for seven years (laughs). They wouldn't throw it away. Then we took the rams back, or they took rams out of it for scrap, and the rest of it went on the tip. It only ever did four venues!

Oh my God! How mad is that!
And we still had, in '75, yeah, still just a two-preset desk. We did have some more different lights; we had some weird lights Jules Fisher had put in.

Who did you use in America, lighting wise?
Tom Fields Associates did The Stones. They were it really, we had no choice in the States, Tom Fields pretty well did everything. There wasn't anybody else.

So, that's the TFA of TFA Electrosound?
Yeah, that's right. It started as Tom Fields Associates. They got gobbled up by Rikki Farr.

Now, in '79 I did a Wings tour for Harvey Goldsmith, and TFA Electrosound was the production company for that.
Where was that, in Europe? I did Wings over Europe before Wings over America.
   The band travelled in an open-top, double-deck bus around Europe. Linda and Paul and the whole band travelled in a bus at 37 miles an hour.

We're all going on a summer holiday! (laughs)
Yeah, oh yeah, that's what it was; I think they had a waterbed up on the top deck.

Now, moving on, tell me about the trucking. Did you come off the road at any time?
Apart from the two breaks, I was on the road in a truck 48 weeks a year from January '74 till '81. Except for '75, I went and did The Stones tour in America, and '77, I went to America again to do an ELP tour. They were going to do a tour with an orchestra, and I was going to indoor venues during the week and then stadiums at the weekends. That's a staging story, which I'll get back to. Apart from those two breaks, I was driving a truck and building up the company all the time. I mean, all the time literally; just came home, turn around and went back again. We still pretty well had a monopoly of the American acts coming to Europe. Part of the attraction, if you want to call it that was, if I went out with it, they got a free stage manager because they were clueless, the Americans didn't know how to deal with Europe. So I would drive the lead truck and pretty well organise the stagehands and the ins and outs, because the Americans didn't have a clue.

So, the word got around, 'If you use Edwin's, everything will go smoothly', that kind of thing. Between Del and myself, we were the lead drivers.

Did you charge extra?
No, should have done though. That was because I was basically working on trying to expand, trying to build the company up. As far as I was concerned, it was all PR work. By 1981, I've finally had enough of that, but all through the '70s, yes, that's what I did. I went out on whoever was the biggest tour and pretty well just tried to make it all run as smoothly as we could.

And how did you decide, 'let's buy another three artics', how did that happen? Did you and Edwin make a kind of plan because you were on the road all the time?
No, not really. We talked on the phone. I mean, as I think I said earlier on, we never had much of a business plan. Right through the '70s, we pretty well ran it by the seat of our pants. If we had a lot of tours coming, we never ever bought that many trucks. We did some leases, these were long time leases, we never owned a large number of tractors. In the '70s, we started with two.

It was trailers that you bought?
We bought a load of trailers because you couldn't rent them. In '74, '75, we ran out of step frames. TIP, the rentals step frame fleet, was pretty limited, and when the Americans started wanting air rides, we couldn't rent an air ride trailer for love nor money in those days, so we decided we had to build some trailers. The first run was I think eight trailers because tractors were relatively easy to rent or lease. Trailers were just like rocking horse shit, so we built them. There's still a couple working actually, I don't know if they are on the road anymore. There were two specific venues that had horrible load-ins because there was a headroom clearance problem on the doors. We built them 3.7 metres high instead of 4 metres high to

get under the doors at the Forest Nationale in Brussels—and I don't know if you remember the Olympia Hall in Munich?

Yes.
You used to not get around the corner because you hit the roof. I'm still rather proud of this. In the old days, okay, I've had a trailer down that ramp, I put a trailer underneath the cycle track in Munich before they rebuilt the load-in dock, the curved ramp that goes all the way down, I've put a 40-foot truck inside there.

It took about 40 minutes to get in because the radius is really too small. There's air vents all the way down the ramp, and of course I was backwards and forwards, and there was the building manager freaking out completely. 'I'm going to do this', I kept thinking, and I did it. I've got a picture of it somewhere, I've got a photograph of an Edwin Shirley trailer underneath where the cycle track lifts up at the back of the stage in the Olympia Hall. Yes, so we built trailers because we couldn't get them.

How did you build them?
No, no, we didn't actually build them. We bought them from manufacturer companies, two or three different people built them, and we just told them the specs and they built them for us.

So, now, you're building it specifically for the music business?
Yes, because they were air ride and no one had airbags on trailers in those days, and the first ones we built were with the low headroom.

Were air bags available anywhere in the UK?
Well, no, none of the rental fleets had them because in those days they weren't that reliable, so you just have to carry spare airbags because they would blow up all the time. Rental fleets wouldn't use them.

Would they have come from America originally?
The first ones we got came from America. The first ones, as I said, we bought some second-hand trailers, and that would have been roundabout the same time, about '74.

Because you would have nearly been teaching the manufacturer how to manufacture them?
Well, we told them how to do it with the airbags because they said, "You know, we don't do airbags in trailers", and we said, "You got to do them, why not?" The company was called Tidd Strongbox that built the first ones for us. We'd spent a lot of time talking to their engineers, I don't know if they are still going but they were quite big in those days, they were based somewhere in Bedfordshire. They built our first order, I don't remember exactly, we bought six or eight, and they were the low-height ones. Then we carried on; I don't know how many trailers we bought, probably 15 or 16 in the end. Now, they've got, I don't know, 60 or 70.

We bought two Volvo F88s in whatever 'N' registration is. GWL is the first one I still remember because I used to drive it all the time. It would have been 1975, we bought two with some finance, when Del joined. We got some finance from Jerry, Uriah Heep's manager. Jerry did a deal with us to lend us some money because the banks didn't trust mad rock-and-rollers. We used Edwin's mum's farm because we had nothing when we started, no capital, nothing to use as any kind of guarantee. Edwin persuaded his mum to use the farm as collateral on our overdraft to get us started. It's where the money came from just trying to get the company off the ground in the first place. The loan, bank loan was guaranteed by his mum.

But you pulled it off.
Well, we kind of did until it collapsed. It got to the point, you know, in the late '70s where the productions started getting bigger and bigger and bigger and more and more trucks. We were all financed by the record companies; all the tours were financed by them.

We did Genesis, '75/'76, they came out and did a TV programme on us actually, there's an interview with me in the truck, we're going through Germany, this huge fleet of Edwin trucks going across one of those big viaducts in Germany, about 10 trucks, it was a great shot, loads of EST livery going right across this bridge. We didn't turn any work down, we should have probably, retrospectively, we probably should've turned some jobs down because it got crazy with more and more trucks, harder and harder. It's all about people, the whole industry's about people, it's not about gear; you've got to grab the right people out there. I started to get woolly drivers that didn't care, it affected our reputation a little bit, I think, in the late '70s. Then it all settled down again and Del, you know, God bless him, and Ollie Kite, when he got more established, did a really good job.

Did you get hit much in the '70s by not getting paid?
We got ripped off a few times, yeah. One thing in our favour was that we got the gear in the truck. We did, on a couple of occasions, I won't tell you which bands, park the truck 'round the corner, until the suitcase arrived. We had a couple of pretty bad debts even in those days.

But for the most part that fortunately didn't happen?
No, no, it really didn't, and we were kind of lucky in that, I suppose. There were never any contracts or anything like that, we just got a phone call. That's started to change now, but in those days, there was no question, just pick up the phone, send 10 trucks. That's kind of like how the staging company started, too.

Well, I wanted to move onto that.
That's about the end of the story on trucking, you know, I drove all the time. I've got no dodgy stories to tell about it, we just drove and it was hard. Nobody cared about the hours. We did stupid overnight runs.

Paper logbooks, tachographs?
Tachographs, I don't think came in until the early '80s and speed limiters. I mean, the first two Volvos that we got, the original F88s came with 240 horsepower. Then, they bought in another version that had 280 horsepower. The first two that came into England we had, Volvo had done something to them, they didn't quite get the back axle diff right, they put a coach diff, or something in the back axle, and the ones we had GWL and KJA would do about 90 miles an hour.

We used to blow up the tyres all the time because we had the step frame trailers with the tiny, tiny wheels, they got the tyres just red hot, so that the tyres just literally blew up. Especially down in southern Europe around the summer. Not that I ever drove at 90 miles an hour.

F88s were the ones that ML had as well, didn't they?
Yeah, they had F88s. Anyway, 1977 I took a break from trucking to go to America. Michael Ahern, who is a connection from the old days at the Rainbow, a John Morris, Bill Graham, kind of good American connection, he was putting together the ELP tour in the States, which had interior shows during the week and football stadiums at the weekends, with an orchestra just at the weekends. So we're doing outdoor shows and we're going to build, you know, a stage in these stadiums, because I've done The Stones in stadiums and stuff like that. Mike Brown Grandstands was the scaffolding company in America, he said, I want you to come over as the, in those days, it was called technical director. Now, it's called site-co.

The one that's supposed to be representing the band, making sure that everything was ready for the production to come in on the Saturday morning. That was all well and good, except that the ticket sales were shit. In the end, they actually scrapped most of the outdoor weekend shows, decided not to do them. Mike Brown, on the other hand, his staging company was really busy, and they said, "Do you want to just stay and work for us?" So I had a crash course in stage building, stayed in America for the whole summer in '77 and built stages all around the States for Mike Brown. We did about half a dozen ELP shows in the end, that's all. The rest of the time, I was going from here to there and everywhere, and I was also driving the truck.

All for music concerts?
Yeah, all gigs, yeah, all in stadiums. Lots of Three Dog Night, Dylan, I'll tell you the Zeppelin story in a minute, Seals and Croft, a whole bunch of acts doing outdoor shows in '77. I was going from here to there driving the truck and building the stages with a Mike Brown roof, called a Phoenix roof, which is the first one we had over here, and it was made for A Star Is Born. There's a picture of a big outdoor stage in Phoenix and that roof, that aluminium structure, was built by Mike Brown. It became the standard touring roof for outdoor shows because it was load-bearing with steel towers and aluminium roofs.

Steel platform?
No, a wooden stage with steel underneath. Steel towers, motors to lift the roof up, put the lights on. I spent the whole summer in '77 doing that. I did a show in Amarillo, Texas, drove back from Amarillo to Los Angeles to reload the truck,

which is 1,700 miles on my own. Reloaded the truck, a guy came in with me and we drove from there, and we were on our way to Buffalo, New York, 2,800 miles away, to do a Led Zeppelin show. Got to about 9 or 10 hours out and we heard Robert Plant's son had tragically died that day, while he was there on a Zeppelin tour in the summer of '77. So Mike said, "Okay, take the truck and park it at Buffalo airport, get on a plane, there's another system sitting in Miami". He was trying to put another system up first thing the following morning. I'd say they were working me to death (laughs). He was a lovely man, Mike Brown, he was our partner when we started the company in 1980.

Where was he based?
He's based in Los Angeles. His company used to do the Rose Bowl and the seating, that's why Mike Brown Grandstands was the name of the company. The Americans somehow found him to help to build the towers and things like that for the first outdoor concerts. Yeah, I'm not quite sure how that happened but he was involved, he was it, really. They designed this aluminium roof grid for the Star Is Born movie, and he built a bunch of them, so he had, pretty well, the monopoly over there. It was just about the load-bearing capacity, no one could put a lighting rig up because nobody's roof was strong enough to hold it, so he came up with a pretty decent bit of kit, but it certainly wouldn't work these days but in those days . . .

What was it? Aluminium truss?
Trusses, yeah. It was folding trusses. There were two trusses hinged at the top so you could flat pack them, they were three feet, pretty deep because it was a 60-foot span in those days, you know, to get any kind of load capacity. They were probably the first ones with maybe two and a half feet, but after the load started going up, they're four feet deep.

So what, 60 foot across just 2 by 30 foots?
No, no, they were just plugged together. They were all in, I think, they were 20-foot sections so 3 by 20s. Maybe 60-foot spans just put together, jacked together. Fleetwood Mac in 1980, came to Europe, Chris Lamb was production manager.

No relation?
No, no relation whatsoever. Chris Lamb, they'd used Mike Brown in America, to do all of the US shows; they couldn't find a stage over here, and they were going to do outdoor shows in Europe. No one could come up with the roof capacity to hold the rig they wanted to use. So Chris called me and said, "There's two roofs in two containers, on their way across the Atlantic", just basically, the aluminium grid structures. "Get two scaff systems and stages together, and we'll see you in Munich in two weeks". That was how we started the staging.

Did you get given drawings?
No. I'd done it in '77, so I just basically duplicated the way we did it in America.
    Yeah, we sent two trucks of scaff out to Munich, set up a production line because then, we built the decks with eight-by-four sheets of plywood, and put

two-by-fours of each underneath to make them strong otherwise they won't be load-bearing.

We just put one on top of the other, and interleave with two-by-fours, so it didn't take up too much room. We had to build a few special bits and pieces to support the decks and stuff like that, but we basically built the whole first system in two weeks for the Fleetwood Mac Tusk tour in 1980.

With John Courage as the tour manager?
Yeah, that's right. So that's how the staging company started.

Fleetwood Mac was paying Mike Brown for the roofs to come over to Europe?
Well, no, we did a deal with Mike Brown. We went into partnership with Mike Brown.

Oh, okay. Sleeping partner?
Yeah, he and I, we had a financial arrangement for him to get a guaranteed amount of rental for his roofs that he sent over to us. He did quite well out of it in the end because those two roofs were getting a bit old by that stage, but we knew that, and of course his goodwill, because anybody he worked with in America would automatically end up coming to us in Europe.

1980, Fleetwood Mac, is that your recollection of the first outdoor arenas in Europe?
No, because we did football stadiums in 1970, with The Stones, we did the same venues.

No roofs though?
No, there's no roof. As I said, we had this aluminium scaffolding set-up, which is highly fucking dangerous, to hang a few lamps up at the back, and it was only 24 lamps in those days, no rain cover, no nothing. If it rained, it was all over.

1980 was the first time, as far as I know, that anyone toured a waterproof, proscenium stage set-up. There wasn't anything. There might have been one-offs in Wembley because, as I said, we did Wembley in '72/'73 with The Who, but I think we just hung stuff off their roof. I can't remember what we did in those days. But that was the first tour, the first touring rig for outdoor shows.

How long did that take to set up?
It wasn't that complicated, you know, because it wasn't very heavy.

Did you have two because you were leapfrogging?
Yeah. We could do it in 24 hours 'round the clock, but two or three days then.

I mean, that's four or five days with the massive new towers systems. If you think back on it, in those days, a 60-foot-wide roof and then two scaffolding towers of PA, so the whole thing was 100 foot wide. Now, it's 250 feet wide and 75 feet deep.

It's monstrous now.
60 feet by 40 feet was the performing area and that was it, in those days. That was the roof, 60 by 40, and everything fits under it. We did use the same stuff, the same roofs in '81, for The Rolling Stones tour, and I did the same thing. I was on the tour, I was actually the stage manager, but I was also overseeing the stage roofing, the staging set-up and the trucks.

Yeah, I was going to say. When you went to Europe, you're in charge of the trucking, putting the stage up and the roof and making sure it's fairly watertight?
Yeah. Well, not really because I mean, by the time I got there, it was all up. I think we had three teams in '81, we were leapfrogging so much. We had two sets of production in '81 for The Stones, two sets of sound and lights, three stages at least, sometimes up to four, I think, because we're doing so many shows.

We were playing darts with Bill Graham and Michael Ahern because it sold so well in '81, I had this great big whiteboard on the wall with a map of Europe and Bill was like, "Right, we want to do a show here". Of course, if you had a show there, you had to look at the ramifications.

Oh, absolutely, the logistics rule.
And if one system, you know, suddenly has to go there, something else becomes impossible to do, so then we can't do it. Unless we bring in another system to cover that one, that's probably why we got four, four sets of stage at one point.

On Supertramp we had Roger Searle site-co stage one, Chris Adamson site-co stage two and Wolfgang Koln the third.
Ahh, Wolfgang Koln, that sounds good.

So that staging really sort of set you up really for Europe after Fleetwood Mac and The Stones?
Yeah, yeah, pretty well, pretty well.

Who was your first competitor?
Stage Co, and they never came along until the towers, till the scaff got outdated. There were a few local companies, always local companies in Germany and Switzerland that did it. Harkins' hippies did it as well, the guys from Somerset, they did the '70s summer Stones tour with us. There's the guy who designed the Roskilde tent, with the tensile roof.

Jon Bray tells me about the '79 Knebworth roof with Bill Harkin.
That was an air-roof, wasn't it? The thing about the tensile roofs and the blow-up roofs is that they're not load-bearing, there's not a lot of capacity. You can't take a normal lighting rig and put it up there because of the way they're constructed.

So explain to me the difference between an air-roof and a tensile roof.
Well, tensile roof is basically a circus tent, so it starts high and then, drops off like the hoop ones and drops off really quickly on the side.

If you're trying to fly up a sensible-sized lighting rig, you can't do it because you hit the sides before you can get it up, you can't get to trim height. Inflatable roofs are okay but they're not load-bearing. Basically anyone can build an inflatable roof and turn the fan on; you don't get wet but you can't hang anything on it.

Basically, an air-roof is just a skin.
That's right, yeah, yeah. The Roskilde roof and the Harkins' hippy roofs were like circus tents, and you couldn't hang anything, really. Their sight lines are terrible because if you imagine these cables almost coming out into the audience and like, that horseshoe, at the front, just trying to look to see from the side, you couldn't see anything because your proscenium opening has closed right down.

So on these original roofs in '80, the roof itself was aluminium and some, sort of, box truss design?
Yeah. Held up by steel scaff towers.

Literally, just like I put up to fix my chimney on my house?
Yeah, yeah. There was some special stuff made because of the motors, we always lifted them up with motors. We had to custom make some bracing for the cantilever system, you know, to get the roof up. We always built the roof, take it up to head height, put the lights on and take it up the rest of the way, and you're ready to go. It changed to towers in, well, I don't know, certainly there weren't any towers holding up roofs in the '80s. Would've been the '90s before towers started.

And what was the advantage of towers over scaff?
Well, it's mainly the load, we started to overload with scaff because two-inch scaff pipe, it's only got a certain amount of load it can take, and the PAs and the lighting rigs are getting heavier and heavier and heavier. We ended up having to put three or four legs all next to each other and put right weird connections to try and spread the load. The other problem is access. The main performance area, okay, it's clear, but the sides, because there's more and more weight upstairs, that means more and more scaff legs. It turned into a wall of scaff all the way down to the side of the stage so there's no access on and off stage, monitor world, guitar world, all the stuff that lives out in the wings, there was nowhere to put it because it was a forest of pipes. Putting in really heavy-duty towers takes all that away because you put a really heavy-duty tower in front, heavy-duty tower at the back, pick the roof up, there's a clear span between the towers.

Then, all of a sudden, you've got space.
So you can build a wing area with complete roll-on access, especially with festivals where you've got risers having to go on and off stage; take away all the scaff and we would increase the access all around for everybody and all the gear. Changeovers now have access to the back and the side and everything on festivals these days.

Back then the loads, the weights just got higher and higher and higher, and the scaff just couldn't deal with it, just couldn't stand it anymore. You imagine the weight on the downstage left and right corners; you've got the weight of a quarter of the lights hanging in the roof. You've also got the whole PA all bearing down on each downstage corner. You know, it just got too much to handle.

Dodgy moments?
Too many. Bent, bending, breaking, all of that, yeah. I'm only going to tell you one story . . . I got a call from Mike Wiseman who was the stage manager for Rod Stewart at that time. It's a bloody miracle, but we were in, I was asleep, 3 o'clock in the morning, right, this is Mike on the phone, "We're in Copenhagen". "Yeah, yeah, what's up?" "Well, we're trying to lift the roof up and it has so much weight on the roof that the motors are running, the motors are going, nothing's happening, and the cantilever beam at the top where the cable is coming down to pick up the roof, in the end it just snapped". So the roof fell. But thank God, no one got hurt. Like it fell and landed on top of a box.

Like, six feet?
The box? No, not that much, two feet, three feet. But the problem was we had a show, you know, the next morning. So what are we going to do? Well, an absolute bloody miracle, we had another system also in Copenhagen that was going to start building the next day. We've got two crews, one to do the Rod Stewart show, another crew to do . . . might have been ELO or somebody like that, anyway, on the other side of the town.

Wow! Lucky!
I said, "Quick, get in the truck, wake the driver up. Get him to come over with these particular bits". Because it wasn't actually faulty, it wasn't overloaded because we did do engineering on it, it was just the gear was rusty, old, you know, some of the bits we had, had been around the block too many times. So we got some brand-new stuff that was on the other system, took it over to the Rod Stewart show and it bailed us out at the last minute.

(Roy has just touched wood. . .)
When I was involved in it, when I was running it from '80, for the first four or five years, '80 to '84, we never had a fatality. We had some accidents, you know. It's a building site, especially in the time of 'everyone wants everything done, yesterday'. You can't expect them to get away with no accidents, but I mean, no one got killed. We had some equipment failure, and every now and again, we just had to stop and say, "no, this doesn't look right, rethink this a bit", we re-engineered everything. At the start of the '70s, nothing got engineered, we just made it up, no one did any calculations on anything.

Were you trained in that sort of field?
Never. No, never. It's all talking, just sitting down talking to engineers and common sense.

Did you cross paths with Richard Hartman on this sort of stuff?
Yeah, Richard physically built the second generation of our roof trusses. The original ones that we'd got from America were of a certain standard, as I said, quite small. When the weight loads started going up and up, we decided to upgrade, build new ones, and Richard physically built them for us. He built two or three roofs, I think it was for Genesis actually because there was no way that we could get the weight of the Genesis rig underneath the Mike Brown roofs. Yeah, so he built them, physically built the second generation of roof trusses for us.

So, where were you keeping these?
Well, we moved at that point, we moved to Crow's Road, so we kept it all there.

So, you just took them apart and they're all in bits?
Yeah, but most of the scaff was rented. The only thing we had was the roof, the physical roofs, and all the special items that we had to design; that's extra supports for the main stage area because the scaff that we were using didn't fit the German specs, it was all right in England but it didn't work with the German specifications, their rules are ridiculous, twice as strict.

We had to have extra, extra support. We had to design a load of extra pieces, widgets, that we could fit onto the existing scaffolding to give us extra bracing to the places where the Germans said we needed it. We had to buy those because we had stock scaff. It wasn't actually just straight scaff, it was system scaff.

What is the difference?
Normal scaff has just got clamps, swivel clamps on it. System scaff has got fittings welded to it and hooks or U-shaped fittings or round fittings and it's a system, so you don't need to put a clamp point everywhere. It just clips in, bangs in with a hammer or something that goes over it, I can't explain it really. There's two or three different proprietary systems, it's like meccano really, every half a metre, there's a fitting.

So where did the guys that put these together for you come from? Were they from the lighting world?
Most of them, a lot of them were, a few were scaffers, riggers because there's a lot of climbing, a lot of high work. On the original stuff, not so bad now when it's all towers, but the original scaff used to be 85 feet in the air.

A lot of riggers came and worked for us. And a lot of the stage crews, the humpers and the guys from Stage Miracles and from Show Stars, they would come and do a couple of builds with us and learn how to do it. Then they'd end up coming to work for us as the team because they'd learnt how the system worked. Quite often, in Europe, a lot of the local labour the promoters would supply us with sometimes were not up to it, so we'd take a busload of scaffers, we call them, you know, from England and they'd follow the tour around and get paid, camp out in the bus. Get paid a reasonable amount of money and follow the tour. Then a couple of other local crews got wind that that was quite a good plan and there were two brilliant crews, one from Hungary and one from Poland. So whenever we went to Europe we would bring those crews in because they knew how to do it and were mad for the work, and they could do it 10 times faster than anybody else and they had a good attitude; the Germans and the Italians didn't want to work.

But, nowadays there must be literally hundreds of crew doing this through the summer—
It's all changed now because it's not scaff anymore. It's much more specific. You've got the towers, the tower systems are all ridiculously heavy, most of the manual labour's gone out of it now. It's all crane and forklift now because the pieces are too heavy, and unmanageable, the weights are going up ridiculously.

How many roofs did you end up with?
I think we ended up with three from Mike Brown and probably had six in the end, six before we went to the tower system, which again was after my time. That was in the '90s.

So, you stepped in really in 1980 with Fleetwood Mac and then stepped out when?
Well, after about '85/'86, I'd had enough of it. I handed off the staging to Henry Crannan and Graham Fleming, those two guys were running the staging side. After that I pretty well concentrated on just doing production. In '88 I started working for Robert Plant, I did all the production management, so from '88 onwards I've concentrated purely on production. I became a sleeping partner, you know, in the trucking company probably in '81; I have nothing to do with trucking whatsoever, just kept my shares.

Built up the staging company till '85-ish, then backed out of that as well. Kept my shares in the trucking company again until '90, I believe it was '91 when Ollie and Del and Tim Norman offered to buy me out. They had to buy Edwin out because he had to go.

Oh!
A bit of a political issue, so I went at the same time. Edwin Shirley sold his shares in the trucking company at the same time as I did. I believe it was '91.

Okay. Did you have anything to do with the setting up of Rock-it?
We started it. I think this is probably public knowledge.

All right, last thing really. I just wanted to ask you how Chris Wright came into your life?
Okay, well, he approached us, because I remember he was working for a company that were doing band gear shipments backwards and forwards to America. Now,

I can't remember the name of them. He was pretty unhappy with the life there, and he came to Edwin and I and said that he wanted to start up on his own, and he said, "Would you be prepared to come in with me and underwrite my wages while I get established?" So we agreed. I can't remember, I can't tell you the exact way the company was broken up, but for the first, I believe, two years we're actually in partnership with Chris in Rock-It Cargo,[3] then after I believe it was two years, he bought us out. The rest is history.

Do you know why he came to you guys?
No, I don't know, to be honest. I know that he knew that we dealt with all the American acts; he knew that we had a lot of contacts, lot of connections. I think he thought that it was a good move, you know, to use all our contacts, yes.

And you guys obviously thought it was a pretty good idea at the time?
Well, we did because you know it was very frustrating trying to deal with the American acts because their gear constantly got held up here and there and didn't arrive, ended up at the wrong airport or whatever.

So, then you are the man on the ground, in the airport?
Yeah, that's right. It would make it a lot easier for us because, as I've said, in those days we pretty well had a monopoly on dealing with all the American acts who would bring their own backline, but never brought production in those days, they'd use local stuff but they'd always bring their own backline.

Yeah. It all comes down to logistics again, doesn't it?
Yeah, it's all about logistics.

You are Mr Logistics. I'm going to ask you three questions I've asked everyone. In your opinion, I'm being specific to 'in the beginning' really, was there any one person who remains in your head as a 'guiding light', somehow affected you?
Yeah, originally it was Chip Monck, no question. And Brian, you know, Brian Croft guided me along the road to get started, but you know that was more of a father figure, I would say. Chip was a guy who had the 'balls' is the word to go to the artists and say "I want you to spend money, I want you to spend some money on production" and basically encouraged the artists to see that it was a worthwhile thing to spend their money on using production, which of course produced all our income, from every aspect, trucking, lighting, everything. If the bands didn't carry production we wouldn't have any work.

Yeah, very true.
And, you know, the figure in America who in '81, a little later, was Bill Graham, he was another one, he had an interest in producing shows of a standard that, you know, instead of just turning up and playing a few tunes, gave the punter something for their money.

Yeah, and he created the Fillmores, of course, both West and East.
And he was the producer of the '81 Stones tour, which probably was, at that time, the biggest production that had ever been on the road in Europe.

This is difficult because you've got a trucking hat on and you've got a staging hat on, to a certain extent, but from your lighting days, can you think of any one particular technical thing that changed the way of touring shows?
Well, from the lighting world, it has to be the moving light, no question. From the trucking side, trucking is trucking, there's no particular thing. I think the only thing that killed us was the rules and regulations, which are ridiculous as they gradually got introduced.

You're not the only one who's said that.
They are not designed for rock-and-roll, they are designed for the drivers who drive non-stop to Spain and turn around and come back again. Our drivers have more time off than pretty well any other drivers in the world, you know, and still break the rules all the time, because they're not allowed to move for half an hour.

The big change on staging was the tower system, but that wasn't until the '90s. And the '80s motors really. Chain motors, when they first came in, that made a massive difference, no one had ever thought about it until, I don't know, I think it came from America. It might have come from Chip or someone, the original concept of hanging up with chain from the roof to pick up a lighting rig; it was like a bolt of lightning. And, of course, the computer, I mean the whole world changed with computers, but that's later.

Do you like computers? Does that make your job easier?
Absolutely. Makes it more productive. I don't know how we survived without them.

It was a diary; it was paper, pencil and a telephone.
Here's another little aside from logistics again. Must have been, can't remember when it was, late '70s. We used e-mail and the Internet before the fax machine. There was a small group of us through British Telecom Gold. British Telecom had a system, and it was linked with a system in America, AT&T. It was called DGS, and it was a closed user group where we had e-mail addresses, we could send each other e-mails including drawings, any file, and it was done with an acoustic coupler on the telephone. It was before the fax machine.

Yeah, yeah, what was it called?
Tandy TRS80s, little computers, it was called DGS, there were a bunch of people, all in the music business, in England, the US and Europe, they were on the system; you could send each other, it was all text based, it was before the mouse. I used to use it, I thought it was fantastic. I used to do training sessions because a lot of the band's management companies started to use it, and I used to go and train the secretaries how to use it. Telex was the only other system. Then the fax machine

appeared, and it died, just like that. And then the Internet appeared. We were on the Internet before the Internet!

Roy Lamb, thank you, very much indeed.
(Laughs) I hope you get something out of that.

At the time of this interview, Roy was preparing production for The Who's next world tour. Roy has continued to freelance as a production manager for many years.

Notes
1 www.nytimes.com/1991/10/27/us/bill-graham-rock-impresario-dies-at-60-in-crash.html
2 clarktransfer.com/about
3 www.rock-itcargo.com/rock-it_history.html

15 Del Roll and Ollie Kite

Del on left, Ollie on right

July 17, 2013 in Royal Docks East London with Del Roll and Ollie Kite

So, Del, where were you born, where did you go to school and when did you leave school? Just a little background, please.
Sure. I was born in 1946 in Edmonton, North London, the same hospital as Bruce Forsythe, one of my few claims to fame, and then was brought up, dragged up in Tottenham. Went to school in Tottenham, and the establishment was called Downhills, which was very appropriate at the time. Yeah, Downhills so much so that when I left at the age of 15 they didn't even have a prize giving because the results were so bad. Left school and already had been taking drum lessons and was in a band.

'61, right? At the beginning of the '60s?
Yeah, yeah. My main aim from leaving school was to work in a drum store, primarily because you used to get drum kits trade price. There were no vacancies, so I took up a job in a saxophone shop called Bill Lewington[1] in London, Soho. After six months a vacancy came up in a nice drum shop right next to Piccadilly

called Footes.[2] I worked there for a couple of years, learning all about drums and meeting a lot of the '60s bands, including, well, from Nero and the Gladiators onwards really, which was a real education, and a lot of jazzers. July 30th or 31st, 1965, I hung my coat up, got on my motorcycle, went to Catford.

Oh, what motorcycle?
It was a Triumph Tiger Cub, motorcycled to the Tiger's Head in Catford, played a gig with my regular band then. and that was the turning point. We all gave up our day jobs that evening, did that gig, Saturday morning, the following morning, got in our van, drove to Harwich.

What was the band's name?
That was called the Chevrons, five-piece band, we had our manager with us called Barrie Marshall, he came with us. It was that long ago that we actually didn't do roll-on/roll-off ferries, they actually craned the van onto the ferry. How did we know Barrie? Well, his wife Jenny, and still his wife Jenny, was the singer in our band at one time. Funnily enough, I was her drummer, and we used to do lots of American air bases and gigs such as like that. She was a superb singer, no question.

Now was Barrie managing more than one act or just you and your band?
Well, he worked for Arthur Howes Agency as well I think at the time. He managed a couple of other bands, one called Mark Four (who became The Creation), who had a bit of a hit with *Making Time*. Anyway, he got us a gig at the Star Club in Essen, Germany. So we all legged it over there on that Saturday, and just changed in the street because the dressing room, which was also the accommodation with another band, was so grim. At that time we were a Tamla-Motown type band, and northern Germany was still rock-and-roll, so to be honest, we didn't go down that well. The club owners said "No problem, we're going to ship you down to Frankfurt", and the following day we ended up in a little town called Giessen just outside Frankfurt that had an American air base, right, and an army base nearby that was more hip to Tamla-Motown music. We played there for nearly a month, and it's the old story really, you're on duty for 9, 10, 11 hours a night alternating with another band. The good or bad thing is that the other band was German but the singer was Scottish, and he knew the ropes. After three-odd weeks it became obvious, we were not going to get paid.

Oh, no!
Yeah. We couldn't report him to the police because no one had work permits or visas. So we agreed that we were all going to leg it in the middle of the night. The Scot said he was going to go north with his band, and we said, "Well, we'll go to Frankfurt, Hauptbahnhof", the main station there. The theory was that people would speak English, and there would be phones and facilities that we could use, even though we didn't have any money. So we drove to Frankfurt, parked up, went into the station and ordered a massive meal between the five of us. We had no

money at all, I rang up our manager, Barrie was our agent at that time, so I rang up our manager, and said, "Well, someone needs to come out because we can't even escape from the restaurant right now". Manager comes out.

Comes out where?
To Frankfurt, he got on a plane, flew out!

While you were sitting in the restaurant?
Yeah, we sat there for 12 hours.

Amazing they didn't call the police.
Well, yes, that's true, we were well behaved. All long-haired Englishmen. Our manager said, "You can now go to an American air base in the middle of France, but you have to have your hair cut". We all went around the corner to the station, had our hair cut, got in the van, drove to the middle of France, and was greeted by the sergeant major or the staff sergeant there who was in charge of entertainments. He could see we hadn't eaten too well. He put us in a car, a big American car, we drove off to a 2 o'clock dive where the only food was snails and bread, which we actually enjoyed, to be honest. Yeah, it was first time I'd ever been over a 100 miles an hour in a car, so it was an experience.

Anyone that stayed in the business?
No, in the nicest way, no, one's died since then. One is in a country western band, the other is a coach driver and the other guy is retired, yeah. No, they didn't last the course, but just to cut a long story short, we played at that air base for a month and was told naively to go to Paris at this address to collect our money.

Oh, no!
Yeah, I know. We end up in Paris, slept under the Eiffel Tower in the van, which you could do then, and of course, there was no Parisian address. Drove to Bologne and we all sold our watches to people in the queue, to get tickets home. We got on the ferry, I dropped everybody off home, got home, had a good sleep, went out the following morning and ran out of petrol. That was our introduction to professional life.

So when did you leave and when did you get back?
Well, we left say, the 31st of July, and we got back couple of months later. It was July, August and September. It was still fairly warm, I remember that.

And pretty much the whole time you never got paid.
Correct. Just character-building stuff. Then to fast forward, we played up and down England, and then we were joined by a saxophone player who was in a band called The Riot Squad which had just broken up. He had the name registered and was looking for a ready-made band. The Riot Squad had had a couple of singles out; their drummer was Mitch Mitchell of Hendrix.

Wow, yeah. Before Hendrix?
Yes, indeed. So he joined us. We took the name, we did a few gigs and then David Bowie joined the band as the singer for about six months.

What year was that then?
Oh, '68, '67/'68. Yeah, he came along, he was looking for a vehicle for himself to progress in what he wanted to do, which was like the Fugs and Velvet Underground sort of music.

He was young then.
Yeah, he was. We all used to sleep in the back of a transit together, funnily enough. Yeah, early days.

Really? He wasn't called David Bowie then?
Yeah, he was. It was after Davy Jones, he was Bowie by then. He used to come around to our house for tea and all that, you know, kind of sweet. Bowie joins for six months. We knew he wasn't going to stay, and he knew he wasn't going to stay. Almost the first gig we did with him was in a youth club in Basildon, Essex, and we were the opening act for Cream of all things.

Wow!
I mean, how bizarre is all that! That's another war story. Anyway, David left and we carried on and we then hooked up with a record producer called Joe Meek, famous for the Tornados, et cetera. We recorded seven flops with him, it got us lots of work, you know, lots of Radio Caroline and lots of Radio London.

Still called the Riot Squad?
Yes, indeed. We toured a lot, which was good, got a bit of a name, you know, as a sort of C or B division band. Then poor old Joe had a bit of a turn, shot his landlady in Holloway Road where his studio was because she had the cheek to ask for rent, I think. Then he shot himself.

Both dead?
Oh, yeah, yeah. The band kind of split up after that. It was a six-piece band, we went three each way. We re-formed and then ended up with a guitarist called Len Tuckey who was married to Suzi Quatro. He was with the Nashville Teens. We went on and on doing Germany.

Len Tuckey, I remember him doing gigs with Cockney Rebel supporting Suzi.
Yeah, there you go, there you go. He's still around. I think he manages Slade now and plays. Long story short, Richard, I end up giving up playing because of being dicked around by band members that didn't really want to rehearse. So this is where it comes in, that's my playing career over. Part two, I gave up playing. I sold my drum kit, needs must! The manager of our band was also the manager of the Keef Hartley band, the old blues drummer.

I saw them as a school kid.
Yeah. He had just left John Mayall with Eric Clapton in the band, I believe, and formed his own band, and they were looking for a road manager, which I absolutely did not want to do. It turned out that I was asked to drive him around to various shops and things, and he was actually auditioning me as a driver. He wanted to make sure as a drummer I didn't get on the kit every time I saw one, which would have pissed him off (laughs). Anyway, I got the audition, I ended up driving the Keef Hartley band around, which was good.

What year was that?
Maybe '69, something like '69, '68. I could check on that.

Yeah, Oxford Town Hall around 1969 when they played there and I saw them. I probably just saw you on stage when I was a school kid coming along to see the show.
Yeah, well, by default, I ended up being the soundman as well. So I become road manager, which you'd call tour manager now, driving the band around, keyboards, trumpet, sax, lead, bass Keef and Keef's dog called Blue, the big Alsatian used to belong to Gino Washington, funnily enough. Anyway, it became a bit of a challenge, much as I loved Keef, he was great fun and great education for me, driving six comedians around wears a bit thin. Because fuel was so cheap, you'd come back from Sheffield, to go back the following day to Leeds, so apart from all the driving and setting up the gear with the roadie at the time, you have a 100-mile trip almost, picking up and dropping the band off, because they lived all over the place.

What were you driving around in?
That was a transit as well, 12-seater mini bus. The gear was in a separate van. I saw what the road manager was doing, which was not very good, in a nicest way, not very organised, and I said to Keef, "How about I swap? I look after your gear, your man drives you around". And he said "Absolutely". So that's when I became a roadie per se, up and down England with the dog. I always ended with the dog.

Driving in what?
Another transit.

Okay. No seats in it.
Correct, yeah, all the gear. Also one of the managers to the band, he had a private pilot's license, so we'd end up doing a lot of festivals where it would just be myself and the band, and maybe, you know, the saxophone, trumpet and a cymbal bag. We did a lot of flying around Europe, which was great, good education.

In a private plane?
Yeah, yeah. Jon Hiseman's (of Colosseum) wife was with the band at one time, Barbara Thompson did a lot of festivals, which was great, Edgar Broughton

band and all those sort. Free were on the bills. Good education. In the end the band split up and Keef, well, couldn't keep a crew on or anything. He didn't know what he wanted to do, and so I started doing a bit of freelancing, driving bands around, for Al Kooper, a duo called Hardin & York, remember them?

Yeah, I remember them, yeah.
Eddie Hardin, who took Steve Winwood's place, and Pete York from Spencer Davis Group. Anyway, Keef broke up the band, I did freelance work and then I got a call from Uriah Heep's management.

Hang on a minute, Keef Hartley Band—did you do the second Bath and West Showground festival? They were on the bill.
With Led Zeppelin on it. Yeah, I did that one, yeah, I think we did a gig that night as well somewhere, maybe in Bristol or something.

1970 Bath and West Festival poster

Because you were one of the opening acts on the Friday.
Yeah, yeah, it was fabulous. I'm still trying to get a poster from that. Anyway, I got a call from the Uriah Heep office. Oh, rewind a bit, the attraction of joining Keef's outfit was that they had just done Woodstock and they were going back to America.

The only thing I wanted in life was to go to America, which of course he never did. Anyway, Heep's people ring me up and say "Come for an interview with Ken Hensley, the keyboard player. If you're okay on this Italian tour we'll take you to America". I said "Fine". The Italian tour, Boxing Day, a 7-1/2 tonner comes to pick me up in Enfield with two roadies in it and immediately they say "You're driving". I said, "Why aren't you?" "Well, we don't have licenses", "How did you get here?", "Well . . . not very easily!"

Well, without being stopped. You had a license for yourself?
Yeah, yeah, yeah, I did by then. So Boxing Day we left.

Why? Sorry, why did you have a license?
I thought it'd be useful! In a minute you'll find out how I got an HGV, which is even more useful. Anyway, we drove to the Belgium-German border, Boxing Day and, of course, it was shut. So we sat there like idiots for 12 hours in the snow to do the paperwork, drove down to Italy, pretty much on my own while the other two slept.

So we're in 1970/'71?
'71, yeah, now late December 71 when we do an Italian club tour, which culminated at the last show (remember no mobile phones, and it was with very dodgy promoters), the last show was a club, I can't remember the town, but I went in there and the stage was full of band equipment. I said to the club owner, "You need to remove this to let the band on". "But this doesn't move", he says "You set up in front", which is only two or three feet. Even then the band had a lot of gear, so I said, "Well, we can't unload until that's moved, end of story". We can't get hold of the band or the road manager, who was Mel Baister by then. I don't know if you know that name.

Del in Rose Morris advert

Yeah, yeah, first time I went to America in '74, Wishbone Ash, Mel Baister's my boss. There's a coincidence!
There you go, he's still around. Anyway, I couldn't get hold of the band, they'd left the hotel and, as you know, it's a pocket full of coins in them days, *gettone*, they used to call it, were coins for Italian phones.

Gettone, Italian telephone coins

Suddenly, a big black limousine turns up in the car park, which was like an ice rink. Obviously it was a "connected" person that got out of it. Johnnie Allen, my old partner in crime on the road, and who is still on the road, bless him, he went to explain, and the next thing I saw from sitting behind the wheel of our truck was this guy got a gun out and started pistol whipping poor old Johnnie. It was at this point I started the engine and drove at the promoter kinda foolishly. The promoter then looked at me, looked at the truck and just fired two shots at the truck and burst a couple of tyres. At that point I stopped and we decided to unload. We removed the gear off the stage, put ours on, Mel turns up, and questions why one of our road managers, Johnnie Allen, has got plaster on his forehead and he's white as a sheet. When he gets told the story, he starts to remonstrate with the promoter and he got threatened, too. So band goes on stage, we're promised tyres, of course, band goes on stage, band play, band fly home. Johnnie flies home, he's not in a state to drive, he needs to get home quick. Of course, two tyres don't turn up, so we drove home in the winter from Italy with just three tyres. I had a spare, three tyres, supposed to be four on the back axle, got home though.

I get the job; following week we're on a plane to Miami, their second and my first American tour. They were supporting Buddy Miles and Deep Purple. From then on they just got bigger and bigger and bigger, and very soon they were headlining America, ice hockey arenas.

I saw them back at Dagenham Roundhouse.
Yeah, there you go, I remember that. I was there probably. Well, yes, it's a small world, isn't it? As Heep got bigger, it became obvious they were going to do venues where you needed to carry your own lighting system. We were already carrying our own audio system.

What was that?
It was Martin bass bins; I can't remember the name of the PA company.

Dave Hartstone? IES?
Oh, IES, yes, exactly, and they were Martin bass bins, but you're right, Dave Hartstone of Covent Garden at one point. Anyway, so I rang up the only lighting company I knew, which was ESP with John Brown running it and Brian Croft. They said, "We could do that, we could do a tour". I said, "Well, we don't need anyone tonight or tomorrow. The Rainbow, they have enough lights, but after that we're in Brighton and then on a UK tour". They said, "We'll send someone down to look at what you've got, you know, what you'd like and everything, and we go from there". So we load into the Rainbow, up at Finsbury Park. The next thing is I get a call, "There's a gentleman at the stage door to see you".

Have you got a lighting designer by this time, got an LD?
No, that was going to be me just turning the lights on and off, really. So we never thought of that, though I used to do odds and sods in America when it was house lights, you know.

Do you remember who did them at the Rainbow? Did someone do them for you?
Me. I used to tell the guy with the machine. It was kind of Michael Ahern era, you know, Roy Lamb was working there at one point. Anyway, the guy from ESP turns up, and I said, "Well, let him in". They said, "Well, we can't because he's banned". So I said, "All right, I'll go out to him", and it's Edwin Shirley at the stage door. He's been banned from the Rainbow for being pissed and obnoxious. So I just sneaked him in.

So what was Edwin Shirley coming for?
He was the representative for ESP Lighting. I got him in and showed him what we wanted. He said, "Right, I'll see you in Brighton tomorrow". We did the Rainbow; success, go to Brighton. Sure enough, not on time but an old North Thames Gas three-tonner turns up belching smoke with Edwin.

He's driving?
Yeah, yeah. He loads in with a lad. We do the show, which is kind of neat, and he turned the lights on and off, you know, at a given time. Very much pioneering and then we loaded out except he kept his truck running, because the battery was flat and we needed it for the tail lift. He filled the Brighton Dome with diesel smoke, which was quite an achievement. I said to him, "Listen, we're only down the road in a hotel, come and join us for a drink". Because he had to go back to get more gear, he said, "Well, I'll join you for a drink in a couple of minutes". And two hours later he rolled in from what was a one-minute drive, and he said, "I'd been pulled by the police". He didn't have an operator's license, so he better get serious. That's when Edwin Shirley decided to get an operator's license.

Sorry, what is an operator's license?
Well, it's a license for you to operate trucks, to run a trucking company, which he was about to do. You need premises to park the vehicles, you needed a contract with the local dealership or repair agents to service it, unless you did it yourself, you needed money in the bank, so you're not going to be Johnny-come-lately, really.

And he needed that because he was running the transport side of ESP.
Well, yeah, he was doing their trucking; that's the way to put it. They would rent out from Hertz or Avis, and they saw that Edwin was more competitive and Edwin was a crewman. So it's double bubble.

Anyway, Edwin and I struck up a relationship. He was doing lighting and, at the same time, with Roy Lamb he started Edwin Shirley Trucking. It was called Edwin Shirley Trucking because as mentioned, you'd need a base for an operator's license, which was his mother's farm. It became easier to use the word Shirley, you know, in paperwork and Edwin being Edwin, it was called Edwin Shirley Trucking. Edwin and I continued touring together with Heep whenever they toured for the next couple of years. He would just not run the trucking company, someone else would. He came to America, Australia, maybe I think Japan with me, we kind of toured the world. It became obvious, amusingly, well, on stage with Uriah Heep, there was always a bar, it was a bit heavy metal, a constant bar for the band to drink while on the stage. The barman was the tour manager, and Edwin and the lights used to be on stage right. So Edwin would help at the bar, if you know what I'm saying.

Help himself at the bar?
Yeah. Which means he wasn't very good at operating the lights. By default, I ended up becoming the lighting designer and Edwin became the pyrotechnics man, which when you are a bit pissed, is dangerous. There's another story there! Anyway, we became closer and closer. It was obvious that Edwin Shirley Trucking was going to become a bit of an entity even in them days because they were the only people in Europe doing it. They were renting trucks; you could only rent really from one company called Dawson's in Leighton Buzzard with sleeper cabs.

And trailers. Where would the trailers come from?
You could rent them as well. We bought a couple of dodgy ones. They needed to buy a couple of trucks, and I wanted to come off the road at that point.

Where are we now, '74, '73?
Yeah, late '73, early '74. So I said, "Here's the deal. I'll come off the road, mostly, although I'll stay on it for a reason to come to in a minute. I can probably get you finance for your first two tractors via the management of Uriah Heep. Okay? In exchange, I'd like to become a director and a shareholder of the company, you know, as I am going to give up a career, a well-paid career on the road". By now, I had a bit of a phone book of contacts.

Did they have the offices at Camden then? Next to the Roundhouse? Bron, that's their record company, wasn't it?
Yeah, Bron, yeah. Exactly that. They hadn't moved there long from Oxford Street of all places, but Edwin had his flat in Bell Street, just off Edgware Road, and his mum's farm where there was a bit of a workshop.

So I went to the manager, Gerry Bron, of Uriah Heep, brother of Elena Bron, the actress, and he said, "Yeah, of course, I'll lend you the money". I said, "Well here is the selling point, A, you can get tax advantage for investing in rolling stock; B, I will stay with the band, but you don't have to pay me during downtime, non-touring time; C, you will get a special deal on trucking". And he said, "Well yeah, except I need you guys to come up with a deposit to show that you are willing". So yeah, Gerry Bron said, "Absolutely, you have to kick up a deposit", of which nobody had no money as they say, double negative. So we gathered up Annie Pocock's brother, Billy Joe Rice, Jimmy Barnett, Roy Lamb, Edwin Shirley, me and a few other people to put some money in, and we met up in Camden at Dingwalls. Everyone got pissed, and everyone agreed to put in a thousand dollars or something each.

Of course, the problem then was, we've now got the two lorries on paper, what colours? As a lighting designer I had a swatch book in my pocket, we got that out and we really couldn't see it too well in the dark. Then I remembered my mum had a shopping bag that was yellow and purple, and I always liked those colours. So I said, "How about that"? Everyone said, "Perfect". So we were off and running. That's it, that's the colours and that's our first two trucks.

And that was in?
Well, the company was formed in January '74.
At this point in our conversation, Ollie Kite (OK) *joins us. Ollie joined Edwin Shirley Trucking in 1976 and is still there today with Del* (DR).

So, two trucks, funded by Gerry Bron. How were you going to pay the money back, by the way? I mean, did you come to an agreement about monthly instalments?

(DR)   Not at all, No, he just trusted us.

What did you buy, by the way?

(DR)   Volvos, they were 15,500 pounds each, I think, and now they are top specs, 102,000 pounds.

Were they air ride then?

(DR)   No.
(OK)   No, they weren't.
(DR)   No such thing as air ride tractors, only air ride trailers, which we'll come to with Clark Transfer at some point in this conversation. Well, money started rolling in. As mentioned, it was based on the farm.

Who else was there? Who else was doing it then, '74?

(DR)   No one. Avis, I guess.

When did ML have their trucks? '76?

(OK)   I didn't start till '76, so I don't know, but they were about when I started in '76.
(DR)   I think, yeah, that would be about right. Our only rival was Artist Services with Don Murfet and Jerry Slater. Yeah, they used to rent some Avis trucks. Maybe the same time before, but they weren't serious about it, and I guess we were.

Have you read Don Murfet's book? I worked for him actually, tour managed Adam and the Ants, when he started managing Adam Ant, in 1981. So who's driving the trucks?

(DR)   Eric Owen, who is still alive today but sadly, not driving, and Mick Conafray was our first-ever driver, sort of, yeah. We nicked him from Curved Air, and he is still driving for us today. Yeah, so still a few originals. A lot of the old faces have disappeared, like Lead Foot. By now, we had three trucks because we bought IES's old Ford D100, as a tractor, as a dog. Yeah, they still went abroad though.

And who was the first band that used all the trucks in one go?

(DR)   Yeah, I suppose it was Santana.
(OK)   I think it was Santana and Earth Wind and Fire together.
(DR)   Yeah. That or Santana and Journey, I am not sure.

In 1970?

(DR)   Well, actually, before '74, wasn't it? '73.

Yeah. But that was rental?

(DR)   Yeah. The first tour for one of the brand-new ones was Uriah Heep as the celebration or arse licking for the manager, either way, and then it just got bigger really. You know, we were just buying and buying trucks and trailers slowly. It was hard to get step frame air ride trailers from manufacturers in England, they just didn't have the capacity or the will to do one-offs really, so at one point, we resorted to importing American step frame air ride trailers from a rock-and-roll company out there called Clark Transfer.

I remember Clark Transfer.

(DR)   Yeah, the Molitch Brothers. To pay for one of them (they always came on the ship obviously), it was filled with McDonald's furniture, allegedly for the first-ever McDonald's in Britain.

(OK) Yeah, sounds about right, doesn't it? That is the story. Anyway, we had these, and the reason that Clark Transfer got rid of the trailers (laughs) is because they'd done their lifespan, but we stretched them for another five, maybe ten years some of them, they'd air ride, that was the attraction.

That was the whole thing; it was all compressed air that kept the suspension?

(DR) Yeah, except they were American threads.

Heavy duty?

(DR) Yeah, well, and getting repairs, I mean, I remember limping to an American air base personally on a tour because the wheel was falling off. An American air base would have those nuts and bolts, which they did.

What happens if one of the things broke? Could you still pull the trailer?

(DR) Well, no—
(OK) You should carry a spare air bag in the box.
(DR) Exactly that, yeah. Spare air bag.

Which would keep the pressure?

(DR) Yeah, or you blank off the pipe, yeah.
(OK) Air bags used to burst occasionally.

Del in an EST calendar

So what's an air bag?

(OK)   The air suspension is a big compressed rubber bag that acts as a spring.

On each axle. Or two on each axle?

(OK)   One on each side, yeah. So instead of a spring, there's a big bag full of air like a concertina.

And why did you want those rather than sprung load?

(OK)   Just bullshit, isn't it?
(DR)   Well, it was bullshit yeah, it was, but it was the magic word in America!

Because roadies thought stuff would get broken?

(DR)   It was half true in a funny way, you know, you're right. We'll come to another buzzword in a minute, but it seemed to give a better ride. I mean, when you were doing (in the old days) Berlin, remember how you got through that corridor?

That's when 'air ride' did come in. It was a popular word. It was like 'step frame trailers' now is a swear word because the buzzword is 'flat floor mega cube trailers', so you don't have a step. Now, actually, it saves four cubic metres at most, and you never fill to the top of the trailer anyway, but it's a buzzword. So sexy to use 'megas' now!

Well, because it's larger?

(DR)   On paper. It cuts out the step.

Why was the step there in the first place?

(DR)   To get the height. Because you need maximum height four metres. Smaller wheels at the back drop the floor down so you've got extra space, so more volume. The front is 13 feet, or a foot taller than the remaining 27 feet as it was.

So right back in the day, at the beginning, what were the trailers?

(DR)   Right at the beginning (they were) York and there was Crane, they were step trailers.

Because of the fifth wheel?

(OK)   Yeah, 13 feet of platform, dropped down about 9 or 10 inches, maybe a foot, and then 27 feet of lower deck to a little ramp to go up on to the top deck. The dance floor.

The sound always went over the dance floor, didn't it?

(DR) Sound, yeah, no, you are right, it was tidy.

And who moaned most about stuff getting broken in the early days because of no air ride?

(OK) Stuff didn't get broken.
(DR) No, it was flight cased.
(OK) It could get a bit shaken up going across crappy East European roads, and connections become loose and things like that, but I don't think things got trashed?
(DR) No, not at all. No, you are right. It was something that you should have air ride really.

And that's because of the American touring.

(DR) Well, they introduced it, yeah. I mean, they were ahead of us with road cases, whatever; they were ahead of us, period. Multicores what an invention that was, right?

Okay, and way ahead of the Australians. They're not even putting wheels on road cases.

(DR) Don't need 'em, cobber (laughs).

Now let me go back to the beginning again. Let me ask about driver's hours and carnets and tachographs. There wasn't any of that, was there?

(OK) Logbooks for drivers.

Oh, tell me about logbooks.

(OK) Well, it's just a multi-page document with a copy, you know, with a carbon copy where you have to write down when you started work and when you stop for a break and when you finished work. You just have to fill in your details on the page for the day, and then write in where you started from, how many miles you had done for where you were going.

Who for, the police?

(OK) Yes, the police, for safety purposes. So you didn't work too much.

Obviously in the UK, but was that in Europe as well?

(OK) Yeah. So there were drivers' hours regulations even then, but there was no way of telling what the truck had been doing unless the driver wrote it down. If you got stopped and you hadn't made an entry of when you started, then that was an offence, and then you have to show when you stopped. You have to keep the logbook up to date on an on going basis. You couldn't just fill it in at the end of the day because you could get stopped.

That's if you're on a long one, you might not fill in the beginning?

(OK)   You had to fill in the beginning even before you started the truck.

Right, no one would know?

(DR)   They would if they stopped you.

But I was going to say if you're stopped, you instantly just put in that you started at midday rather than 9 in the morning.

(DR)   You've got to be quick.
(OK)   Yeah, you got to be quick.
(DR)   The answer could have been, and was in America, is that you run with three logbooks. You had to be on the case. One from what you've just done, one on what you're doing, and one for when you get there. And you got to juggle it, depends when you get stopped. That's a lot of thinking! Logbooks are the answer. I don't even know when tacos came in, do you?

No double drivers really then?

(DR)   If needed, yeah, because physically sometimes you can't actually get there. You need to keep going. You're right. So yes, double drivers was the answer frankly.
(OK)   Yeah, it was a bit hit and miss. Of course, Carnet, ATA Carnet for everywhere.

From the beginning?

(OK)   Always, yeah.

And obviously ATA Carnet came out of any trading not specifically for rock tours.

(OK)   Yeah, temporary imports. That was the only way that you could get a rock-and-roll show from anywhere to anywhere.
(DR)   And as you know it was before the EEC.
(OK)   You have to do a Carnet into France, a Carnet out of France into Belgium, out of Belgium into Luxembourg, out of Luxembourg and so on.

So every border you had to talk to two different customs agents and get two different stamps?

(OK)   Yeah. So it took forever.

Sundays was a strange day in Europe, wasn't it?

(OK) Still is. Driving bans everywhere, not everywhere but Germany, Austria.
(DR) Certainly France, yes, Switzerland, yes, much more . . . what's the word. . ., litigious these days?

So what do you do say on a Milan to Madrid, with a one day off and the day off is a Sunday.

(OK) You've got to get a permit.
(DR) Because music or showbiz is an exempt commodity. Yeah. Like milk.
(OK) Yeah, but you still have to have a piece of paper to say that you're exempt.

And back then?

(OK) Yeah, you've always had to have a piece of paper.

And the promoter used to get that? Still does?

(DR) Carnets, you're familiar with anyway, you've heard that a million times. It's improved because obviously there are a lot of countries in the EEC or the common market.

Therefore, less borders to have to stop it.

(DR) Yes, and only last week Serbia joined.
(OK) Croatia.
(DR) Croatia joined the common market. That's yet another country.
   Ollie will get on to the Russian saga in a minute because he's good at that and as to what happens there. Tachographs, we know about them, but they've been superceded by the digi cards.
(OK) Electronic tacho. Which is completely tamper-proof, and so it's actually much better because it is what it is. There's absolutely no debate.

And your arm can't be twisted by a tour manager or anyone!

(OK) There's absolutely no debate that the tacho can tell exactly what you've been doing and when you've been doing it, and they don't even have to stop you to find out. They could read it electronically.

They all say in the music business, you know, the show must go on, it's a bit of a cliché, but 99 times out of 100, it does.

(DR) Yeah, yeah.
(OK) I guess it got more difficult as productions got bigger. So instead of, you know, one truck full of instruments and the stuff to make the music, they had lighting, more complicated sound, videos, set and gags. The Rolling Stones went from three transit vans to a hundred trucks.

Yes. As more and more of the whole stage infrastructure was purpose-built.

(OK) So as shows scaled up, the logistics of getting it to where it is supposed to go got more and more complicated, and more and more crucial. I am sure with a transit full of backline you can turn up an hour before the show and get something together, but you got 27 artics full of stuff. You got to be there when you got to be there.

But they also plan for longer set-up times.

(OK) Yeah, they plan it, yeah, because it's so expensive to keep those massive shows on the road, you've got to get the maximum number of shows a week that the artist can actually dare to do. You can't do one a week, you got to do four or five a week at least. It makes for long drives and it's a lot more complicated, lots of people, and it's like a monster self-feeding itself.

'83 for me was a massive year because I got hired by Supertramp. Do you remember '83, the Supertramp Football Stadium tour across Europe? It was 'Famous Last Words'. You guys, I think, did it. I think it was the biggest logistical tour to date, and the Serious Moonlight Tour with Davie Bowie was going around Europe too, at the same time.

(OK) We did both.
(DR) Yeah, we did Serious Moonlight.

Do you remember with Supertramp, we leapfrogged stages, I think it's the first time, but tell me if I am wrong.

(DR) You may be, to be honest, because Bowie leapfrogged as well.
(OK) Okay. When was the first big Stones stadium tour? What year was that?
(DR) '82 I thought.
(OK) Yeah. So I think that was probably the first.

Did that leapfrog stages? We had three different stages and we were leapfrogging stages into football stadiums around Europe.

(OK) It's all about the same time. But I think the Stones, that Stones' tour, was the first.
(DR) Well, I was going to say about logistics side. For me, one of the turning points was in the old old days, you could do a tour of Germany quite easily. There's so many venues and towns, you know, whether it's Peter Frampton or Wishbone, you could do so many gigs. Then, as you know, the American army and Air Force started pulling out of Germany, and thus less punters, and therefore we were finding that the bands were looking for further and further venues. So, suddenly Belgrade was on the map, the Eastern Bloc. Even doing Lisbon was a bit of an unusual thing or Porto, in the early days. So going beyond the boundaries, I mean you'd never do Istanbul in them days but—

I suppose you guys probably did the first Russian one. Was that Elton?

(DR) Oh, that was ML. Yeah, Pete Burke and Rivett, I think.

(OK) I think you are probably right. I think it was Elton, then we did Billy Joel.

When you got your first two trucks, you said that you were working out of Edwin's mum's farm and where else?

(DR) Bell Street was our London Headquarters. Just off Edgware Road. That was an upstairs flat, which meant the hall was always full of ramps, not popular with the other residents.

You're running it out of a flat?

(DR) Yeah, at one point. Now remember there were no credit cards in them days, so drivers carried cash, and they would come park their artics in that area, and of course one artic used to equal something like four parking metres. We used to have to keep a little map on the wall to see who was parked where, so you could feed the metre. It was a bit Mickey Mouse, to be honest.

So, Ollie, why did you come into this?

(OK) I came into it towards the end of late '76 because I lived in Cranbrook, where Edwin's mum's farm was. I was aware of this trucking company, and a mate of my sister worked for Edwin. I used to see Edwin in the pub occasionally.

What were you doing?

(OK) I was training to be a lawyer. That didn't work out particularly well. What happened was I must have run into Edwin in the pub or something. He said, "If you're looking for a job, we could have one so come to see me", so I went to see him.

How old were you?

(OK) Twenty-five maybe?

Had you gone to University?

(OK) No, no, I've just been to night school. From school I joined a firm of solicitors, and I'd been going to college at night. Edwin decided to add another string to his bow, which was wine delivery. He got a job with a wine distribution company, who would send a rep to your house, having made an appointment, and turn up with a briefcase full of wine. You would do an in-house wine tasting and, at the end of the evening, you'd fill in all the forms and four weeks later a case of Beaujolais or whatever would arrive.

Anyway, Edwin did a wine tasting and got talking to the guy, and it turned out they had a real problem actually delivering the wine to the punter. Edwin in his wisdom decided that as we were a trucking company, he would be that wine delivery. He got the job, and I got the job to deliver the wine. So I used to go up to Camden, funnily enough, where the wine warehouse was, pick up a truck full of wine, bring it back to the yard, put it into a smaller truck, and then distribute it all around Kent.

Was that up in Camden in a warehouse, customs cleared?

(OK)   Yeah, it was a bonded warehouse that you used to go to, up there and pick up the delivery notes, and they just pile a load of wine in the back of the truck. Take it back to Kent and then try and sort out all the individual orders. I eventually ended up hiring another guy to help me, called Frankie Enfield, who is now tour manager for The Rolling Stones. Yeah, so that's how we both started. I literally started the wine division and then Frankie took over, and I started working in the office.
(DR)   As our legal eagle.
(OK)   Yeah, legal department. So that was my job delivering wine. Then moving into the office one or two days a week to do stuff. Then I just ended up staying in the office.

Where did the office move to, from Bell Street?

(DR)   To Battersea, Chelsea Bridge. The Battersea depot wasn't big really, maybe half a dozen trailers. By now we are getting the famous Saviem seven and a half tons on the road. Remember them? With sleepers, that was a big seller. We still had to keep the farm on for doing our servicing.

At some point we ended up working in the office in the day then shipping the seven and a half tonners down to Kent at night to be serviced, so you could come back with another truck. It was a bit full on to be honest. We were young and foolish then, so yeah we gave up Bell Street, moved to Battersea. Stayed there for two or three years.

You never did buses did you?

(OK)   Buses, yeah, we did.
(DR)   The first ever.
(OK)   The first ever in Europe.

Really! What year was that?

(DR)   '81.
(OK)   Probably.
(DR)   Yeah, because we did it for the Stones in '82. Yeah, I'll bet you're right.
(OK)   Yeah, we invented European rock-n-roll bussing.

Before Len Wright?

(OK) Yeah, yeah, way before Len Wright. Yeah, a year or two before.

(DR) Well he basically copied our idea. Made it more financially viable.

Did ML's two sleepers come after yours?

(OK) Yeah, Robin and Mike drove in ours.

(DR) Yeah, we definitely had the first European custom-built rock-and-roll sleeper.

Who built yours?

(DR) Lambourne, yeah. Horsebox builders.

(OK) They look like a horsebox.

That's interesting because ML's was designed by a canal boat builder.

(OK) Theirs was probably better on water than ours!

(DR) Our theory was that horses travelled better than people in Britain. So get a horsebox dealer to do it.

That's a good theory.

(DR) It is actually.

But you didn't stay in it. How many buses? Berryhurst came into it then as well.

(OK) Yeah, we did the first one, and then we did a second even better one, by which time Len Wright was getting into it. We carried on with the two for a while, and it just got too much grief. The trucking company was getting bigger and bigger and the staging company was getting bigger and bigger.

Well I was going to go on to staging.

(OK) And the bussing was just too much grief. Wasn't it?

(DR) Yeah. High maintenance.

High maintenance, just two buses?

(DR) Yeah.

(OK) Constantly busy. We sold them to Berryhurst.

(DR) Did we really? I didn't know that.

Is that how Berryhurst started?

(OK) Probably.

(DR)   They probably had some buses already, but they bought our two.
(OK)   I remember doing the deal with Mike Levy.

From '77 through to '86 every diary I owned was a Berryhurst diary.

(OK)   Right.
(DR)   And they personalised it with your name on the front.

Okay, staging. How did that come about?

(DR)   Fleetwood Mac. We went through this the other day.
(OK)   We did, yeah.
(DR)   Somewhere in Germany. Was it Monchengladbach or something like that?
(OK)   Loreley or one of those famous festivals. They needed an outdoor stage. And so, Roy Lamb, in fact, had worked for somebody in America, Mike Brown, building stages, so we got Mike Brown to send over an aluminium roof from America, which we then put onto a scaffolding structure for that Fleetwood Mac show. That was the start of Edwin Shirley staging.

What year was that, Ollie?

(DR)   Was it '80? Yeah, I'd say '80. Also the deal was that Fleetwood Mac couldn't or wouldn't afford to ship a stage over because it would have been airfreight at the time and ship it back. So the deal was they'd pay for it to come over and we would pay for it to go back; that's part of the deal. See what I'm saying? That's how we got Fleetwood Mac trucking and the staging and the use of that stage for the season, except the season turned out to be about ten years.
(OK)   Yeah, it did.
(DR)   That roof lasted quite a long time then.
(OK)   It lasted until it got nicked, didn't it?
(DR)   That's true, there's the story.
(OK)   Someone nicked the whole trailer. Nicked a trailer from where it's parked outside the yard.
(DR)   West Ham.
(OK)   Yeah, I mean for scrap. We never did see that again.

Why didn't Fleetwood Mac call a promoter? Were there no stages in Europe? Outdoor stages?

(OK)   Not really, no.
(DR)   They were all scaffold with an unsuitable roof. That's the thing.
(OK)   They presumably played under a Mike Brown roof in America. That was what they wanted, so we took the Mike Brown roof and then we developed our own.

What was Roy doing at that time then? He used to stage production manage all the time, didn't he?

(DR) Yeah. He did some driving that whole period.

(OK) He did some production managing.

(DR) He's an electrician by trade, a rock-and-roll electrician, and production management. Yeah, he was The Stones' stage manager or production manager in '82.

Then he learned rigging as well. He definitely made himself invaluable, and he still does to this day. But, by mutual agreement, by his own admission, he's not very good in an office.

(OK) Yeah, I don't think he wanted to be constrained by having to turn up at the office 10 o'clock every morning. He's much better off production managing.

(DR) So that was the deal, really.

Was Roy there at the same time as you?

(DR) Before me.

So it was Edwin and him and then you came along?

(DR) Yeah. Yeah. The deal was when I came along, well, you know half the story, we agreed that between the three of us that Roy and I would stay on the road for a while to make more contacts because I was going to and from America. A lot of our clients are Americans, so it was good to build up the phone book, you know, which Roy did as well.

(OK) Talking about buying trucks, we got Manfred Mann to buy us a truck, didn't we? Remember?

(DR) That's right. Bloody hell, Ollie. Bloody hell.

(OK) Because capital purchases like trucks are tax deductible. So Manfred Mann, also on the Bron label, he bought a truck, which we then leased from him.

(DR) Should have been a MAN, shouldn't it?

(OK) Should have been a MAN, but it wasn't. It was another Volvo.

(DR) I had forgotten that one. Wow.

(OK) And he slept in it one day, didn't he, because he wanted to get his money's worth. Lead Foot was driving, and he slept in the bunk.

Who did?

(OK) Manfred because he was doing Manfred Mann's Earth Band tours at the time, so he used to ride in his truck.

(DR) Funny. But then you get rivals, then, of course.

Who was the first rival?

(OK) ML, really, I suppose.

(DR) There's a company called Thompson's.
(OK) But they never expanded, ML did though.

We digress but that's allowed. So staging, go on, there was the Fleetwood Mac rental...

(OK) Yeah. It kind of went, it just went nuts from there.
(DR) We did Donnington. Remember that? And then, once again, pioneering, because there was no other staging company.
(OK) Yeah. We did the first Donnington.

For MCP?

(OK) Yeah. Yeah.
(DR) Maurice Jones and Tim Parsons.
(OK) Then it started to become tours and we needed more than one, so we then copied the Mike Brown one, didn't we? We had somebody build a copy.

Did Richard Hartman get involved in this?

(OK) No. I think he was around, though, wasn't he? He might have worked for us. He might have built some towers for us or something. We had to build more roofs, rent more scaff.

Who was running this?

(OK) Roy was the main element of the staging, wasn't he?
(DR) Yeah and Henry Crallin?
(OK) And Henry Crallin, who was one of the drivers at that time. We all had input into it. Graham Fleming got involved.
   Scaff was always and it still is yellow, and of course rock bands didn't want the bright yellow scaff, they wanted black scaff. Pipe and clamp is silver, silver poles, but quick-form scaffolding was always yellow because you wanted to be able to see it, and on a rock-and-roll show, you didn't want to be able to see it! We had to get into painting the yellow scaff black, so we did a deal with the scaff company to produce black scaff that we would then rent. It just got more and more complicated.

And, of course, you want it in the summer but you don't want it in the winter.

(OK) Yeah. Then you've got all this scaff sitting around.
(DR) That's when you get into opening an Australian branch because their winter is our summer.
(OK) And an American branch.
(DR) That's true.
(OK) And then a Japanese branch.

Your company got big.

(OK) It got very big.

So the staging got bigger than the trucking from a business point of view?

(OK) Yeah. Yeah. We had ESS Limited, then ESS Australia, ESS Japan, ESS Inc.

With the same gear, moving it around?

(OK) A lot of the same gear being moved, yeah, because people did international stadium tours, and they wanted the same stuff. So we did on occasions fly scaff.

Sounds insane!

(OK) Sounds insane, yeah. We tried to use local scaff, just fly the roofs.

Now we're talking the late '80s here, aren't we?

(OK) Early to mid-'80s.
(DR) Initially, we did, when we went to America we did a partnership with Mike Brown with UPS, United Production Services. Over here, it was called UPS UK.

Yeah. I don't know anything about the history of staging from an American point of view at all. I suppose the original festivals here in the UK made it out of scaffolding. I mean like Reading and so on.

(OK) Yeah.
(DR) Yeah.

But you were the first company to actually take a grip of the onward-going increasing demand in outdoor good quality. So rivals again, did you get involved with Glastonbury?

(OK) No, because they built their own stage right from the off didn't they, really.

Do you remember Hav-a-Van? Rickmansworth?

(OK) Yeah. Yeah. We used to rent from them.
(DR) Don't have a van.

They were pretty much the first one, weren't they?

(DR) Yeah. There was Wheels, out of Barnet, Elton John used to use them. Yes, we've got rivals. As I said there's Thompson's out in Manchester, green and

white, and they mainly did 10cc by coincidence. There was a company right by Olympic called Music Haul, which I thought was a great name.
(OK) Keedwell.
(DR) Keedwell, yeah. Yeah. Very good.
(OK) And Euro trucks. Latterly.
(DR) Oh, Up Up? Out of White City, wasn't it? Yeah. Who was that?
(OK) It was Bill somebody.
(DR) Bloody hell, Ollie, you scare me.
(OK) I will think of it in a minute. It was a hand with a cloud on it, something to do with a cloud.
(DR) Yeah. Pete Edmonds is there somewhere. They weren't that serious, to be honest. They took work from us.
(OK) Redburn worked for them.
(DR) Stardis? Yeah. David Steinberg.
(OK) Yeah. David Steinberg. There's a lot of rivals in the end.
(DR) Well it looked like an attractive way of earning money.
(OK) Yeah.
(DR) And groovy. So staging, we've kind of done bussing, haven't we? Bussing didn't really last long. Two or three years, maybe four.
(OK) Maybe a bit more, about four or five years. Yeah. It was good while it lasted, but then people like Len Wright started to get into it in a big way.
    Buses weren't for us; it was good while it lasted, but it just got to be a pain. Two's never enough. You need ten.

How mad and how big is it now?

(OK) It's crazy. These things cost 750,000 or something to build. It's just crazy.
(DR) Who was it that shipped some Beat the Streets down to Australia? Dolly Parton, wasn't it? She shipped Austrian buses down to Australia because she likes 'em so much. How bizarre is that? And then they didn't use them all the time because "maybe she'll take the limo to the hotel and not the bus".

And going on to the staging, that just grew and grew.

(OK) Just grew and grew and grew. I'm pretty sure it was The Stones did the first multi-stage stadium tour because it then got into really serious big business. We did the trucking and the staging and the bussing. We did everything on that tour.
(DR) That was '82, yeah.
(OK) That was '82. In fact, we were the main contractor for sound and lights and everything, with Showco. We did a joint project with Showco. They did the sound and lights element; we did the trucking, bussing, staging.

That would have been the first one then?

(OK) Yeah, and that really opened everybody's eyes up because they're doing a stadium every other night and 40,000 people or whatever it was, big-ticket sales, everybody thought this is the way to go.

Who booked that? Do you know?

(OK) Harvey, wasn't it? In those days.

(DR) I guess so, yeah, but then he'd partner up, as you say, with Mama or Andy Bechir or people like that.

(OK) I think we had three stages on that? One going up, one being used, one coming down.

(DR) The big logistic thing was obviously Michael Jackson because he didn't leapfrog town-to-town or country-to-country; it was continent to continent. That was big. They had to use aeroplanes to ship certainly the production, sound and lights—

(OK) Yeah. Yeah. Two Antonovs.

(DR) Three. Forty tons they carried, each.

(OK) Three Antonovs.

Who hired them?

(DR) Rock-it Cargo.

Rock-it hired them?

(DR) Yeah, which we formed.

(OK) We started Rock-it Cargo as well.

Oh, blimey.

(DR) Going back to when I'd done these tours with Uriah Heep; almost the end of the first tour, which would have been 1972. We flew the backline equipment back into Heathrow, and you can't just clear the carnet at Heathrow, you need an agent. British Airways cargo suggested they knew a little back street airfreight agent that would deal with rock-and-roll bands called Dateline Forwarding.

We used them, and I went around there and the owner said, "Oh, I'll give you one of my boys (who was Chris Wright) to look after you". Chris did a great job. It was obvious that he was a big music fan, and him and I became real close friends and our families became friends. That went on for two, three or four years, and eventually he got itchy feet and had the idea that he wanted to start a rock-and-roll Air Sea Freight Company. So he came to us in whatever year it was. Ollie, remember that?

(OK) It must have been '80, '81?

(DR) Yeah, so early '80s, and said to me, "Would you guarantee me a year's wages?", would we gamble a year's wages to pay him and keep his family stable. Chris was living in a caravan at the time, because that's what he could afford. Of course, I took it to all the guys in EST. Everybody said "Yes" without hesitation. So we guaranteed him a year's wages. He took off. It was the same girl that did the now-famous Rock-it Cargo logo that designed all our calendars and note paper and artwork, so that still remains, and Rock-it got bigger and bigger, worldwide now, of course. It came to a point where it was obvious that Chris, EST and Rock-it couldn't work together because it was a clash. How would you put it?

(OK)   Well it was Chris and Edwin who couldn't work together.

(DR)   Oh, that was nicely put. Yeah, yeah. Chris and Edwin.

(OK)   They just clashed. I mean, just different mentalities and personalities, so they couldn't work together really. In the end, I helped them kind of broker a buyout, so Chris then bought EST out of Rock-it Cargo. It was getting to the point that it was starting to become fairly acrimonious.

Why?

(OK)   Just personality really. Chris's way of doing things and Edwin's way of doing things. I'm not saying one was right and one was wrong, but two different ways of going about things.

(DR)   Two different ways, no doubt about it. We're still very close with Chris, funnily enough. I'm a godfather to his son anyways. Chris is not now a major shareholder in Rock-it anymore. He doesn't steer the ship completely. It's run by Americans now, but yeah, that started with EST so trucks, buses, stages, airfreight.

(OK)   Yeah. We didn't let anybody know because we thought the bands like . . .

(DR)   The Who.

(OK)   The Who wouldn't use Rock-it Cargo because it would be like putting money in Edwin Shirley Truckin's pocket 'cos they had ML. ML did their trucking, and if they wanted to use Rock-it, we felt like they wouldn't want to if they knew it was us, so we completely de-noticed it so nobody ever found out. People who didn't like us could still like Rock-it Cargo. I remember a famous speech or famous address by Harvey Goldsmith, some conference where we're all attending . . .

(DR)   ILMC (International Live Music Conference).

(OK)   ILMC, where someone had a go at him about ticket prices. He turned it back and said the reason ticket prices were so high it's because you so and so's, on production, are developing such complicated shows that we've got no choice but to put the ticket price up. He singled us out as a staging company for particular abuse because I think probably about the time we were doing Michael Jackson, we've got moving platforms and we've got all sorts of stuff, cost a fortune. He blamed us for the escalating price of tickets. I actually pointed out to him there and then that he was the one driving the Rolls Royce. We were the ones driving the second-hand MGs.

Did that go down well?

(DR)   No, great line though.

(OK)   You can't shut Harvey up once he gets on the soapbox, if he's got something to say, he says it.

Yeah. That's how I really started off as a freelancer, working for Harvey, he was great with me, he helped me up the ladder. Did you drive artics a long time, Del?

(DR) Yeah, yeah, yeah. Got my license in maybe '73 and then did a lot of EST trucking and did a lot of driving in America for Heep and Showco a bit. Mike Brown, I've done a lot of driving. There was a turning point back to the formative years when it was getting bigger and Ollie was around and Tim really.

Were you driving, Ollie?

(OK) No, I never drove. I decided really early on not to get a truck license, because I used to have to phone people up in the middle of the night, and they had to get out of bed as so and so needs an extra truck. If I'm phoning up those people and I've got a driving license, they will just say, "Well you do it". I could phone people with a pretty much clear conscience. I was asking them to do something I couldn't do.

(DR) Yeah. It got busier and busier, and in the end Edwin said to me, "You've either got to shit or bust really, I mean Roy's never going to come back to the office, bless him. We need either you in the office or we get a replacement and you stay on the road, either way". The turning point was Uriah Heep again. They stopped being the headline act. We flew back to America one more time, and maybe by now it's '78. I was doing two jobs and we rolled up to the first gig, and this time Kiss was headlining and Heep were the opening act, and I thought this is a sign of them going downhill frankly. We got abused on that tour as the opening act by Kiss's people, abused. It's awful. So I thought I better get involved with the lorry company more seriously. That was the turning point for me stopping driving really.

(OK) You still used to dabble with it?

(DR) Yeah, yeah. If we run out of people and there's a tour to go on—

Have you kept your license up?

(DR) Yeah, yeah. I haven't driven for maybe 10 years.

(OK) Once a driver, always a driver.

(DR) Yeah, you can't . . . it's the best thing in the world. But not these days, that's the problem, and because of litigation, because of speed limiters, 56 miles an hour's not everyone's idea of fun. You see, I'm sure you've seen it, when you go up the A1, the A34 at night, all these truckers tacking behind, because they can't overtake. They all got their foot down and doing 56. You look at the back of a trailer all your life. Probably the same trailer every night. People do that for a living, but not for us. That was it and we really started getting seriously into it as a trucking company. We was only kind of dicking around up to then.

(OK) It was a bit of a laugh, wasn't it?

(DR) Yeah, it was better than working.

(OK) Better than a real job. I think for Edwin Shirley Trucking the major game changer was stadium shows. When it stopped being one, two, three trucks on a tour and suddenly it became eight, nine, ten trucks on a tour plus staging, plus then the logistics side of it started to get really complicated. Up to then anyone with a truck could have a go. They'd probably fail. People would come along for a year, have a go at being Edwin Shirley Trucking, not make it, disappear and then another one would come along. But once it got really serious then it was good, wasn't it? It was rewarding in many ways, and financially it was rewarding and it was a real challenge.

If you could mention one person in your careers that stood out?

(DR) Michael Ahern. Yeah. I'd say that for me, because he was loyal for a start. He was realistic, he was great at his job, he was a pioneer for too many reasons. He was loyal, no question.

(OK) A good man.

(DR) He was on The Stones' payroll, wasn't he, for a long time?

(OK) He must have worked for someone before.

(DR) Yeah, he was very tight with Bill Graham. Day on the Green (held in California from 1973) and all that sort of thing.

(OK) Yeah, he was a pioneer, a bit like Chip Monck in the old days.

Brian Croft said Chip Monck.

(OK) Yeah, Chip Monck is kind of credited with upgrading rock-and-roll to a bigger, more theatrical thing.

(DR) Visually.

Is there someone that sticks in your head, Ollie?

(OK) Ah yeah, Michael Ahern, Jerry Stickles. You know, Chris Lamb, all those guys that you kind of look up to, you'd look up to respectfully. As far as I was concerned they were older than me, so they were the kind of leaders when I was just starting out. You could see that they knew what they were doing and they were driving things forward. You could have a laugh, that it was all good fun in the '70s and early '80s. The fun started to be bled out of it because the act realised, probably quite rightly, that they could make an awful lot of money. So they stopped giving it to the rest of us and started keeping it, which is fair enough, I suppose!

Del, Ollie, thank you very much.

Del and Ollie continue to this day running Edwin Shirley Trucking Ltd at www.yourock-weroll.com.

Notes

1 www.bill-lewington.com/
2 www.footesmusic.com/

Part 7
Pioneers of Bussing

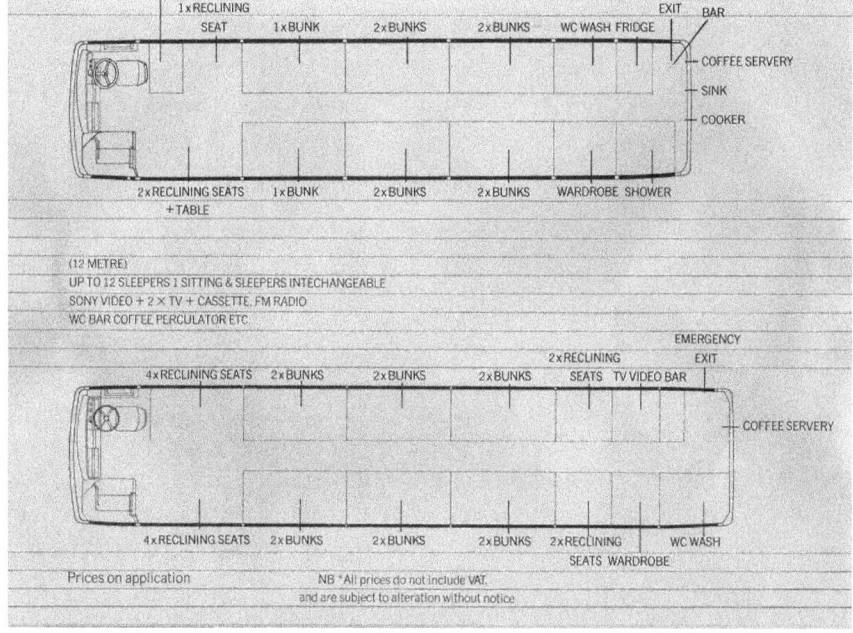

Plans to ML Executives sleeper buses

Having something in common with these 'pioneers' is indeed a great treat for me. It wasn't until I sat down with Len Wright at his home in 2010 that I discovered we had, in the past, a mutual employer, that employer being Supertramp.

When Supertramp hired a bus from National Travel back in 1975, it just so happened that Len Wright was the guy that got the job of driving them around the UK, and that gave him the urge to go out on a limb, to start a rock-and-roll bus company.

Initially, crews hired cars to get from A to B with no sleep for days on end. It wasn't until lighting became part of a band's performance that the crew started to outgrow the transit van or car and the bus started to come into mainstream touring requirements.

There were sole owner/operators that rented day buses in the early '70s, but as Len tells, if you wanted to hire a coach in the UK you had to deal with one of the nationalised coach companies more often than not.

I must not forget Mike Levy and his company Berryhurst at this juncture, a London-based company that owned a fleet of limousines, used by the rich and famous and record companies, who spent fortunes ferrying bands around. When Len began his company, Len Wright Travel, Berryhurst followed suit pretty soon after and were buying coaches to hire out as well as their fleet of limos. The coaches were day coaches though, and it took the likes of trucking companies, ML Executives and Edwin Shirley, to build the first UK sleeper buses before other coach hire companies took the plunge and invested in sleepers for the future.

16 Len Wright

Len Wright

November 30, 2010 at Len Wright's Home in Wiltshire

Where did you go to school and what were you doing before music?
I was brought up in Essex, went to Brentwood School, failed my A-levels, but had an interest in transport. I was always interested in buses and coaches.

Why?
Just had an interest, which developed as a kid, you know, your classic bus spotter rather than train spotter, I suppose.

I was a bus spotter. You used to be able to get all the books with ...
... Absolutely! And mark off the numbers.
   Well, I've still got them from those days, in the '60s. I've still got loads and loads of photos that I used to take, but I'd always wanted to do that as a career. I ended up with a management-training job at East Kent, the famous East Kent Road Car Company, and I was there until the early '70s, probably about '71. Then

I worked for Grey-Green Coaches in London in a management job, so I did have a proper job at one stage of my life.

And what did East Kent Road Car Company do?
Well, they ran a regional bus company. They ran all the local buses around Canterbury, Folkestone, Dover, Margate area.

Double-deckers and single-deckers?
Yeah, everything. They were unusual for that sort of company, because even though they had a big coach fleet, they also served all the holiday resorts, and did quite a lot of continental tours.

So it wasn't just regional?
No, it wasn't just buses. About a third of their fleet were coaches, which was unusual for that sort of company and purely because of where the openings were; when I finished the training scheme, I ended up working on the coach side of the business.

Did they teach you to drive the coach?
Yeah, I took my PSV driving test on an old Guy Arab double-deck bus. I like to remember going out with the instructor.

Where would he sit?
He sat behind you. They took the window out behind the driver's cab. So he'd sit in the lower deck and instruct you from behind, over your shoulder. I can remember going around the corner when I was out on a training run, and being tapped on the shoulder, "Pull over". I thought, what have I done wrong now? Had I done anything wrong? And he said, "What do think happened when you went around that last corner?" "Well", I said, "I didn't hit the curb, didn't see any other traffic or anything". He said, "Everybody on the top deck fell off their seats. Slow down next time". So anyway, yeah, it was an old double-decker with no synchromesh on the gears.

Was it usual to take one's test on a double-decker?
Yeah.

Was it a different test to single-decker?
It wasn't a different test, there wasn't so much of a difference between single and double deck. It was simply that they tended to use the older vehicles in the fleet, which tended to be the double-decks for training. The key thing was that it was a manual gearbox because that made a difference to your license. In fact, then, there weren't that many automatic or semi-auto. With the buses, the big thing in the '70s was the introduction of semi-automatics. You still made a manual change, but you had a little lever by the steering wheel when you didn't have a clutch. That was a key difference, it was completely different from driving a coach. The first

time I took a coach out, which was a lot longer, a lot faster, I can remember nearly taking out the traffic lights on the first corner.

But yeah, ridiculous as it is, having taken the test on that old double-decker with a top speed of 40 miles an hour, if you were lucky, immediately you can drive a coach, which is twice as long and at least twice as fast.

I then ended up basically in charge of private hire at East Kent, and particularly developing their incoming tourist market because we had coaches on the channel port side. I was given the job of developing business from France, Belgium and other incoming tourist markets.

Incoming?
Yeah. In those days, not everybody crossed the channel with their coach, so you would get, for example, French school kids, we used to have contracts with various education authorities.

In France?
In France. They would charter cross-channel ferries, you know, passenger ferries, and they would bring a thousand kids over at a time, and then we would take them up to London, to Canterbury and various trips.

Wow, did you get those contracts in the first place?
Yeah, because nobody was really doing much of it, you see, the mid-'60s was when the incoming tourist market in this country really took off. Prior to that, it wasn't really that organised or any big volume. It was one of those situations where I came to the end of the training scheme, and one of the managers was (a) given the job of doing something about the incoming tourist business and (b) had to find a job for this trainee who'd just finished his training course. So that was it, I was just told to go off to find Thanet depot, where there was a telex machine, and do something about the incoming tourist market. It was quite a steep learning curve.

Did you not go over to France, knock on the door and introduce yourself?
Yeah, that was what I used to do. But I had no training for it, nobody told me how to do that! So off we went. Then during that job, I came into contact with people from various other coach companies because they used to go off on British tourist authorities or trade mission-type things. I'd met various other people, one of whom was the director of Grey-Green Coaches, which was one of the bigger London companies. He offered me a job there. So I went there in, I think, about '71, and I was there for about three years in various management jobs. I got very frustrated for all sorts of reasons. I mean, I was given the job of developing the volume of business that they were getting, and I did that, I got huge amounts of additional business for them. But the service delivery was rubbish. I spent one year getting all this new business, and the second year, going around apologising for the quality of the service.

What sort of business were you getting from them?
Again, a lot of it was dealing with incoming traffic and just general private hire in London, all sorts of corporate stuff, no music business stuff though. I mean, nobody really thought about coaches for the music business then.

Everyone's still in the back of the transit.
Absolutely.

Football teams, did you do any football teams?
Yeah, we did one or two. We used to do Spurs, who were on our doorstep. Not many others, I think they were using coaches, but nothing like the sort of facilities that they would have now.

So I suppose Spurs in those days, you would have gone off to an away game in a . . .
. . . Fairly regular coach. I was driving a lot of weekends because it was a very seasonal business, as you can imagine, you know, East End Coach Company. You get busy from May to September, at every weekend, you'd have all the Londoners going down to the coast, tours going away, all that traditional market. I worked in a management job during the week and drove coaches at weekends. Then I found the management job getting more and more frustrating. I enjoyed driving, I was in my early twenties, so I just said, "Sod this, I am going to go and get a job as a coach driver". So I left Grey-Green.

In a way though, going down a peg.
Oh yeah, definitely. God knows what my parents thought. I mean, they didn't say anything about it, but I went and got a job at National Travel, based out down at Victoria, which is now National Express.

Earning more money?
Yeah. Well, debatable, but similar probably, if you take all the tips and overtime. I worked there and had a great time for two or three years; it was one of those situations where if you had anything about you, you could very quickly get on to the interesting work.

Were you talking over the bus PA system at the same time?
Sometimes, yeah, that was a good example actually because I was quite able and willing to do driver/courier tours where you had to look after the hotel check-ins, be the tour director, tour manager, as well as the driver. A lot of the drivers that were there either didn't want to or weren't really up to doing that sort of work or didn't want to be away from home.

For that reason, you know, I got on to interesting jobs quite quickly, but also that obviously creates a bit of resentment because you end up with people who have been there for 30, 40 years who don't want the job themselves, but don't actually want to see the new kid come in and do it either. Generally, I got on pretty well with them.

So I was working for National Travel. Of course, in that sort of business, in those days, your season, you know, if you are a tour driver—was sort of May to September, and then you'd be scratching around on school contracts and all sorts of odd jobs during the winter.

National Travel, tell me a little bit about them.
Well, the National Bus Company was formed in '68, and they then created National Express. All the separate regional companies had their own express coach services, and when it was all nationalised, they put them all into National Express. You know, the idea was to create a single brand, a sort of American Greyhound approach to running the coach services. The chairman of National Bus Company is a guy called Freddie Wood who had originally run Croda International, he'd been in various sorts of big corporates.

I never knew they nationalised the buses.
Yeah. Well, part of it, the regional companies used to be either Tilling Group[1] or BET Group, most of them. Now, the Tilling Group, the government acquired in 1947. They were for a long time, nationally owned, but run very much at arm's length. Then in '68, Barbara Castle was threatening to nationalise the rest of the bus industry. The other big group was the BET Group,[2] who owned the company I worked for, in Kent.

BET stands for?
British Electric Traction; they were originally a tramways network. Anyway, they campaigned vigorously against nationalisation, until in the end the government offered them a huge amount of money. We'd have our buses in Kent running around with these anti-nationalisation posters all over them. I went into work one morning, and there was this mad panic to get these all stripped off because overnight the business had been sold to the government and had been nationalised.

That must have taken thousands of jobs?
Well, no, if anything, it created them, and it took away a lot of the commercial accountability, as tends to happen in any nationalised business. So anyway, they then formed National Express and they then formed a few regional coach companies. Now, there had always been a coach company called Samuelson's, which ultimately had a connection with Samuelson's Lighting in the film business.

Right back there was a connection between the two; it's the same family. But this company called Samuelson's was based just across the road from Victoria Coach Station. That then became part of National Travel Southeast, which is who I was employed by.

Then in, must have been, October or November '75, Supertramp booked a coach with National Travel for their UK tour, which was just at the point where they were making it big. It was after they'd released *Crime of The Century* and they'd done well on the back of that, and they just brought out *Crisis? What Crisis?* This tour was to promote *Crisis? What Crisis?* and they'd booked a coach with somebody else, a one-man band, owner-driver, who for some reason couldn't do it at the last minute. Noel Millier was the guy, he lived in South London and did a bit of show business /rock-and-roll work in those days. It was due to Noel letting Supertramp down at short notice that I ended up on my first tour with them! They came to National Travel with 24 hours' notice for about a seven- or eight-week tour. I was the only driver who was prepared at a moment's notice to go off on a tour of that length, so I got allocated it. I remember going into the depot one day from whatever job I'd been doing in London, and I'd thought it was the end of the touring season. You go in and look at what you're due to do the next day. There was a big board up with all the work allocations. I got called into the office and they said, "Don't take any notice of that; you need to be back in. You can go home now, don't worry about the other job you got this afternoon because you need to get ready; you're on a seven-week tour. You're going out tomorrow morning". "Oh really, who is it?" I'd never heard of Supertramp. You had sort of tiers of drivers, so because I was a tour driver, I had a different uniform. I turned up next morning in this blue blazer with the National Travel logo on the pocket and smart grey trousers and everything, probably I picked them up from the A&M offices that were then in New King's Road, and this gang of fairly scruffy people got on the bus and off we went to Bristol.

Long hair, beards . . .
Yeah, off we went to Bristol. Got out on the M4, and we'd got somebody else to pick up in Maidenhead. Whereas a lot of drivers would have said, "It doesn't say anything about that on here", I said, "Fine, you tell me where we've got to go". I was reminding John (Helliwell) the other day. I can remember pulling up at the back of wherever, because he and Christine lived in a flat over the shops in Maidenhead.

I can remember pulling up 'round the back, and Christine coming out with Charles, he's a babe in arms, to wave goodbye. That's when I first met up with John. Off we went to Bristol. We got checked in the hotel, and right then it was, "We need you to take us down to Colston Hall" at whatever time it was. So I did that, went back up to the hotel. Then it was band and the crew. After the show, I brought the band back to the hotel, went back down for the crew and then somehow got involved in loading the trucks. They had got a huge problem about how to fit the equipment in the trucks. I got involved in sorting that out with them.

Do you recall what truck it was?
They were Edwin Shirley trucks. The two truck drivers were Dave Connor and Pat O'Doherty. Now, Dave was a tall, thin guy, who then from there on did all the Supertramp tours. What I do remember about him was he operated about 24 hours a day for about seven weeks, yeah.

What were they driving then? Seven-tonners, ten-tonners?
No, no, these were artics. They were Volvo F88s, they had two Volvo F88s. It was the first tour I think, that they got onto that size of production.

Now, at that point, we're talking '75. Was there anyone doing buses in the music business?
No, not on a regular basis. Not that I came across. That's how I got involved. That was my first rock-and-roll tour.

Had you ever been to a concert before that?
I don't think so. Might have bought the occasional Carpenters record or something like that, that was about it. Anyway, as it sort of evolved, I got on really well with them and really got involved. I think it's about the second or third day when this official National Travel tour driver's blazer went, and I started wearing the tour jacket. I mean, you know what they are like, anyway they're just a really nice crowd, particularly Dave, who was managing them. Of all the bands I've ever worked with, they were the ones that really sort of brought everybody involved in as part of the family. You know, you had to do your job and had to do it well, but you were involved. They were just really nice people to be able to work with the band and the crew. Through them, I got to know some of the A&M Records people as well, because obviously they were around on the tour.

Tell me about the bus. What was the bus?
Oh, it was a Bristol RE, which was the pride and joy of the National Travel London fleet. It'd won the British Coach Rally that year. They didn't have any work for it, but it was a great ego trip bus. It was fitted; it had a loo, it had tables and it had a servery.

Oh, it had tables? Tables where you are facing each other?
Oh yeah, which is why it was booked for the tour and had a little servery area, very basic servery area at the back and a loo and so on. As I said, it had won the British Coach Rally in Brighton that year.

What colour was it?
White with the arrow and National down the side, bit like the National Express coaches, same livery as National Express Coaches. That caused a bit of friction because the sort of top driver was allocated that because it was the top coach, but he didn't want to go away on a seven-week tour, so his coach, his pride and joy, got taken off him and given to me to drive.

How old were you then?
That was '75, I would have been 27 and had been working there for two years, yes. Not the most popular move when that happened, but that tour went on virtually to Christmas '75. Then they had a big European tour, going in the New Year, and they booked that with National Travel on the condition that I was allocated the tour. I then had Christmas off and then went away for about another two or three months, all over Europe.

The first Supertramp tour was in this National Travel Bristol RE
© Steve Thoroughgood

Did you look forward to that?
Oh, very much so, yeah, because I was single then.

Happy days.
Oh, absolutely. I was in a job where you didn't normally expect to get any tours away between autumn and spring, and suddenly I'm away, virtually all winter, which was great, and with people I got on well with, and going to interesting places and having some fun.

Had you been to those places before?
Some of them, but by no means all of them. Obviously places like Paris and Rome are on the regular tourist circuit, but in those days, places like Berlin weren't. I mean, going through the corridor routes to Berlin for the first time was quite an experience. You know, taking three hours to go through the border and having these people push mirrors under the bus and so on. Also, I think it was on that first trip with them; yes, it was, because I was still in a National Travel Coach; we were in Berlin, they'd done a show in Amsterdam, and then there was a day off, it was either one or two days off, and then a show in Berlin. We drove to Berlin, and several of the band and the crew wanted to go into East Berlin, so I said, "Well, I'll find out what the deal is", and found out you could take the bus over, you just had various bits of paperwork.

So I arranged that and went through Checkpoint Charlie to East Berlin, which was interesting partly for what it was then, but also because one or two members of the crew had bought some interesting literature in Amsterdam to read on the bus trip to Hamburg, and I hadn't had the chance to clear the coach out. So we arrived at Checkpoint Charlie, went through the Western control, got to the East German control, and they promptly searched the coach.

You are talking about adult magazines here, aren't you?
Yeah, and all this Amsterdam porn got confiscated, and I got pulled off the bus and lectured about this kind of material not being allowed in the German Republic. Anyway, that was quite fun, and going through the corridor routes was quite an adventure.

Did that teach you 'search the bus before going across a border'?
Well, I mean, I'd already learnt that. Obviously then, you had frontiers all over Europe, and I'd already learnt that 20 minutes before you got to a border, you let the tour manager or somebody know, and he'd go 'round and clean the ash trays.

How did you find your way?
Just looking up on maps, plan it in advance. We didn't have Sat Nav, but I had quite a big collection of maps and street maps.

Was anyone in the band particularly interested and assisted you?
Sometimes, yeah, I mean, that's how I got to know John better than any of them because he would often sit up front in the courier seat, chatting when we were on a journey. Certainly, if it were somewhere where it was a bit difficult or new to navigate, he'd help. I also got to know him better than the others because he was the only one that used to get up for breakfast!

And he still does!
But I mean navigation wasn't easy in those days. If the worst came to the worst, if you are driving into a French town, you'd simply hire a taxi to lead you in if you were really stuck.

That's what I always did as a tour manager, if in doubt.
Yeah. You don't want to be driving 'round and 'round in circles. Anyway, after that, they were due to go on a big American tour. They wanted me to go with them, which is one of those great things that you sort of look back on as the biggest missed opportunity in my life. I'd never been to America. I didn't really think about the implications. I was asked, "Will you come and drive the bus?" "Yeah, sure I will". I'd never been to America, didn't know where anything was, but was quite up for driving the tour bus. Dave Margereson found somebody in the States. I think, generally with the band buses, it's much more the case that you rent the bus and then rent the driver and put your own guide, whomever you want, goes on the bus. You actually rent the bus on a self-drive basis.

I was totally up to that, but then because Supertramp were doing so well at that time, they decided the tour could not be done with one bus, there were sections they had to fly, so it didn't happen, which I was quite disappointed about at the time.

Also during those two tours, I'd always at the back of mind wanted to start my own business. I suddenly realised that nobody else was doing it, you know, and there were these bands touring and looking for drivers. I was getting feedback from people at the record company and other promoters, various people I met on those tours. When they did hire buses, they just got a lot of aggravation because they'd get drivers who didn't understand what was going on or tune into it, were just difficult and awkward and unreasonable. Both the drivers and the bus company owners assumed that it would be a problem because it was a rock band.

So you had this sort of confrontation, this sort of cultural confrontation, before you even started the tour. If you've got that, then of course, things do go wrong. There was nobody out there that the customer was satisfied with, there was a general tendency to treat road crews as the scum of the earth and so any basic old banger will do, whereas actually they're some of the most discerning travel customers that you're ever going to get. You know, they travel; they know the difference between the bus that rides well and one that doesn't, because they are in them all the time, all over the world, and so there was that obvious gap.

At one stage, there was a suggestion that A&M Records were going to buy a bus or buses, that I would then get involved with them in running the business, but probably quite rightly, that didn't happen. A&M Records were saying to me, "Well, if you get a bus, we will give you some more tours, give you some work to get you going". I'd worked at Grey-Green with a guy called Simon Budden,[3] whose family ran a coach business down in Hampshire, and I was quite friendly with Simon.

So one evening, I phoned him and said, "Do you think your dad would be interested in backing a new coach business maybe for the rock-and-roll industry?". He said, "Hold on, I'll go and ask him". He comes back, he said, "Yeah, he thinks he might, can you come see us on Tuesday". So it was set up, Len Wright Travel was set up. You know, I never owned a controlling interest, I had 49% and the Budden family had 51%. It was set up that way with a second-hand Ford coach that was in their fleet, which had reclining seats and tables but nothing else in it. A&M Records paid me upfront for a Gallagher & Lyle tour, a Chris de Burgh tour, Nils Lofgren, and a Joan Armatrading tour.

Wow, okay.
Which got me going with the support from the Buddens.

That's the yellow and brown livery I know so well?
No, no, not at that stage. That was green and cream; that was in the Buddens fleet, still in their livery. Yeah, so I did those tours and a few other bits of work.

All UK?
UK and Europe, I think. What I'd already realised is that, in those days, the music business touring was really winter, you know, it was sort of autumn to

spring—which was great if you're getting a coach business started because you can always find work for coaches in the summer. That must have been spring '76. I think that happened after I came back off that Supertramp European tour. I did a couple of tours for A&M, then I was into the summer doing general coach work for various people. I remember we had no intention then of trying to fund another vehicle, but Supertramp came along and wanted to tour that autumn, and I'd already got commitments then. So we ordered the first new coach, the second vehicle in the fleet, totally on the back of a Supertramp tour because I couldn't say no to them. We went out and got it on HP (hire purchase), got this brand-new, state-of-the-art AEC coach. That was the first one in the Len Wright Travel brown and yellow livery.

There is a great piece with you and a bus on the web; Len Wright Travel, what is that about?
That's a bit of old cine film. My nephew put that up on YouTube because my father, who is dead now, used to take a lot of cine film. I mean it's pretty poor quality, but that was copied by my nephew coming over here to go through his grandfather's old cine film, and projecting onto the wall probably in here, and then recording it with a digital video.

That first AEC, actually I am not sure if I've got pictures of it, for the time, it was a real state-of-the-art high-spec coach. It still wasn't a sleeper bus, but we did a load of tours with that. It was like all these sort of things; for the UK, it was a real

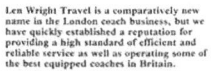

Len Wright brochure circa 1978

top-of-the-range coach. If you took it to Scandinavia in the winter, you'd gradually lose all the gears because it's a manual gearbox, and the transfer box at the bottom of it was quite exposed to the airflow underneath. In January, February in Sweden, Norway, the lubricant would start to solidify, and you'd suddenly find you've only got two or three gears left. So we developed a technique.

Oh, I was going to say, you set up a fire underneath to warm it up?!
Oh, you can—yeah, that's another story, but you can't when you are driving. What you did then, you would take one of the ducts that led to the demisters out and take the sleeve off around the gear lever, and shove the hot air blower down there, instead of the windscreen, so you could still drive!

Just going back a bit, the very first European tour with Supertramp when I was still driving the National Travel Coach, I didn't actually make it to the end of that tour because the last shows were Oslo and Stockholm. We were due to go overnight from Oslo to Stockholm, and the temperature dropped to about minus 30. Okay, it was a Bristol RE, there are hundreds of them, if not thousands in this country, but the radiator was at the front and the engine at the rear. So, all the coolant went down these long pipes, the full length of the vehicle, and froze up. I can remember to this day, I'm surprised I didn't get frostbite lying under it at about 1 o'clock in the morning, after the show, trying to get the coolant to circulate, with a blowtorch lying underneath. . . (laughter).

In the end, we had to give up. Dave Margereson said, "Look, this isn't going to work. We will stay here tonight, and I'll book them on a flight in the morning".

Len Wright brochure circa 1978

Anyway, at that point we'd now got two coaches, the original Ford and then this brand-new AEC, which had been bought just because we've got an overlapping tour with Supertramp. I was still living in a flat with three other guys, in Stamford Hill in North London, pretty rundown flat. The depot was the street outside, and there were a couple of coaches occasionally outside our flat taking up all the parking spaces.

One of the other guys who lived in the flat also worked for the coach company I'd worked for previously in London. He left there and drove the second coach.

His name was?
Bill Lear.

So he was your first employee?
Absolutely, yeah. The way we used to manage it was the phone number for bookings and anything businesswise was Simon Budden's office in Romsey. I had the office on my bus with a series of files and folders, and a portable typewriter that I kept in the side locker. I'd be away on tour and whenever we got to wherever we were that night, you know, we go down for soundcheck, I'd get on the phone to Simon to deal with any messages, and then I'll get the office out from the side locker and I'd be typing quotes and invoices and things, sitting at one of the tables in the bus, you know, until soundcheck was over, put the office away and go back to the hotel. I worked that way for quite a while, you know, just using Simon's office in Romsey as the message point.

Well, it just sort of gradually grew, because once you started getting a reputation, you know, the market itself was growing and nobody else was in it, initially.

Who was the first person to come up with another one, was it Berryhurst?
No, Trathens. We suddenly were getting more work; we ordered another new coach and so on, but the work was expanding at a greater rate than we could really accommodate. We were starting to subcontract work, and we used Trathens, who were one of the few other companies that had similar standard of vehicles. We subcontracted some work to them, and they then decided this was a good market to be in and got into it.

Who were Trathens?
That was a family business down in Devon, based at Yelverton, just outside Plymouth. The original business became huge in the '80s, they were all over, doing everything, and then went spectacularly bust in a very big way. It was a family business, and the two sons who were a similar sort of age to me, Dave and Mike Trathens took it over. They started Trathens Star Riders, which were for the music business. After they went bust, what was left of the business, we took over the Star Rider business, and a company called Park's of Hamilton, which is a big Scottish coach company, took over the rest of the business. I think it's still the Trathens brothers behind it; you start to see Trathens Star Riders again now, so I think they've gone back into it.

So Trathens, you started subletting to them?
Yeah, and then the Berryhurst connection came in, because Berryhurst were quite big in chauffer drive in the business, and they used to give us a lot of work; they gave us the coach work they were asked to provide, and they contracted it out to us.

Mike Levy.
Yeah, and then of course, they decided, why give the work away? You know, we might as well do it ourselves. They then started to buy coaches. Also Edwin Shirley used to give us a lot of work, and he would get asked for a complete transport package, so yeah, he'd contract the coaches out to us. So would Roger Searle at ML, which again, then I think they also thought, "well, actually, we might have to run them ourselves". It suddenly became a lot more competitive. One of the lessons that I learnt then is as long as you keep your service standards up, you come under price pressure, but you don't have to undercut. We had a situation for a while where you'd have tour managers coming on saying, they get a quote from us, they get a quote from various other people, and then a lot of them would come back and say, "Well, Len, we'd like to use you, but you're sort of 20% adrift. I can justify 5 or 10, but I can't justify 20", so you'd end up doing a bit of haggling, and you'd still get some sort of premium over the competition. Of course, you get all these growing pains that you get with any business, and we then had to start recruiting drivers and making sure that they understood what was going on.

How did you do that? Throw them in the deep end.
More or less.

Did they have the Len Wright chat?
Yeah, oh yes. I mean, you got to make it clear because the first time you go out on a band tour, it's very easy to think 'this is a great party', you know, 'we're going to have a great time'. What you have to understand is that everybody on that tour has got a specific job to do, and yours is to be at the right place, right time with a fuel tank full. Do that and then you can party. If you've got an overnight drive, you're going to be fresh, ready and sober to do it, you know, and you pick your time and place to party. The first time that I put a coach out without driving myself on a Supertramp tour, we had a problem. One of the Buddens drivers, who'd later on became sort of world famous as a rock-and-roll tour driver, Graham—did you ever come across Graham Hewitt, Turnaround Ted, he was one of my first drivers. The first tour he did with Supertramp, I had to go out and take over because it was one of those situations where he probably was never going to win anyway. He failed to win spectacularly and was constantly getting lost everywhere, and they were turning up late at places and so on. That was a lesson learnt for him, as well as for me. Later on, he became a very popular driver.

It was a little bit difficult for me with Supertramp; it was, you know, with the growing business, I couldn't go off on a three-month tour, so that was a bit difficult. Then, what really pushed us on sleeper buses was people like Edwin Shirley and ML getting involved with those. I had resisted for a long time, having anything that was purpose-built as a band bus because of losing the flexibility to use it during the summer on other work.

So it'd been discussed?
Oh yes.

And why was it discussed?
Well, because they were available in the States.

The band and the crew used to go over to the States and tour by bus.
Travel on a sleeper bus.

Travel on a sleeper, come back and say to you, "Why haven't you got one then?"
Yeah, absolutely. We did have some purpose-built ones, but a lot of the work that we did, we priced it differently; it was to have demountable bunks.

Yes, I remember that because I'm sure I used them.
We would fit out a bus for the winter as a crew bus, with bunks in, then come the spring, they would come out, all the seats would go back in it, and it would spend the summer doing shuttle runs up and down to the south of France and Spain and tours around London, that sort of thing.

When did that start happening, Len, and who designed and put those beds in for you? Were they already there?
No, they weren't; it was early '80s. Well, certainly late '70s. It was probably maybe 1979, I think.

Do you think ML had done a purpose-built one before you were rigging and de-rigging?
I honestly can't remember now what the sequence was, but certainly there was a general pressure to get sleeper bunks in on the vehicles. Actually, they must have been a bit earlier because the first purpose-built one we had, I think, was 1981, the first Nightliner.

Now, all of the fitting out for the purpose-built ones, we used to buy the shell from a regular coach manufacturer, and then it would be fitted out by a company called Fair Stage Conversions. We had a depot in what used to be the old goods yard at Richmond Station in London, it looked like a scrapyard, this place; we had a corner of it for our coaches and a portacabin office. There was a haulage company on the other side of the bit that we were in; there was a second-hand London bus dealer down the far end; there was a local coach company that was pretty scruffy, pretty basic.

Len Wright Viewmaster coach at Bristol's Colston Hall and called Nightliner One
© Terry Jones

Pretty big yard.
Oh yeah, big yard. There was also a company that used to tow in accident cars for the police. You can imagine the state, you know, what the place looked like, but in part of the old goods shed is this company of about three or four guys, who used to fit out exhibition vehicles, and because we were in the same place, got to know them. They did the fit outs, they built the bunk units for us, the demountable ones, and they also did the fit out for the purpose-built ones.

I remember curtains, and you must have built in some sort of internal scaffolding?
The curtains would be there anyway because one of the options when you buy a regular coach is having curtains in it, and then what we would do—aahh okay, you're thinking of the curtains on the gangway side, inside. What we would do for the bunk unit, there would be bulkheads that would fit into the floor channel because all the seating on coaches is in a channel, so it gives you flexibility to fix tables, seats or whatever.

We'd have feet on the bulkheads that fitted into those channels, and then we'd bolt them into the parcel rack structure above. Then you'd have brackets and the bunks would bolt into the bulkheads. We also had panels that fitted in on the window side. So you have a blank, you know, you wouldn't just be next to the glass, you'd have a panel inside the glass and then curtains that went across the bunks. But the first purpose-built one we had was Nightliner One, which I think was 1981. That was the first. And I can't remember whether we had the demountable ones before or after that, but it may have been a year or so before.

I got a feeling it was before.
Yeah, but not that much earlier. I don't think it would have been before '78 at the earliest. Then, of course, Edwin Shirley was famous for his horsebox, because his first coach was actually built by Lambourne, who made the big horse transporters.

Rather than go to a coach manufacturer, he went and got a Lambourne horsebox shell and fitted that out as a coach. Yeah, I think he had two altogether, but they sort of gave up on that and decided to concentrate on trucking; you know, one reason we did not go into running trucks is that it is actually quite different. There is a tendency to think, well, they're both vehicles, all the issues are going to be the same, and they are not.

I think ML were reasonably successful with their coaches. I think Edwin Shirley had quite a lot of aggravation with theirs, and in the end, came back to working with us. Even when they had their coaches, it didn't cover all of their requirements. You know, we still did quite a bit of work with them and with Roger Searle. It all sort of evolved and then, of course now, it's all about double-decks, more and more facilities.

Well, you know what, that's the whole cycle, you know, because Simon Budden, who I mentioned earlier, his family sold their 51% to a guy who ran a coach company in St. Albans who I knew, because it had got to a point where they felt it was getting too big for them to support.

Oh really? You didn't want to buy the whole thing?
I couldn't, I wasn't in a position to. Anyway, their shareholding went to a guy in St Albans, and then by the mid-'80s, the National Bus Company was being privatised, and it got more and more complicated.

Len, thank you very much for this fascinating story.

Len Wright was something of a pioneer in the UK bus industry; he started London's first tendered bus operation—London Bus Lines—in 1985.
Today Len runs a web travel site called www.choosewhere.com.

Notes
1 www.petergould.co.uk/local_transport_history/pioneers/people/tilling.htm
2 www.rediffusion.info/bet.html
3 www.countrybus.co.uk/buddens.htm

Part 8

Pioneers of Catering

I can only retell my personal experience of touring and eating in the '70s, and the quality of the catering provided would depend on which country you were in. When the band 'rider' was created, it became part of the contract between the band and promoter; the promoter was legally bound to keep his side of the deal, right? Well sometimes!

In the UK it was the 'takeaway' that fed us most of the time, both band and crew, and the promoter would have a variety of local menus to choose from before sending out his man to go for burgers, fish and chips, Chinese and Indian food. Over in Europe though it was more effectively organised. The promoter would have a local restaurant open its kitchen early in the evening, and after completing the soundcheck, band and crew would all be escorted to a convenient, cozy restaurant to eat dinner.

Initially, hot food served in the actual venue didn't happen, not here in the UK or for that matter in Europe. There were lavish plates of sandwiches, crisps and nuts, but that was about the scale of it, along with the soft drinks, beers and spirits. In the US they were ahead of us, and they did have venue catering.

All the takeaways and curly sandwiches stopped once tour caterers joined our entourage, but that didn't really start until the very end of the '70s. My first experience of this way of touring was actually in the US in '78, with the Grateful Dead. They had a catering team they hired from the New Jersey venue, Capitol Theatre Passaic, that went on the road with them and it was AMAZING . . .

It wasn't until 1983 that I got to work with an awesome team under the leadership of Debbie Bray and her company Just Desserts, feeding over 100 people every day on the Famous Last Words Supertramp European tour. In all weather conditions, these guys and girls created a tented space of calm and comfort, as we played to crowds in football stadiums.

Here in the UK it was Chris Adamson, with the help of Richard Dale, who started his own production company, CA Touring, and brought in Debbie Bray and later on in '78, Val Bowes. I'll never forget the in-house caterers at the Hammersmith Odeon from the early '80s and their founder, Val Bowes, tells me her story of the creation of Flying Saucers.

Sandi Grabham created a very early catering company in the UK called Home Cookin'. Her amazing rise as a pioneer in tour catering came via a strange but interesting business opportunity, way back in 1977, but I'll let her tell the rest.

17 Sandi Grabham

Sandi Grabham

March 24, 2016 with Sandi Grabham by Skype

Let me ask you, where did you go to school?
I went to many schools because my mother became a widow when I was only six, so she shortly after that got housekeeping jobs in different places. I started off school in a place called South Shields, and I ended up at school in London in Notting Hill, all over the place, yes.

But you still have that distinctive Northeast accent.
I have, but believe me, by accident, I employ a lot of northerners. They're hard workers, and I don't get any grief from them. When I am with them, I have to say, "Say that again", because I don't sometimes understand what they're actually saying to me. So, yes, you never get rid of your accent!

Okay, the formative years were in South Shields. You came down to London at what age?
Oh, I think I would have been about 12.

And finished off your education, in London?
Yes.

And did you go on to do A-levels, O-levels?
I didn't, I left school at 16, and my aim was to become a nurse, and I went into nursing, in I think 1964.

Okay. Tell me the story as you remember it from there then.
Well, my husband was—well, he still is—he was a musician, and he's a northerner. I left nursing because I was in love, and I would have had to live-in for three years, and I didn't want to do that at the time, but I love nursing.

So hang on, this is Mick Grabham,[1] isn't it? How did you meet Mick?
A local coffee bar at the time in Sunderland. At that time, he was gigging. He worked, but he also did gigs locally, that's how I met him, really.

How come you're back up north, meeting him in Sunderland?
My mother stopped being a housekeeper then, and we had owned some flats, and so we went back to one of the flats that wasn't rented out. Then she got a normal job. Actually, she was a cook, they called them cooks in those days. I had moved back up north and was completely depressed because when I left the south, it was the Mods and the Rockers. Can you remember that period? In Newcastle it was just so behind the times then. I was completely depressed being back up there, it wasn't 'happening'. It is now, of course, but at that time, it wasn't. London was the happening place to be.

Yeah. And so you'd started your nursing training in Sunderland?
It was only about a year that I was a cadet nurse, I started at 16. When you got to 18, you took your exams to decide what kind of nursing you were going to do. Then in those days, you lived in for three years whilst the training was going on.

What hospital?
Oh, it was called the Ingham Infirmary in South Shields.
   Then my husband got a break. He joined a band called The Plastic Penny, and they were in the charts at that time. They needed a drummer and a lead guitarist, and through a friend of theirs, both Mick and Nigel Olsson (the drummer, who is still the drummer with Elton John) came down and joined The Plastic Penny together. It was a few years, three or four years maximum, and when that happened, we moved down south.

How did Mick find that gig, an advert in *Melody Maker*?
No, it wasn't anything like that. A friend of theirs was a bouncer at a nightclub in Sunderland. The band had told him about them not having a drummer or a

guitarist, and he said, "I know two people that do that and are very good at it". And that's how it happened.

Oh, wow. And so did Mick know Nigel?
Oh yes, they were in bands together, up north, yes.

Our friend, the bouncer, he used to run the Seaburn Hotel, I think it was called; it's hard to remember. He used to have bands play there years ago like Cream, all different bands like that.

Okay, so did you come down with him to London?
I followed him down a month later. We lived in Kensington. Then he joined many bands after that. He was in a band called Cochise, and he ended up in Procol Harum. Nigel ended up with Elton John, and he's still there now.

Instead of me touring with Mick, he realised one day that we were able to buy a home, and he also realised that what he earned, me going with him, it came out of his money. So he toured, and I wanted to do something because obviously, I didn't have my daughter then. I wanted to start a little business. So through contacts, through Mick, I heard about Gus Dudgeon and his studio in Cookham. It was a beautiful place, The Mill,[2] but they didn't have catering. When they had a band in, they couldn't get an evening meal, and that's how I started. Cooking an evening meal at home, actually half cooking it, and then driving to Cookham, finishing it off in their kitchen, serving the band and then going back home.

How long did you do that for?
Well, like anything, there would be a band in and then there'd be a gap, and then there'd be another band in, so probably a couple of years, starting in 1977 or early '78.

The Mill at Cookham
© Rod Duggan

The studio at The Mill
© Rod Duggan

The following note and the two photos are from Rod Duggan, who was at this time the technical manager and later to become the studio manager of The Mill.

Hello Richard, The Mill (aka The Sol) was the late Gus Dudgeon's private recording studio built in a converted water mill in Cookham by Eddie Veale around 1975. It was intended for quadrophonic mixing of Elton John's back catalogue, but quad never survived that long. It was arguably the finest and most beautiful studio ever built and utilised an American MCI console and tape machines all in custom wood veneered housings. I was the technical engineer until Gus's financial issues forced the sale, when the new owner Jimmy Page took over. I was then promoted to Studio Manager.

Artists using the studio included Chris Rea, Mick Fleetwood, George Harrison, Elton John, Boomtown Rats, Bill Wyman, and Led Zeppelin (the Coda album). They shot the video for "Je Suis un Rock Star" in the grounds of The Mill.

Do you remember any of the bands?
Well, Chris Rea was one of them. Elton John came in, ones that you probably wouldn't have heard of, a German band I can't even think of the name. Everybody who worked at the studio went to Chris Rea's first gig, which I think was at the Hammersmith Odeon.

So, this was keeping you quite busy. Where were you living at this point?
Well, we had moved from Kensington, and we lived in a place called Hatch End in Middlesex, it was only about half an hour from Cookham.

Hot-boxing it.
Yeah, and finishing it off when we got there because there was a kitchen.

Did you give them a menu or did you just do what you wanted?
Oh no, it wasn't like that in those days. You just cooked what you thought would be correct. If there was a vegetarian, you'd cater for the vegetarian, nothing like it is now where you've got to have a lot of choices. But you know, I'd only be cooking for, say, 10 maximum.

Yeah, lovely. Go on. Did that in itself turn into an artist saying, "Well, will you come and do. . .?"
Someone that both Mick and I knew found out that Al Stewart needed caterers, they were already on tour, Time Passages Tour Europe 1978, and they had the two Welsh guys, Bubble and Squeak, out with them. For some reason they got paid off, and I turned up. I rented a Range Rover and turned up with a friend of mine; there were only about four or five gigs left to do, not a lot. We turned up with a hostess trolley, some electric rings and some paper plates and that. Well, as funny as that sounds now, at the time, there was no guidance on what went on because the bands Mick had been in, the promoter just put some fruit and shit in the dressing room; that was it. So it was successful!

Was it in England?
Yeah, the Manchester Apollo. But you could imagine driving; in those days there was no Sat Nav, no mobile bloody phone or anything. You had to just find your way to the gig, which now of course is a piece of cake but it wasn't then.

No, the itinerary was all typed on one page!
If we had one! We didn't get anything, I don't think. I did these few gigs, and I am guessing Harvey Goldsmith promoted it. Years ago, I really only worked for people like MCP in Walsall, and Harvey Goldsmith, they were the promoters around at that time.

Yes, that's right.
So we did that Al Stewart tour, that was successful. Actually they gave us a tip, which is very rare. Now, you don't get tips. They were that pleased with us. I can remember saying to Rachel, who was with me, there was a rider for the dressing room and everything, and I can remember this American saying, "You're going to have to set the bar up", and I went running downstairs and said to Rachel, "Oh, they want me to stay in their dressing room and run the bar", completely green I was! But they didn't, they just wanted me to set it up, and Rachel said, "Well how am I going to manage to feed 25 people on my own?" We were panicking because I'd never done this kind of thing before, but obviously, your knowledge grows the more you do it.

Who paid you?
The promoter would have paid us. Oh, it would have been Harvey Goldsmith, yes.

Okay. And you said Rachel. Rachel was your assistant?
Rachel Hawkins, yes, she did many jobs with me.

What happened after that? Were you still doing the studio?
Well, I was, but there was nobody in at that time, so that was good. I don't know if I went back to the studio. Rachel and I took it in turns to go in, but not for very long after that because I think the second tour we did was Frank Zappa, Sheik Yerbouti tour.[3] We were learning all the time.

Okay. So again, did the promoter hire you?
Yes, well, you know what? They didn't have an option. There weren't any companies.

And was that Harvey again?
It would have been Harvey Goldsmith.

Was that just the UK again?
Yes. Oh yeah, we never went into Europe at that point. Never. Remember, we were still self-driving in a van, you were still loading your own equipment in, nobody helped you then, you still washed up and had to do the shopping.

Are you on electric or are you on gas?
Then it would have been on a household oven, Calor Gas.

Did the truck carry your equipment?
No, you put it in a transit van; I only had about one flight case. You know, it was in baskets and things, it was awkward.

Yes. And what were you cooking? Did you talk to the tour manager or did you make it up?
No, nobody asked! Nobody said anything, because they had been used to getting a few quid to go to the pub and have a sandwich or fish and chips or something. When they got different food, they were grateful, like Sunday, probably do a Sunday lunch with Yorkshire pudding, the whole business. Nowadays, I've got the bain-marie you know, there's six choices of all kinds of food now.

So that's how it ended up. Now, you've got to be a restaurant really.

It's extraordinary what you guys can do nowadays. Absolutely extraordinary. Frank Zappa, was that a long tour? Do you remember?
No, it wouldn't have been long. It would have been a week or 10 days. I have most of my passes laminated, but I have got a hell of a lot that aren't. They're in the office, so I am just looking around to try and find dates on any of them. I've only found Band Aid at the moment, that was 1995.

So Sandi, Al Stewart, you went and just took over, were you Home Cookin' then?
Well, we kind of had that name. But we only had that name from going to the studio. We just used to say, "well it's home-cooked", so the company became Home Cookin'.

Who were you dealing with at MCP? Maurice or Tim?
Well, Tim Parsons mainly. I knew Maurice very well, but he was like the boss, and he just came to different shows.

We're in the mid- to late '70s then, it's Home Cookin', and you're purely working for just two promoters in the UK? You'd be doing one-offs?
We did one-offs, yes.

More than tours?
Mmm. Probably equal.

Do you remember the first time you did a gig where the band had hired you?
I don't believe any band hired us. I think where the band's concerned, the powers that be, the promotion, or maybe the production put it to them; do you want to use the same company and the answer's either yes or no. Fortunately, you know, these days, I've done 12 years of the X Factor so I must be doing something right. Every one of Boyzone's tours, I've kept the acts I've worked with.

Yeah. So nowadays, you are working for the artists, not the promoter?
No, the production—as you know, these production companies now get everything together; well, they also get catering together. Once that's sorted, the promoter goes along with their decision. That's how it seems to work these days.

Towards the end of the '70s, there came a point where bands used to take catering companies on tour with them.
Not many. They took like a personal chef with them, that's what happened. We did a few AC/DC tours at the time. One of the chefs that worked on that, they took him to Nassau while they recorded an album there.

I am looking at all their passes and, you know, nowadays, you put dates on, don't you, but there's no dates back then. Oh '88, AC/DC, Back in Black. So I did one, two, three, four, five tours then.

Do you remember the first time you went to Europe?
I don't think we did in the '70s. It was becoming a big deal for the band and production to start taking caterers with them. If you ask me what was the first tour in Europe we did, I haven't got a clue. We did Pink Floyd. Then, of course, other companies started to spring up then, catering companies like Debbie Bray's Just Desserts, I think.

That's right, and Val Bowes. Yeah, do you remember Val got the contract to do the catering at the Hammersmith Odeon?
Yes, I remember that.

That's what really catapulted her into the touring world. She formed her company called Flying Saucers and is still operating. It would have been well after you'd been working with these promoters in the mid-'70s.
Yeah, there wasn't anybody. Maybe Bubble and Squeak still did it. The '80s is far more clear than the '70s for me!

Yeah, it really kicked off properly in the '80s. I remember tours in the '70s, if you went to Europe, the promoter would just take you to a restaurant after the sound check. In England, it was, as you said, packet of crisps and a hot dog, if you were lucky.
It did become quite a big deal taking caterers; suddenly they could have tea and coffee whenever they wanted it, water, soft drinks, you know. It was on tap for them, and a breakfast. Hey, I remember one day being out of my brain because the crew used to roll joints in catering; that would never be allowed nowadays, but it was far more laidback then. Didn't seem that unusual, did it, at the time? I never liked it, so I never smoked it, but the air was thick, and some of these small dining rooms . . . if you can call them a dining room!

Did you ever do a festival back in those days?
Well, we did, and I was trying to think on that because they weren't really called festivals then. I remember doing one at Knebworth, Pete Wilson was promoter's rep at the time; he isn't anymore, he was one of the partners in 3A Entertainment.

Pete was still with Harvey at this point?
He was. Well, I remember Pete calling us up and saying if we did a good job on three warm-up gigs in Scotland and we did them well, there was a chance we could do some outdoor shows with these people. It was The Rolling Stones. Our first gig was in Aberdeen, really small tiny little theatre in Aberdeen, Aberdeen Capitol Theatre. Glasgow Apollo[4] and Edinburgh Playhouse were the other two warm-up gigs.

That must have been part of the Tattoo You European Tour in 1982.[5]
There was more than one big show actually. We definitely did Bristol, and we did that big park in Leeds, a huge place, Roundhay Park. We did the shows successfully, but it was a bloody nightmare doing the three little gigs.

When did you get flight cases going?
Well, they gradually came in. I am guessing the more work I got, the more I could afford to get a flight case. I guess I had got them by the '80s. Not as many as you would need, but they were building up.

You must have gone to Europe eventually?
We did go to Europe. I don't know when it was the first time we went, it might have been AC/DC. I remember we employed a driver, because we still weren't on the bus or anything.

Really!
Yeah. I think our equipment may have gone on the bus, and we hired some vehicle to take the caterers. We got a driver because none of us had been to Europe.

Believe it or not, buses, especially sleeper buses, came really quite late in the day.
Well, now everybody's got it easy. The gear gets picked up from your warehouse. You're on the bus and you're in the same hotels as everybody else. So completely different. When I tell the people that work for me, "Oh, we used to wash up". I remember somebody asking if we had a special kind of mustard, and I said, "I'll just go and have a look in the van and see, I'm sure it's in there"; it wasn't in there, so I drove to the shop and got one and came straight back. I said "I've found it, I am sorry I've taken a while". "Oh, that's all right, you've got it". It wasn't just like Colman's mustard. It would have been a French mustard. Even in those days, herbal teas weren't fashionable, they didn't have any, really.

Did you work for MAM, for Barry Dickens?
I think I did one or two gigs. I must have worked with them on the Diamond tour because I did Neil Diamond. I am sure he was the promoter.

Have you kept your diaries?
No. I just got loads and loads of paperwork and swag from going back donkey's years, but it's all in the warehouse. I think, right, when I do retire, I am going to get a ghostwriter to write my book, because it's all about things I shouldn't be telling. I mean, it's not incriminating anybody, but it's things that happened with the catering while they were busy doing the show.

You should do it, definitely.
There are some hilarious stories really in there. I just shouldn't say them at the moment.

Could you tell me one?
Well, we did a David Bowie tour. I don't know when that was. We must have done a few outdoor gigs with them, and we were doing Sunderland football club, the old club, and Bowie wanted sushi. You could get sushi from London, that wasn't a problem, and Birmingham because I know the sushi came up on the train, but we struggled to think how we would give them sushi in Sunderland. I mean, people would think this is stupid now, but it wasn't then. So we rang the Nissan factory because we knew they had sent English chefs to learn how to do sushi in Japan, and he did it for us, for the gig!

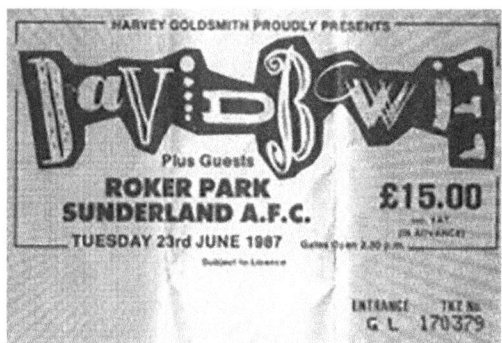

1987 David Bowie concert ticket at Roker Park
vintagerock.wordpress.com

The other thing we did, terrible thing, Bowie suddenly wanted about 8 o'clock at night, peppermint tea. We were in Sunderland, remember, you couldn't get it. I don't think we'd put herbal teas out because there weren't herbal teas then.

Earl Grey, maybe.
Earl Grey, yes, but certainly not peppermint. If we'd been in London, we would have got it, yes. So we made a pot of tea and put Trebor mints in, and he drank it!

Perfect.
And we were trying to get these bloody mints to dissolve, and they don't really dissolve. In the end, we just said, "We've got to put them in", but nobody said a word. And he had a cup or two.

So it worked.
Yeah.

I remember on Fleetwood Mac in 1977, on their rider were six limes. This is the same story, really. You could get them in London, but you couldn't get a lime anywhere else in the country, so I had a case of limes in the back of my car.
(laughs) Well, I think people these days would find a lot of this hard to believe, but it's just the truth. You couldn't just get things just like that. Yeah, well, that was a nightmare, we didn't think so at the time, but knowing what I know now, I think how the hell did we manage that?

Just amazing those early days, what you just got on and did.
Well, you had to, but I have to say, from my point of view, it's a great grounding in those early days.

Yeah, did you have a favourite venue?
I can't say I have. You say a good venue, I have these days it's the O2 because of the space!

Lovely talking to you, thank you very much.
Thank you, Richard. Thanks a lot now.

Sandi continues to work in the music business as a freelance catering executive at her company Gig-a-Bite. They provide a total catering solution for music tours and corporate events, and have worked with many world-famous names.

Notes
1 musiciansolympus.blogspot.co.uk/2011/10/mickgrabham-guitar.html
2 www.themillrecordingstudio.com/sessions.html
3 www.zappateers.com/fzshows/79.html
4 www.glasgowapollo.com/index.asp?s_id=1&m_id=2&page=1
5 www.timeisonourside.com/chron1982.html

18 Val Bowes

Val with Barry Manilow

March 16, 2015 with Val Bowes by Skype

Where did it all begin for Val Bowes?
Well, Richard, I grew up in North Yorkshire, a daughter of a North Yorkshire farmer in the middle of nowhere on the North York Moors. I had a really amazing upbringing. We had a home farm, owned by the local Lord and Lady, and we had everything at the farm. We used to milk our cows and make our own butter. I'm not that old, but it's wonderful to remember my mother (I am that old, actually!) making butter every Tuesday in the yard from the cows that we milked, which is amazing, really.

Yes, I remember my grandmother doing the same, you know. It was, yeah, lovely. And you still have that tinge of North Yorkshire accent to your voice.
It's true. I left there when I was 18, and my family is still there, but there's still things I say like 'bath' and 'path' and 'grass'. I will never not say that!

You went to school up there, too? Did you go to university?
I went to school up there, and this is kind of interesting because this is one of those cases of 'life is what happens to you when you're planning something else', as we

know. I did my A-levels, and I only got one A-level because I had a broken heart at the time. The only course I could do that was left, virtually, was in London, and it was a Hotel Catering and Management course at the Polytechnic, an HND (Higher National Diploma). So I moved to London from North Yorkshire, and it was the best course I could ever have done in my life. I didn't even know I could cook until I got there, so that was great.

Perfect. What Poly?
It was the Poly of North London at the time. I don't know what it's called now. It could be some university or other, but it was in Holloway.

Oh, it was in Holloway, okay. I went to North East London Poly. Where were you living?
A variety of places to start with, but ultimately, I moved to Greenwich. Do you remember a person called Dafydd ap Rees? He was a mutual friend of mine and the guy I was living with, do you remember a band called Charlie?

Yes, I do.
Well, I lived with Julian Colbeck, the keyboard player, for eight years.

Okay. Boyfriend?
Yeah, yeah, yeah, and we lived above a bookshop in Greenwich which belonged to Dafydd, he rented the flat and we took over the flat. To cut a long story short, we bought the lease in the end, but it was a second-hand bookshop called Book Lovers. We started a cookery shop called Cook Lovers and just changed the 'B' to a 'C'. That's clever, eh? At that point, I was an assistant domestic service manager at Greenwich District Hospital. I was only 21 or 22 or something. This was 1978. It all kind of coincided with getting the shop, me being at the hospital. And then, my God, what was his name? The guy who started a lighting company, he worked with Chris Adamson.

Richard Dale?
Yeah, him, he was stuck for someone to go on a tour. Styx, it was, he begged me to go and replace the person who couldn't do it. So I went, and that's how I started doing catering, because someone dropped out.

Just roll back a minute here, how did you know Richard?
He was associated with Julian who I was living with, just through the industry. That's all I know, really, as I can remember.

When you were in Greenwich, were you a big music fan? Did you go along to lots of shows?
No, not really. Luckily, I think, I'm not totally star-struck or anything, I've got really eclectic musical taste but I didn't really go to that many things. I went to the Marquee because that's where Julian used to play, not when he was in Charlie, but

when he was in a strange band called Greep. Anyway, I used to go to the Marquee in Wardour Street quite often.

Okay, so through Julian and Richard, you got asked to step into a tour to do the catering, and at the time, you're working as a manager in a hospital?
No, at that point, I'd left the hospital to run the shop, the Cook Lovers shop. It was a shop that sold lots of different things for the kitchen, but many of them were locally made by potters and local artisans, you know, screen-printed table cloths and pottery, kitchen pottery, and all that kind of thing, as well as cookery books and stuff. It was a really sweet little shop, actually.

What road was that in?
Trafalgar Road, opposite what was the Bricklayer's Arms where Squeeze started life.

Who was running the catering then on that first tour with Styx?
Debbie Bray and another person, someone called Kate, who moved into lighting, I can't remember her surname. It was a long time ago. Anyway, after that, Richard asked me to do a Hall and Oates tour, Daryl Hall and John Oates, on my own. It was only supposed to be 12 people, so I said "All right then". Self-drive, washing up, none of these bloody helpers or anything, you know. I did it, and then the band got wind of the fact that everybody rather liked the food, and so they all wanted to eat, and suddenly it turned into 25 people, and I was still on my own. I was like, "Oh, my God. This is just ridiculous". I did the tour and I thought, "If I'm going to do this, and it's this hard, I'm going to do it for myself", and that's when I decided I would start Flying Saucers.

Yep, yep. Initially, on that first tour that you did, I'm going to assume that Debbie Bray was already a member of CA Tours?
She was, yes.

Was all the catering equipment very organized with CA Tours? Was it all very basic?
Oh, no. Everything was in flight cases then. And when I started Flying Saucers, which was about the same time, I automatically had everything in flight cases. I wouldn't have done that if it wasn't the norm, I guess I just followed the format of having flight cases.

What do you remember about that first tour?
I don't remember anything at all, really, about the Styx one! I remember how lovely the Daryl Hall and John Oates people were though, and that good feeling you get from finding out that you can do something like that, really well. It was an amazing feeling. You know that feeling. It was very intense, it's very physically intense, really mentally intense. It's all-consuming and that gives you such a buzz, doesn't it?

Yes, absolutely. Of all the people that work from dawn until dusk, it's you guys up at the stoves, sometimes in terrible conditions. When are we talking about? We're talking about 1980?
It must have been. I started Flying Saucers in October 1980, so it was the run-up to that, maybe potentially, even a year's run-up to that. It might have been the beginning of that year. It was all quite quick, my decision to do it for myself.

Had Debbie, by this time, started Just Desserts?
No, she stayed with CA for a while, then it wasn't that far after though, maybe a couple of years later. There were only about three companies when I started, and one of them was definitely Bubble and Squeak. I think the other was Home Cookin'. The other one was CA Tours, and I think I was about fourth and now, there's really rather a lot, which obviously shrinks the market share. I've spawned a few because, of course, Tony Laurenson used to work for me.

Is that where he started?
Yeah. And so did Debbie Sharpe and Kim Davenport. They both worked for me, and they started Eat Your Hearts Out!

So off of the back of the Hall and Oates tour, why did you decide to do this? I mean, yes, it was fun, but it was incredibly hard work.
Again, something just rose up in me, that "if I'm going to do this, then I'm going to do it for myself because it's too bloody hard to do it for somebody else and not actually get the best you can out of it", you know.

I was really lucky because Kennedy Street did that tour, and they took a big liking to me. Then within a couple of years, two years, maybe, I was absolutely lucky enough to be given the outbuilding at the back of Hammersmith Odeon, London, to have as my office and base and to cater for everything that came into the Hammersmith Odeon. It was just like the most amazing thing because I got to know so many tour managers and bands and stuff. You know, in a really short space of time, so that was an absolute blessing.

How did that come about?
I'd been catering for a band there, and the manager then approached me and said, "Look, we've got this building, and I'd really like you to take it over, and we'll put as much as we can your way, to build it up". That's just how it happened. He didn't ask me for rent or anything, for three years, which was awesome, of course. He just gave me the building because it needed to be done up and everything. I did all that. Then I was given no notice whatsoever to leave. I had to find somewhere to go, and so I woke up in the middle of the night, one night, and a little voice was telling me to go to Nomis Studios, Nomis Studios. I rented a place at Nomis Studios for an office, but I had to do that in a nano moment, really.

They had their own catering at Nomis there, didn't they?
Oh, yeah, we didn't do their catering, we just had storage and an office there.

Going back to Hammersmith Odeon, how long had you been on the road before that opportunity came up?
I think it was within about a year or so. Then I was there for three years. That was just so lucky.

Did you have an addition to the building? I seem to remember two bits to it.
Yes, the other bit was Stage Miracles, I think, where they used it as a crew room. I did have a little upstairs part, up some really steep stairs, where my little office was. God knows how I never broke my neck, quite frankly.

We all went out there to eat, didn't we, actually out in that building? Yeah, because those dressing rooms were tiny, I remember.
Yeah, yeah.

Back in the early days, correct me if I'm wrong, you used to carry around gas everywhere?
Oh yeah, gas was perfectly acceptable back then, but we don't carry gas at all now. We only use it if we're in a festival situation or something. Actually, we don't even send it on the road anymore.

Why was it gas then, and not electricity?
Well, the power resources were potentially not good. At least, you were self-contained if you had the gas, you know, in terms of cooking power.

I can't remember who it was, but I swear I saw a caterer once having to actually do it all in the back of the truck, did you ever have to?
Very occasionally, but that would be utterly impossible now in terms of health and safety, obviously, because we have to comply with the same hefty legislation that any catering operation has to.

But, in those early days, no health and safety?
There must have been health and safety, but it certainly wasn't the same in any way. We used to get the odd inspections but now, it's massive, obviously massive.

When you were doing the Hammersmith Odeon, did you also go touring?
Yeah, well, I had somebody called Jude Kelly. She used to work with me there so that if we had tours going on and I was off on tour, then she would look after the Hammersmith Odeon, and it was at that point that I was building up more teams so that I had more than one tour running at one time. That lasted about three years, and then I suddenly had to leave really quickly, that's when I moved to Nomis Studios. Yeah, I moved out of London in 1985, and up to Suffolk, and so that was all within that five-year period, '80 to '85.

Did it get to a stage where you thought, 'This is getting too big'?
I purposely kept it small, actually. You know, it's always been this, the maximum of four tours at any one time so that it's properly manageable with proper attention to detail, really.

Have you had staff that stayed with you for a long time? Apart from the bastards that went off and set up on their own (laughs).
Yeah, the bastards! There's been a switch from the original set of people that I had, but that's obviously going to happen as people get older. Even now, I've got people like Wilf, for example, now when did he start? . . . On those early U2 tours, and that's a long time ago. It must be almost 20 years ago.

You mentioned Bubble and Squeak; do you remember the people that worked for them?
I remember there was a husband and wife team called 'Paul and Muffet'. I can't remember the surname at all. Paul and some other guy owned Bubble and Squeak. But then Muffet was Paul's wife, and they used to tour together.

That's exactly my memory. I seem to remember Bubble and Squeak being two guys.
Yeah, and Muffet was pretty highly involved just because, obviously, of being married to Paul. I can remember a Police tour in the early '80s when they were kind of trying to come out of it. We did this Police tour with Paul and Muffet because they couldn't handle it on their own, it was quite big. Then it's all a bit woolly. . .!

That's with Bubble and Squeak. What about Home Cookin', do you remember anything about them?
Us caterers, we try to have quite a kind of nice vibe between us because we come to each other if we have a panic or we suddenly have to replace somebody. Sandi was someone we used to talk to quite regularly at Home Cookin', Sandi Grabham. I believe she still is at the helm, I'm sure she is.

I was going to ask you to just talk about the beginning, the limited amount of equipment, therefore, to a certain extent, limited provision of menus.
Not really related to equipment but the norm at the time. When I started, people were delighted when you had cottage pie on the menu, "Oh, that's it, that's what we're having for dinner tonight guys, cottage pie", "Oh lovely, cottage pie, yeah great". They'd just happily take what they were given, and there wasn't any choice. I don't even remember that we had very many vegetarians in those days either. I was the first person to introduce multi-choice menus because we did a big tour. Big tours weren't much the norm then, but we did a 'Kids from Fame' tour and we had to feed a hundred people, and that's a lot because it was normally only 20 or 30. Me and my team thought, the best way of tackling this is if we give different choices because then you're not cooking a hundred portions of one thing. That's when multi-choice menus came in, and everybody followed me after that.

They certainly did, and we're all eternally grateful. So it was because of the volume of people?
Yeah, that's why I brought it in, and everyone went, "Oh, this is a really nice idea. We can have choice now". Back then it would be white meat, maybe fish, although

fish wasn't necessarily that popular then although it is now, vegetarians, or pasta or blah, blah, blah. There'd be at least three choices of main dishes on the menu.

Do you recall any memories with the early American tours coming over and the difference in our national taste buds?
Yeah, definitely. I guess that still is the case, to a degree, but you very quickly learn to understand and adapt to the needs and the expectations of people from somewhere else, and that's not difficult. It isn't difficult at all, really. We find now that bands are less likely to have lots of stuff in their dressing room because it's hard to keep it nice in there. In catering, we've got refrigerated salad bars and all that kind of thing all the time.

How did your crew cope? Was there much difference between, originally, tours with 20, 25 people and now up to 200?
The biggest we've ever done was when we were involved with U2, we were their caterers for 14 years, which was a lovely one, but everything comes to an end. We had a touring situation of 300, something like that. It was like a military operation. Leapfrogging, as well. Leapfrogging teams.

Leapfrogging teams because you couldn't set up quick enough on the day. You had to set up in advance? Was that also because the leapfrog production team also needed catering for?
Yeah, yeah.

On the U2 tour what was your maximum staffing?
Gosh, I'm trying to remember, but I think it was about 14. Also I recruited about three or four Americans when we did the Worldwide U2 Tours, because I thought that was a clever thing to do. We had a wonderful American cook called Nedra who was fabulous and completely understood, clearly, what Americans wanted to eat and she'd make grits for breakfast and just do all sorts of stuff that we wouldn't have thought of, she was fabulous.

Now, that's an interesting one. You raised an interesting point here because my recollections of catering in the industry is that in the UK and Europe, we hired caterers to come on tour with us, they travelled with us. When you went to America, there wasn't any catering on the road with the band at all, and it would all be individual caterers attached to, more often than not, the promoters.
Yeah, well, attached to the building, actually more often than not, because it's all unionised, so you couldn't go in because there were specific people in that building who did the catering.

The only reason it worked for us with a big tour like U2 was because the catering was split and we would do U2 and their immediate entourage—ha! 300!—and then someone else in the building would take care of the local catering crew so they were still involved and were earning money out of it, if you see what I mean.

Okay, but you're talking about baseball stadiums and football stadiums for the most part, rather than arenas.
Yeah, absolutely. Even in arenas, I think they still potentially have some big in-house unions where you cannot take that job away from that person.

Now, you've got non-union and union arenas all over that country. Was U2 the first time you actually ventured outside of Europe?
Yeah, I think it was actually. I think it was the early '90s, it was the Zooropa Tour, so about '93. We were with them for 14 years, and that was fantastic.

Let me ask you a question from a catering point of view for production. You kicked off in the very early '80s. Obviously, everything was flight cased, but each day you probably didn't know where you were going to cook, or what you were going to cook and how it was going to be set up because pretty much in those early days, you were going into those venues for the first time?
Yeah, the venues have predominantly improved over time. Especially as we all, as companies sang from the same cue sheet for power requirements and refrigeration and all that kind of thing, so it was just better for them to just put those things in. Now we're experiencing a pretty similar problem with the Eastern European venues that we're going to, which haven't been around that long, you know.

At least, I guess, with power, it's easier than gas nowadays.
Theoretically, yes.

Theoretically. Well, if they can get the power for the lights, they should be able to get the power for the kitchen, shouldn't they?
Yeah, you'd think so. Yes.

Was there anything that sort of just made the whole thing a lot easier?
The only thing actually is that gas is a really fantastic medium because it gives you that real intense heat that you can't get off electric, but now that we can have induction hobs, they're just like gas. They're fabulous. I've got induction hobs in all my rigs now, and they do just the same. You can cook fantastic steaks on them, and it's just similar heat, instant big heat that gas has got.

Anything else that is very much different 35 years on than it was then?
Well the only stuff we do have a lot more of now, in terms of people that we look after, is their intolerances, you know, food intolerances. I don't recall anyone telling us that they were a coeliac, which is gluten intolerance, or allergic to anything. I think the only thing that anyone might have been allergic to before was nuts, which is quite common, but now the EU in their wisdom have brought in legislation where we have to identify the 14 most common allergens which are in any dishes that we cook, which is a nightmare. It's all right if you run a café and you do the same things all the time, but we cook different things every day in foreign countries but, by law, we have to have a food allergen policy and, if someone asks

us about a food allergen in there, we have to be able to tell them what is in the food that we cook and serve them.

Before you go, Richard, I'd like to ask you, didn't we do Duran Duran?

Yes, we did. I did their two biggest tours as tour manager, The Rio World Tour and the Ragged Tiger World Tour.
I've got, if I can find it, I have a picture of all of us with the band and the whole crew in a centrepiece, I think it was Smash Hits, but I've still got that.

I'm looking at the itinerary for the UK Duran Duran Tour in 1983, and it was Kim Davenport, Lynn O'Neil and yourself.

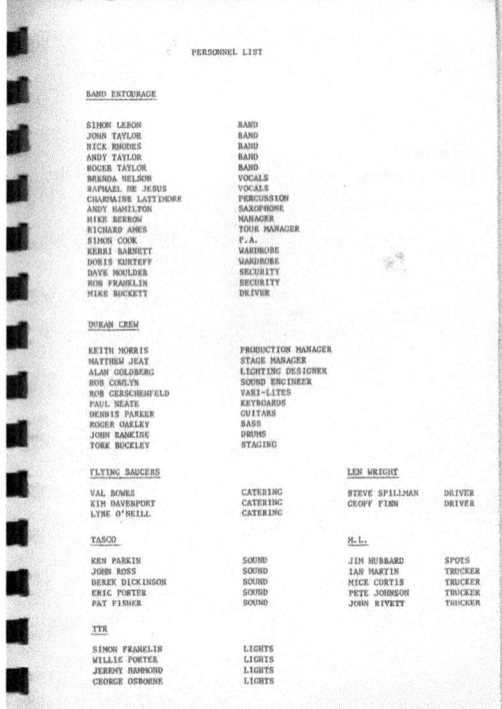

Personnel list of the 1983 Duran Duran UK tour

Well, Val, thank you very much.

Established in 1980, Flying Saucers is still going strong in the entertainment industry today with Val at the helm.

Part 9

Pioneers of Travel Agency

Back in the day as a tour manager, one of the first meetings held would have been with the band's agent, principally to confirm that the routing and distances between towns and cities was physically possible to achieve. Calculating the time required for the safe transportation of the equipment included time to rig and set up the production in the venue plus a soundcheck before the doors opened for the show.

The second meeting was always with your travel agent. On paper, the trucks could get from A to B in time, but could you also get the band and crew there? The trusty travel agent held that information in the 'bible' that was the worldwide guide to airline schedules, and with that you could proceed to pull the rest of the tour together.

What evolved from those early days of the mid-'70s was a circuit, not only of venues, but of hotels—those that welcomed rock-and-roll bands and worked with us to accommodate the little differences required; after all, we were a team working highly irregular hours compared to your average businessperson. I'm talking about late breakfasts, late opening times for the bar and 12-hour laundry service; these were all the little things that improved our stay.

It was a learning curve for all involved, and I for one sat down on many an occasion at Trinifold Travel, with Mike Hawksworth or his right-hand man, Alan Newing, to set up one tour after another from the mid-'70s to the mid-'80s. Mike Hawksworth, the founder of Trinifold Travel, tells his story of how his business grew out of booking rock-and-roll tours.

19 Mike Hawksworth

Mike Hawksworth standing on right

November 10, 2015 with Mike Hawksworth by Skype from the USA

I was going to ask you to start off with where you were born, where did you go to school and how did you get into this business?
I was born in 1946 in Guildford in Surrey, and I lived most of my life between Mitcham and Tooting in South London. I went to school at Pollards Hill Secondary School, Mitcham in Surrey, and I came out with just normal qualifications and that was it.

O-levels?
Yes. I always had a love of geography and of football, and this is how I got my first job. I didn't quite know what to do, I didn't know what to do as a living, but I used to follow European football, and I'd look at where, say, Madrid was on a map—like Real Madrid—and I'd look at another team from, say Italy, Roma, and I had built this love of geography, and from that I built love or the interest of the different cultures. Anyway, I didn't know what to do, I had no idea.

Had you done A-levels?
No, I wanted to get out and work. Previous to that, I'd worked for about five years when I was a lad. I used to help a milkman, so I've been working since I was 11. When I was about 11, I used to take my money each week and buy a record, okay, which was late '50s, early 1960, so late '50s era, Buddy Holly, Presley, Eddie Cochran, and then on to the early '60s, kind of Lonnie Donegan and that stuff, cast your mind back; I used to love music, just love records. I love music and I love the idea of travel and everything else like that. With the milkman, on some Christmases, I used to get six pence as a Christmas tip from some of the people. I thought I was earning serious dough.

With the same milkman? What was his name?
Yes, I knew him over years, dead now, of course, but lovely man, his name was Dick Peck.

So when I left school, I saw someone with a Thomas Cook traveller's cheque and I thought, that's interesting, and I said to the person, "Where are you going?" "Oh, we're going on a holiday". "That's good", I said. So I wrote to Thomas Cook, and I got a job as an office boy, right? That was my first job. Alan Newing, funny enough, worked at the same place as me about two or three years later, but we didn't know each other. That's pure coincidence.

And you'd moved on by then?
Yeah, I went out to work with an attitude of, I'll learn a bit, then I'll move on. It was in '63, '64, '65 time. I had about as many jobs in as many years.

Where was the first one, the Thomas Cook shop?
Thomas Cook's in Barclay Square, I think it's still there. So I moved around and moved around, and in the late '60s I worked for a company called JD Hewitt, in Jermyn Street, London, and they had an office out in, I think it was Pinner, Middlesex, and the office was like a normal travel agency front.

I went over there one day for something, and I saw all these names being written down for flights like P. Townsend, K. Moon, R. Daltrey, and then they used to send these over to the office where I worked to issue the tickets. I couldn't believe it! You know how you get first impressions, and I was really impressed that I was issuing the airline ticket for The Who, sounds crazy now, doesn't it?

They (The Who) had an office in Old Compton Street in London under Track Records, I think. I got friendly with a guy called John Field, who was the accountant there, and another guy called Keith Swallow. We went out a couple of times and had a drink. I decided that I'd had enough of travelling to London so thought I would work locally and started with Frank Hepner of Renowned Travel, later Hep Travel. They used to do a lot of record company work, they didn't do touring as such.

Where is local? Where were you living then?
I was living in Kent, Rainham in Kent.

And so I went down there and, anyway, Sally Arnold rung me up one day and said, "Oh, I don't know if I'm supposed to call you, or I'm supposed to call someone else?". So I was like, "No, you're supposed to call me", and she said, "Okay, I want you to book this tour". Well, I'd done airline tickets and I'd done a bit of hotels but hadn't really done a tour as such.

Who is Sally working for?
Sally was working for Bill at Trinifold Management, Bill Curbishley. This is where it'll all link in. Trinifold was The Who's management in those days, it was originally Track Records with Chris Stamp and Kit Lambert, and then the band, I believe, changed managers.

So then there was Bill, and they sent me down a tour. I'd done some work for Humble Pie, but it was only kind of booking hotels, and then they sent me a Lynyrd Skynyrd and Golden Earring tour. Golden Earring were the headline, and there was about 25 people in Golden Earring and about 20 people on Lynyrd Skynyrd in those days, I thought, 'this is a lot of business'! I was all excited, you know, and I must have done a good job because Lynyrd Skynyrd came back again. This is about early '74. Bill calls me up and he said, "Have you ever thought of opening your own travel agency?" I said, "No, no, no. I'm married, I just bought another house, I earn £2,000 a year, I don't want to take any risk". He said, "Well, we could do a lot of business", because at that time he was looking after The Rolling Stones in Europe as well, along with Peter Rudge; it was Rudge who looked after The Who in America and Bill looked after The Stones in Europe. He said, "Look, I'll give you £3,000 a year, we'll go into a partnership and I'll pay your travel expenses to London, so you know, you'll get a 50% raise". In those days, that was big money, an extra grand a year, like huge. So I went home and I thought about it and concluded you only get a chance like this once in your life, so I said, "Yeah, let's do it", and that's how Trinifold Travel started, but fundamentally, the people that pushed Bill to make the approach to me were Sally Arnold and Bryan Walters.

That's where I first met you was with Cockney Rebel, I think? That's where I first knew you. I'd done that tour and Steve Harley had that huge hit with 'Come up and see me', didn't he?

Yeah, that was in '75.
Okay, well in '74, we'd done that and then in '75, I think, it was just doing really Lynyrd Skynyrd and Golden Earring, and I think there was a couple of others that we'd done. I can't remember now, maybe Genesis.

At that time, Harvey Goldsmith started booking hotels through us for the shows that he was promoting. Bill had obviously said, "We've opened a travel agency, if you could throw him a few bones", and we were off to the races.

So at Trinifold Travel, you became partners, and the directors at the beginning then were you, Bill Curbishley and Bryan Walters, correct? And Sally was working for Bill?
Working for Bill, correct. She was Trinifold Management staff. She was doing all the support/secretarial work for Bill.

Where were the offices? Where did you start working?
Well, we started working in Bryan's old office, in Kingly Street.

Oh, okay. I used to go there because Bryan was Steve Harley's accountant.
Bill occupied another office down the other end of Kingly Street; I think it was number 12 or 21. Bryan took the upstairs and we had a couple of broom cupboards basically, downstairs, and then there was another office where the tour managers used to come in, like Dave Clark—not Dave Clark the singer, Dave Clark the old tour manager who worked with Bill. Of course, the band came in all the time, and it was fun. There was me, and a young girl called Margaret Mortlock came along to help me as we'd got busy.

I remember Margaret.
Well, Margaret was great, Scottish accent, she was really good. Around that time, '75 into '76, we met Rudge and Sally, and Sally said "The Rolling Stones are going out, we're going to put you in the frame to do it". I think, Genesis went out before that in '75, and that was a big, big tour then. It was the tour where Tony Smith had taken over their management, and they went out forever. They were the two bands that really broke it for us, and then people at Harvey's started liking us, and we got other bands coming on via various sources, tour managers used to recommend us, you came into the picture, you started recommending us . . . you know how it worked in those days.

1975 Backstage pass to the Genesis show at Wembley Arena
© R.Ames

And Brian Croft was great, Brian Croft was a real true friend to us; he recommended us to anyone that he could, he used to put his crews out and book them through us.

Still in Kingly Street or had you moved?
No, still in Kingly Street, we didn't move until—wow! Now, let's think. We were doing a tour for Chicago when they had that big hit, so when was that? '78?

Yeah, I've got you in '78 on a phone number of 01 437 2491.
No, that was still Bill's old office.

And then later in '78 or '79, I've got you in Artillery Road?
That's the one. So it's '78 or '79, that's for definite, and that's when Alan (Newing) joined us. So, it was me, Margaret and Alan.

That's when he came in and he, of course, was instrumental with his personality. He started doing the more individual stuff. I remember Jerry Stickles came to us at that time with Queen. A guy called Doug Smith, who used to be a heavy metal manager. He gave us lots of bands like Girlschool and all those sort of bands. Yeah, heavy metal stuff, loads of them. Neil Warnock, who was then an agent, started off at Bron but then went out on his own, he gave us loads of stuff.

Then Harvey Goldsmith and Tony Smith came into the business under John Smith's Entertainments with a piece of the company. That's what happened, they came in as a company investment.

Okay. Let me just go back a minute. So, still at Kingly Street, the contact is Michael, Margaret, Pat or Beverly?
There were two offices; we had an office down in Kent because in those days you couldn't issue airline tickets unless you had a front office. Like, you know the old travel agencies in the street? I didn't want to do that at London prices when I could do it in Kent for like no money at all.

There was nothing down there that was in Kent, it was all these developing housing estates, and I thought, well, you know, if they book holidays and we get our airline tickets done out there for no money, that's a pretty smart way to go, and it proved so, it proved a good investment.

Yeah. In fact, I seem to recall that when you were in Artillery Road, the air tickets used to come up the next morning.
That's right, absolutely right, yeah.

I did three tours with you in '78, The Cars which Kenny McPherson brought me in to do, the band Sailor, and a European tour and Egypt for The Grateful Dead (the European leg got cancelled) that was a Harvey Goldsmith introduction.
Yeah, I remember The Cars. Tell me—well, how is Kenny now, because we used to do Billy Connelly with him, of course? They were pure magic.

Do you remember where Kenny came from?
Kenny came out of Harvey's.

No, before that he was the assistant manager to the Apollo in Glasgow, or the Green's Playhouse, as it used to be called. That's where he started.
I never knew that. You know, he was one of the people—like yourself, those people that really supported us—he was one of them as well, and as I said before Brian Croft. I had a lot of time for Brian Croft.

Trinifold Travel Ltd

5 London Road,
Rainham,
Kent ME8 7RG

Telephone: Medway (0634) 368441/2/3/4
Accounts: Medway (0634) 371626
Telex: 896691 TLX 1RG
Attention: TRINITRAV
Admin. Office: 112 Wardour Street
London, W1V 3LD
Tel: 01 437 2491

Invoice/Credit No: A5439

Grateful Dead re Richard Ames

7/9/78

Passenger(s): Mr. R. Ames

Date	Details	Cost	VAT	Total
24/8/78	TO:- Air ticket London/Rome/ Venice/Rome/ Cairo/London Alitalia Inv. No. 65311	£257.00		

DIRECTORS: MP Hawksworth WG Curbishley HA Goldsmith B Walters, A.C.A. JA Smith
Registered Office: 21 Kingly Street, London, W.1. Registered Number: 1185995 England

Trinifold Travel Invoice, 1978

I've a fantastic interview with Brian for this book.
Yeah. He's just a lovely man.

So in the early days with Margaret I recall in '78 you were on a roll. There were loads of people in the office, everyone had two, three, four tours on the go at the same time, and you were building, as a travel agent, special rates for the music business really.
Well, that's right. What happened there is when we started, the promoters were getting these local rates, you know they were getting them in Germany, but what was happening was Fritz Rau[1] was getting one rate, another promoter was getting another rate and we're all using these different hotels. So I spoke to Fritz one day, he'd come over to the office and Bill introduced us, and I said, "How is it that the hotels will give us the same rates as you but they won't give us commission?" He said, "Well, I don't know". He said, "I don't want to do it, I don't want to do the rooming list and all". I decided what I'd do is I'd approach each hotel on the basis that I could get Fritz Rau's business and probably double it if you gave me a lower rate, right, and Fritz said, he didn't care. So, really it was kind of, not a scam, but a pressured sell to the hotel.

Everyone said, 'they can't do it, they can't, and I said, "Well, how do you know that until you've tried?". So I called each hotel and said, "Look, give me rooms at 75 Deutsche Marks a night, we want 10% commission but we'll give you 500 rooms a year, okay?". And they said, "500 rooms, okay, we'll do that for you". I did that in the main territories in Germany. Then I went to the main territories in the rest of Europe and in England and said the same thing.

Before doing these deals I remember going to a hotel in Manchester when I was setting up a small tour for The Who in 1974 or '75—it was 1975, just looking at that poster on my wall now as I speak. I rang the hotel in Manchester and I'd say, "I'll have five suites, 10 king singles and . . ." all the rest of it. They said, "That's great, what dates?" So, they said, "Yes, we can do that" and I said, "Okay, well, can't you give me a break?", they said "No, no, no". So they said, "And who is this group?" I said, "Well, it's a music group". They said, "Okay, we don't like music groups but who is this one?" So I said, "It's The Who", and the phone went, click.

Hung up on you?
Hung up on me, hung up on me, true story. I think they had four shows, five maybe, and the trouble that I had getting hotel rooms was like unbelievable. But once you broke that barrier, like in those days with the Albany in Glasgow, if you remember it, once you broke those barriers and you were in then, you were in. Of course, as we got more and more revenue, we could demand more and get more from the hotels. We had a very exclusive lock on it in those days.

I remember you had a special book if I remember, right?
Yeah, we printed a book; we printed a book so that we could give it to people and say, "Look, these are the rates we can get, have a look" and that does a lot of

favours, that was kind of a big financial investment in those days, but it paid off handsomely, it really did.

Did you get many hotels like with The Who in the early days that didn't want to take your bookings?
Yeah, because it was very early days, not so much with a band with reputation. They used to just kind of say, "Well, do we want The Rolling Stones because it's The Rolling Stones, it'd be good for our marketing" or "Do we want The Rolling Stones because it's going to be a pain in the ass, you know?" And that was the same thing with The Who. Once you got a relationship with someone it was okay, the problem was getting that relationship, if that makes any sense.

· It's like once you got a hotel in France you gave them all the business, once you get a hotel in Brussels, you gave them all and so on and so forth. Thomas Johansson was great with us in Scandinavia, he used to do stuff himself, and he recommended hotels to us and so, you know, people were good because they didn't want that kind of responsibility on their shoulders, you know?

Okay, did you have to do a lot of that in the early days with bookers at hotels?
No, I never kickbacked. No, I'd take them out for a drink, if we're going to use as a property in a major way, I used to go and visit them, and I always used to take them offsite, I never had a drink with them in the hotel. I always used to take them offsite so that they weren't being looked at by their superiors or their staff or whatever.

I felt that that was the way it would work, and it did work because people would say, "Yeah" offsite, much easier than they would if you're sitting negotiating in an office, you know. And then, of course, you always used to say, "if you want a couple of tickets to the show, let us know", the normal stuff.

Yeah, yeah. Now, again in the early days, when you were booking hotels as I recall, one of the first meetings as a tour manager would be with you, to sit down and plan. I mean I used to literally plan the tour with you, sitting down with the airline books and seeing what times we could move from place to place, and also choosing the hotels in each city. Did you get asked to stay in specific hotels at that time or were you actually saying, this is the best place to stay?
Occasionally we got it, but normally we went for a good-quality hotel that wasn't going to be pretentious towards the client. In those days some of them were very pretentious towards the clients. It's not so much now, for example, The Four Seasons will take anyone within reason, but in those days they were very selective, so we had to be careful. I mean, in Copenhagen, it was the old Plaza where they always used to love rock bands.

One of my favourite hotels back in the day.
Then you had these chains started, like the Scandinavia hotels started as a new group thing in Oslo and Stockholm, so they wanted your business, they had heard

Brass plaques of bands and artists that have stayed at the hotel, on a wooden column in the lobby of the Plaza Hotel, Copenhagen
© Richard Ames

of you by then. It became a different thing, so whenever a new hotel opened in any territory, they called us and said, "Can you give us your business?" and of course, the shoe was on the other foot now, instead of us pleading with them, now they were pleading with us.

About '76 to '78, it really exploded. We brought in Alan, who was great. We brought in Eva, who was great, Eva Ström, she was fabulous; then we had a girl called Claudia Bennett, she was great, and Sue Hall was just fantastic. Sue formulated all the rooming lists and letters of introduction everyone uses today.

It was a young company, everyone loved the business that we were doing. You had Eva who was married to Harvey Baker, a roadie. Sue was an out and out rock-and-roller, you know. Alan played his guitar and loved the music. It's a whole wonderful combination of personalities. Those people made the business. You're only ever as good as your staff allow you to be, no matter what you do; the company would not have made it without those people, no question, no question.

So, tell me how Alan came into the business?
We were needing someone else. We were needing a third person, Margaret and I. By that time we were doing Queen, Rick Wakeman and Manhattan Transfer, quite a bit more stuff, and we needed another pair of hands, and I went to an agency that dealt with people in the travel business. I called them up and I said, "I'm looking for someone to come and work with us", she said, "Okay, okay, I'll send you this guy, he is very good but he's got long hair, very long hair", so I said (laughs), this is true, you ask Alan this, you ask Alan, this is true, I said "Perfect". Me and Alan met, we got along, and I said, "Right, start as soon as you can", so he did, and the day Alan started, we were moving to Artillery Row, and we left him with a phone and a set of airline manuals and said, "Get on with it, come to this address tomorrow", and Alan thought "what the hell have I got myself into here?", you know, he was great. He's still great, love him to bits, and I'll always love him to bits, smashing, lovely man, love him.

Yeah, that was a long working relationship you had with him.
Then, we had Clive, looking after the accounts at that time as well. He went down to Rainham and looked after all the accounts for both offices. He was another one that was priceless, priceless in the development of the business.

So, you say the other office, by the time you moved to Artillery Row, you've moved from Kent?
No, we kept the Kent office for ages. Rainham was the Kent office, kept that for years, and then we sold it. We decided to get our own license for the London office, and so we sold Rainham and we got good money for it, so that was back in the '80s. The laws had changed by this time and you did not need to have a front, a shop front, but we got one in London, which made it all much easier.

Yeah, absolutely. Going back to Artillery Row, you mentioned earlier that, at some point Harvey and Tony Smith came in as business partners. Why?
To develop it further, Bill wanted to develop it further.

You said 'develop it' so, take on more staff, bigger offices, more turnover?
I didn't particularly want that.

But that's what happened.
That's what happened. You're on such a roll, it's like you're getting business in from all angles, you know, from tour manager recommendation, from Neil Warnock, from Doug Smith, from Harvey's people, people that knew Bill, it was just like American agencies were recommending us, Donna Dietz in Los Angeles in—what's the name of that travel agency? Air Apparent.

There was Starflight Travel in New York?
Yeah, Starflight Travel, Ron Erickson, yeah.

And that's where Brooks Ogden started.
Yeah, I thought she did. Then in the early '80s, we opened an office in New York, and our first two clients were The Who and Genesis, but it proved to be a disaster, it was absolutely catastrophic, putting it politely.

It was a bit of a mess-up. So, before that 'experiment', just call it that, occurred. were you booking outside of Europe?
No, no, that was the point of it.

Right. So, you'd book the flights across to America, for example, but you wouldn't book internals in America.
On some we did, on Genesis tours we'd sometimes book the internal flights, but we wouldn't do the hotels. Then, of course, in those days everyone was flying commercially, no one was really chartering like they do now. Very occasionally they would, like Des Moines one day; get on another plane, go to Pittsburgh; get on another plane, go to Philadelphia, it was all that sort of stuff.

Oh, I did loads of tours like that.
Yes, you know the ones I mean? That's the way it was then, but now we mainly concentrate on Europe, we knew where our strength was and concentrated on that. Then around the early '80s, Robin came in and there are a couple of other people, it made it even stronger. I would say we were really strong between '83 and '88, that five-year period, we got bigger beyond that, but that was our peak strength because we had a real asset of staff that were 100% behind the project, and that made all the difference. Then it got too big and you couldn't control it.

How was it with American bands coming over to Europe?
Sometimes it was bad, man. I remember Jefferson Starship, when they toured with Genesis, you know, do you remember that? When Grace Slick refused to go on in Loreley and they burnt the stage down?

Oh, yeah, I remember that.
They rang me up and said, "We may come into London a little earlier than we anticipated". Well, I've heard all these fire engines and horns and everything going off in the background, and I was talking to Bill Thompson (the production manager), I said, "What's going on there Bill?", he says, "We have a slight problem", I said, "What do you mean you got a slight problem", he said "Well Grace doesn't want to perform and they're actually breaking up the stage, no, no, now they're setting it on fire", and this guy was so, so calm about it.

Me and Alan Newing went to Heathrow to get them, as we couldn't get them into the hotel, so we met them at Heathrow. They were all coming through in satin shorts and skateboards and all this sort of thing. I mean, Alan looked at each one of them thinking, "what have we got here?", so we put them in this coach and they said, "Is it all right to smoke?" I was "Yeah, cool, you know, we'll have a cigarette with you", both of us smoked in those days, and of course they have got all their joints and the pipes come out (laughs), it was great.

In '83 with Supertramp was the first time I was involved with a tour that had so many people, we played in football stadiums everywhere, the crew was massive, in total there were over 100 personnel, just from Supertramp's side of things moving around. We had someone on tour from Trinifold.
It might have been Pete Collins. I can't remember if it was Pete or Alan. I know we had Bowie and Supertramp at the same time.

Yes. So that Serious Moonlight tour. I was asked to tour manage that tour as well by Wayne Forte, however I was already committed to Supertramp.
Yeah, we had half the bloody office out on the road I remember that. Now who tour managed that in the end?

It was Frankie Enfield. Now was that the first that you actually had to move your personnel onto the road? Did you realise that tours wouldn't happen smoothly if you weren't actually on the ground there with it rather than just booking hotel rooms and flight tickets from the office?
Actually, I never particularly endorsed that theory; I didn't really see the point of that. I thought that if you are doing the job in an office you could do it just as well, but of course you've got to remember in those days there was no e-mail, fax was just coming in, and all those sort of things.

It was a phase when bands thought it was a cool thing to do rather than it being a real practical thing to do; yes, you had someone going and advance the hotel and yes, you had your keys there on the counter when you got in and yes, you could change your flight but in real terms, I don't think it was of any great benefit. In fact, I think it was detrimental in some ways because the agent who was dealing with it got embroiled in the rock-and-roll side of it rather than the travel agent side of it. You got to remember these are still pretty early days of it. So, Alan for one did, he had issues with it. We never did it with The Stones. I never did it with The Who. I never did it with Michael Jackson, never.

The other issue with that was that you had so many people giving so many instructions instead of it coming from just one person that called the office.

Aah, yes, too many chiefs.
You had one person on the road who was dealing with all these people going, "Oh, my wife wants to do this, oh, can you do this, can you do that, I want to upgrade my room", instead of there being a management person who said, "Okay, I'm going to call Trinifold and say this is what we need". So, you had too many fingers in the pies.

That was a problem with that and, of course, in those days no one had ever done a leapfrogging multi-stadium tour. It just hadn't been done in those days in

Europe. This was when you were doing it with Supertramp, you, Bowie, Queen, they were all doing these multi flip-flop. So, you had all these different teams. You had all these red, white, blue teams. Then you had all of Edwin's, who didn't know their arse from their bloody elbow. They knew their jobs but travelling . . . We were all in the same boat.

Okay, great (laughing), when you look back at it, I mean, man, we were geniuses really. All of us were absolute geniuses to make those things happen with the facilities that were available in that particular moment in time. I was listening to an interview of Edwin Shirley the other day, and he was asked, "How did you start?". "Well", he said, "I got a bread van. I thought I could put stuff in it. I thought it would be a good idea". Who would have thought of that? Well, it worked, and look how successful they were.

And I've got the story from interviewing Roy Lamb and Del Roll, too.
Del, Roy and, of course, Ollie Kite, these are the classic people, the great people. They all made it together and all of us we were in the same boat half the time, but we knew what we were doing, but we also knew that we had to think twice about it because we didn't quite know what we are doing! We did in the end, it didn't take long, but you had great people out there, you know, yeah, great, some great tour managers; Stickles was a great tour manager as well. He was wonderful.

Yeah, yeah. He became the first super tour manager. I sometimes thought that it would be cool if I could clone myself so I could actually do more than one tour at a time, and he was the first one to actually be able to do that. He just had someone else to go on the road, but he became the de facto Tour Director.

Before we finish, tell me about the relationship between the music business travel agent and the airlines?
Well, that was something that progressed, okay, as we built the business up and progressed more and more, that's when you could negotiate better deals. Some of the time you used to go out to Pan Am, to say to Freddy Lloyd—he was great, Freddy at Pan Am—we used to say, "Fred, we've got so-and-so on this flight needing to upgrade", and on a pretty regular basis you'd get it from him, you know, that sort of stuff.

And excess backage?
Sometimes not necessarily, most of the time, that was more a case of an airport contact.

Okay, did you get involved with moving equipment at all?
No, that was always Rock-It Cargo, Alan Escombe, David Bernstein.

You didn't know at the time that you were pioneering, did you?
It was blazing a trail, man! When you look back, you know, I have such affection of the memories and how we were all kids in a sense, you know. We just

struggled through it, then we got there, but equally we couldn't be fools, otherwise we wouldn't have got through it. I mean, none of us were fools, those were wonderful days, and I've told Robin several times, those days will never, ever be relived, it's impossible, they won't ever be relived again. It's too bureaucratic. It's too systematic. It's too . . . It's not personal anymore.

Can you tell me a story?
Okay, Bob Marley, okay, or Pete Tosh, one of the reggae bands. Alan will tell you this story. So they are out on tour in Europe and first date is up in Newcastle, middle of January. They got to the Holiday Inn in Newcastle, they've gone up there and checked in. The next morning Alan called them up, just a courtesy call to the tour manager. He says, "Can I speak to, say, 'Mr Ames' you know, the tour manager", and the hotel says, "Oh no, sorry, Mr Ames has checked out", he says, "What do you mean he's checked out, they were supposed to check in last night".

"Oh, they did check in and it started snowing, okay, and they thought it was a sign, a Rastafarian sign that it was dangerous to stay there, and went back to Jamaica". That's a true story, you ask Alan Newing, that's a true story. There's thousands of them, when you cast your mind back, there's so many of them. Oh, Lord above, Alan's got a great sense of humour as well. He could see the funny side of it. I mean, we used to have some real laughs, and I mean real situations where you had to laugh at it because it was so unbelievable.

I always remember another one, which is stupid, but I remember, I was doing a Queen tour, okay. So I sent this telex to a Paris hotel for the crew, and I said, "Could you please book 60 king singles for Queen crew 20th July three nights", and the message comes back, it says, "We would love to accommodate Her Majesty but we are sold out on those particular days". It was the Holiday Inn Republic. You remember the Holiday Inn Republic in Paris? Yes, there's just thousands of them, Alan's got a catalogue of stories.

What's the best Moon story you recall?
From what I recall he was lovely, I mean, he was funny, but I didn't know him very well. I'll tell you a story about the office. We were sharing it with Bill, and one day we'd gone in, it's just me and Margaret, and the phone started ringing about 9 o'clock. The receptionist wasn't in, so we went to pick up the phone, and it was one of those old phones where the receiver was, you know, kind of laying over the phone, those old phones where you have to take the receiver off to dial a number, you remember? So I've gone to pick up the phone receiver and the whole unit comes up, the receiver hasn't come off; the whole unit has lifted off the desk. I said, "What the hell", so it kept ringing and ringing, so I went over to the next one, try to pick up the receiver and the whole unit again comes up.

So there's six phones in the office like this, and there was seven phones, he'd left one unstuck open, right, so I said "Hello", and Keith said, "Took you long enough to answer the bloody phone, didn't it? If this is the service I'm going to get, then I'm going elsewhere", and he put the phone down.

That was him?
Yeah!

One last question. At the time that Trinifold Travel was growing and you were starting, who else was doing it? Do you remember Mark Allen Travel?
Mark Allen was doing it, Mark Allen, but again they were more record company. They were doing some touring that we didn't do. I'll tell you who else gave me a real big break in those days when I reflect back now. There was someone in Warner Brothers, a lady called Sarah Radclyffe who I think has now become very big in films.

She worked in WEA, and another guy called Peter Ker, they used to bring all these acts in like Seals and Crofts for promos from America, and we were using the Montcalm hotel in those days, if you remember. They gave us a lot of business, and they always wanted to stay at the Montcalm, so again I went to the Montcalm, got a deal for everybody. When all the bands come in, they all used to stay at the Montcalm. Most of that stuff with all the record companies like CBS or whatever, it was dealt with by Mark Allen.

There was a guy there called Terry Austin, I think, and he had done The Moody Blues and a couple of other acts, but we really did most of the touring stuff. Then you had Ian Wright and John Giddings and those guys came on-board, they had, around that '83 to '88 time, they all had a lot of business, Barry Dickins, another truly great character.

One last question then. On your rise, building Trinifold Travel, was there anyone you particularly looked up to that you'd like to give a mention to in helping you in your career.
Obviously, it has to be Bill; if it wasn't for Bill, it would never have happened. It has to be him. Along the way, I think there are five people that really made business: Alan Newing, one; Sue Hall, two; Robin Hawksworth (nee Rhodes), three; Clive Lawrence, four; and Eva Ström. Eva Ström had a big influence. That will be my big five, without them it would have never of happened, simple as that. It wouldn't. I'm very grateful to those for the life they gave me, or helped give me.

Of course, the tour managers like George Travis, who I worked with for 40 years, and Andrew Zweck, who has been very good to me over the years, Jerry Stickles and let's not forget the Edwin Shirley's guys. There were so many people who have been kind and recommended us, too numerous to mention.

Lovely, Mike, thanks very much.

Mike and his wife Robin continue to book travel for artists around the world, but based now in North Carolina, USA under the business name of Tour Company.

Note
1 www.dw.com/en/fritz-rau-who-brought-the-stars-to-germany-dies/a-17035692

Appendix
Richard Ames' Early CV

I started in this business in 1972, first as a roadie, for Cockney Rebel, but then as their sound engineer and, within a year, their tour manager. It was a quick rise up the touring hierarchy, but I've got to say it was out of necessity from my point of view as I wasn't a very good roadie, an even worse soundman, but thankfully for me, I was a good organiser.

Tour Manager (unless stated otherwise)

1973	: Cockney Rebel—UK & Europe
1974	: Wishbone Ash—USA Tour, **Asst Tour Manager**
	: Cockney Rebel—UK & Europe
1975	: Cockney Rebel—World Tour
1976	: Steve Harley & Cockney Rebel—UK & Europe
	: Harvey Goldsmith, **Production Co-ordinator**
	: Miles Copeland, **Production Co-ordinator**
1977	: Fleetwood Mac Rumours—UK Tour, **Tour/Production Co-ordinator**
1978	: The Cars—UK & Europe, **Tour/Production Co-ordinator**
	: Sailor—UK & Europe
	: Rich Kids—UK
	: The Grateful Dead—Egypt & Europe, **Tour/Production Co-ordinator**
	: Steve Harley & Cockney Rebel—UK & Europe
	: Frankie Miller—Europe
1979	: Paul McCartney & Wings—UK, **Tour/Production Co-ordinator**
	: Kate Bush—UK & Europe
	: Sniff & the Tears—USA & Canada
1980	: Paul McCartney & Wings—Japan, **Tour/Production Co-ordinator**
	: Steve Harley & Cockney Rebel—UK
	: XTC—World Tour
1981	: Icehouse (Flowers)—Australia & NZ
	: Adam & the Ants—UK & Europe
	: Richie Blackmore's Rainbow—South America, **Tour/Production Co-ordinator**
1982	: Duran Duran—World Tour

1983 : Supertramp—World Tour
: Duran Duran—Australia & UK
1984 : Duran Duran—World Tour
1985 : Supertramp—US & Canada
: Paul Young—World Tour
1986 : Supertramp—European Tour